Insect Diversity Conservation

This groundbreaking book is an up-to-date global synthesis of the rapidly developing and important field of insect conservation biology. Insects are by far the most speciose organisms on earth, yet barely known. They play important roles in terrestrial ecological processes and in maintaining the world as we know it. They therefore present particular conservation challenges, especially as a quarter may well become extinct in the next few decades.

This book first addresses the ethical foundation of insect conservation, and asks why we should concern ourselves with conservation of a butterfly, beetle or bug. The success of insects and their diversity, which have survived the comings and goings of glaciers, is now facing a more formidable obstacle: the massive impact of humans. After addressing threats, from invasive alien plants to global climate change, the book then explores ways in which insects and their habitats are prioritized, mapped, monitored and conserved. Landscape and species approaches are considered. Restoration, and the role of conventions and social issues are also discussed. The book is for undergraduates, postgraduates, researchers and managers both in conservation biology or entomology and in the wider biological and environmental sciences.

MICHAEL J. SAMWAYS is Professor of Entomology at the University of Stellenbosch, South Africa. He is internationally known as a conservation biologist and policy advisor.

Insect Diversity Conservation

MICHAEL J. SAMWAYS
University of Stellenbosch, South Africa

CAMBRIDGE
UNIVERSITY PRESS

PUBLISHED BY THE PRESS SYNDICATE OF THE UNIVERSITY OF CAMBRIDGE
The Pitt Building, Trumpington Street, Cambridge, United Kingdom

CAMBRIDGE UNIVERSITY PRESS
The Edinburgh Building, Cambridge, CB2 2RU, UK
40 West 20th Street, New York, NY 10011–4211, USA
477 Williamstown Road, Port Melbourne, VIC 3207, Australia
Ruiz de Alarcón 13, 28014 Madrid, Spain
Dock House, The Waterfront, Cape Town 8001, South Africa

http://www.cambridge.org

First published 2005

Printed in the United Kingdom at the University Press, Cambridge

Typeface Swift 9.5/14 pt. *System* LaTeX 2_ε [TB]

A catalogue record for this book is available from the British Library

Library of Congress Cataloguing in Publication data
Samways, Michael J.
Insect diversity conservation / Michael J. Samways.
p. cm.
Includes bibliographical references and index.
ISBN 0 521 78338 0 (hardback: alk. paper) – ISBN 0 521 78947 8 (pbk.: alk. paper)
1. Insects–Ecology. 2. Wildlife conservation. 3. Biological diversity conservation.
I. Title.
QL496.4S364 2005
333.95′5716 – dc22 2004048882

ISBN 0 521 78338 0 hardback
ISBN 0 521 78947 8 paperback

Contents

Preface

Some say that 'the cockroach' will be the last species on Earth to survive. Then it has been calculated that one gravid aphid, left to reproduce with zero mortality, will, after one year, cover the globe with an aphid layer over 140 km thick. Not forgetting too, that flies and fleas vector disease. So, why should we even consider conserving insects? Quite simply, without insects, the likelihood is that the world as we know it would be radically changed in a matter of days. Besides, it is only a tiny minority of insects that harm our lives. These two faces of insects, friend and foe, are just one of the several paradoxes that characterize insect conservation from other facets of taxon-based conservation biology. Our impacting on landscapes can turn a benign insect species into a pest, while, on the other hand, it may cause an extinction of another species. Focusing on the land mosaic, its composition, structure and function, is thus central to insect conservation.

We have no idea of the outcomes from our modification of the biosphere. Blindfolded, we are turning the many faces of the Rubik Cube of biological diversity conservation in the hope that all the faces will match. It is not that we are incapable, it is just that the world is so complex. A thousand species, for example, in the same community (not an unreasonable figure) potentially produces 0.5 million interactions. In addition, strengths of those interactions, and hence outcomes, also vary.

Bleak as this may seem, we are beginning to lift the blindfold and make rational decisions for conserving biological diversity. Insects and their interactions are a major component of that diversity. Indeed, insects are virtual ambassadors for biological diversity. These little animals, by their great variety and abundance, play an unheralded yet pivotal role in our, and many other organisms', lives. We now have a major challenge before us: How do we go about conserving this largely unseen, unknown majority?

The ambassadorial status of insects for terrestrial and aquatic ecosystems is the reasoning behind the title 'Insect Diversity Conservation'. 'Biological' is simply replaced by 'Insect'. This is not in pursuit of entomological chauvinism but rather to emphasize that insects are central, yet with many special features, to biodiversity conservation.

The aim here is to overview and critically appraise the conservation of insect diversity. It focuses strongly on the variety and differences among insects, and links these to landscape and other large-scale conservation initiatives. After all, insects do not rule the world alone. This is not, though, to ignore special cases where a particular insect species requires particular conservation attention.

Conservation cannot be done without clearly defining our feelings and motives for why we are doing it. This goes beyond simply the utilitarian value of insects for us. This field of environmental ethics in relation to insects is therefore addressed in the first chapter, and is a foundation for all that follows. In Chapter 2, the special case for insects, in comparison with and in contrast to other organisms, is argued. This is not to say that insect conservation is tangential to mainstream conservation. Rather, it is central, especially as insects play so many keystone roles in non-marine ecosystems. These roles and others are discussed in Chapter 3. These first three chapters together are the launching point for the rest of the book, and address why there is the need for insect diversity conservation. Part II (Chapters 4–7) addresses threats to insect variety, and emphasizes that many of these threats are multiplicative, with one threat compounding another. Part III (Chapters 8–12) then reviews the options that we have to ameliorate these threats.

As insects are now featuring much more strongly in biodiversity conservation, the field of insect conservation biology has grown enormously in recent years. It is clear too that there are many varying, even conflicting conclusions when various studies are compared. These differences seem to arise mainly from three different perspectives: differences in spatial scale of the study, differences in biogeographical regions, and differences in the focal taxa used. This is healthy and indicative of a rapidly growing field of study. Nevertheless, some general principles are beginning to emerge, and in terms of management, these are synthesized in Part III.

Insect conservation has been a rapid growth area in recent years, often with intense debate. In response, I really do appreciate the stimulating feedback from Jonathan Ball, Andy Beattie, Steve Compton, Eduardo Galante, Henk Geertsema, Justin Gerlach, Jeff Lockwood, Melodie McGeoch, Tim New, Paul Pearce-Kelly, Andrew Pullin, Nigel Stork and Stuart Taylor, as well as the lively minds of my research students over the years. Many authors kindly made available text figures, and these are acknowledged with each figure.

This work would not have been possible without the amazing support of Colleen Louw (processing text), Stuart Taylor (processing figures) and James Pryke (processing references). Anni Coetzer produced the beautiful cover and text illustrations, so rich in symbolism. Ward Cooper and Jo Bottrill saw production of the book to its completion. My warm thanks to these friends for making this such an enjoyable enterprise.

Cover picture

In an ever-changing, human-transformed world, we often overlook the importance of each little creature in our earthly ecosystem. Conscious and moral Man holds the future well-being of the world in his hand. Yet the race is against time, that these noble efforts are not in vain. Insects are an ancient, ecologically significant and beautiful component of the world, symbolized here by the dragonfly. The variety of insect life around us, be it in our garden or city park, are a constant reminder that these small but numerous animals are part of the fabric and health of our planet. By destroying these living creatures and their habitat, the delicate glass of our ecosystem will shatter, leaving us with a transformed, bare world, devoid of colour and life as we once knew it. Time is now short for ensuring the future of this amazing insect diversity.

Anni Coetzer

I The need for insect diversity conservation

Never use the words higher and lower . . . Certainly, they are difficult words, not only descriptive but value laden . . . while bald eagles are an endangered species, so are 129 species of American freshwater mussels . . . Is it more important to save the eagle than ten dozen species of mussels? . . . Perhaps eagles and mussels are just there, and neither is higher or lower. Of the animal biomass on our planet, 90 percent is invertebrates, who account for 95 percent of all animal species.

Charles Darwin (a penned note)

The diversity of insect life today is, as far as we know, the richest it has ever been. The variety is so great that insects make up three-quarters of all species. Insects have radiated into so many diverse forms that we have been able only to describe a small fraction of them. They are a major component of all life we see around us. Out of simple beginnings, the earliest life forms continued to radiate through the process of variation/selection/retention to endow the earth with a fantastically rich tapestry of form and colour, of development and dispersal, that has enriched every corner of terrestrial systems with insect character of some sort. Humans are a latter-day arrival who hold in their palms the future of the insect mosaic. This insect variety is losing its spatial and compositional integrity as we enter the new era, the Homogenocene, which is a mere Blink of a geological eyelid.

1 Ethical foundation for insect conservation

We may notice . . . that the tree-hopper, called by the Greeks *Tettix*, by the latins *Cicada*, received also from the former the title of "Earth-born," – a title lofty in its lowliness, because it was an implied acknowledgment from men of Athens and of Arcady of a common origin with themselves – an admission that the insect was their brother, sprung (as they fabled) from the earth, their common parent, – whence, also, they wore golden tree-hoppers in their hair.

Acheta domestica (1851)

We feel our world in crisis.

David Rothenberg (1989)

1.1 Introduction

Conservation action must have a sound philosophical and ethical foundation. This gives the action meaning and direction. It is the 'why' we are doing it. At the most superficial level, that of utility, nature is at our service to be used, ideally sustainably. In this philosophy, humans have complete dominion over nature, and this is the language of most international agreements and conventions.

Deeper levels require more wrestling with thoughts and ideals. Among these is one philosophical approach where humans and nature are still separate, but nature is to be admired and enjoyed. An alternative view is that humans are part of the fabric of nature, and nature is used sustainably yet respected deeply.

In recent years, a more profound environmental philosophy has emerged, where organisms, including insects, have the right to exist without necessarily being of any service to humans. A powerful epithet to this deep ecology view has emerged: that we should appreciate and love other organisms without the expectation of anything in return.

Different world religions have recognized the environmental crisis and have made declarations that bridge their faith for the future well-being of the world. While philosophy is an essential foundation for how we approach conservation activities, religion is a spiritual complement, which in some countries such as India can play a significant role at the local scale.

Insect diversity conservation has received an enormous upsurge in recent years, principally with the recognition of the major role that insects play in maintaining terrestrial ecological processes. Yet there is recognition too, that insect individuals and species are being lost at an enormous rate. Stemming this loss of diversity is a vast task. A philosophical base helps decide on which value systems we should use to approach the challenge. Religion then provides spiritual recognition that what we are doing is a good thing. These lead to the scientific pursuit of insect diversity conservation, which is the subject of the following chapters.

1.2 Environmental philosophy and insect conservation

1.2.1 Ethical foundations

No conservation effort can meaningfully begin without a firm foundation of human value systems or ethics. Such ethics are the language of conservation strategies. Without some moral guidelines, it is difficult to define our goals and hence the expected outcomes of conservation activity.

There is little to separate insects from other organismal aspects of biodiversity in environmental philosophy. A noteworthy exception is that not all insects are good for each other, or for us. Insects can be parasitoids or disease vectors. Indeed, we exploit parasitoids as biological control agents.

1.1 A ramification of the Resource Conservation Ethic. Mopane 'worms' (larvae of the emperor moth *Gonimbrasia belina*) are harvested and dried, and considered an important source of protein and fat to people in Africa.

At the arguably lowest level of ethical consideration, insects have utilitarian or instrumental value for us. This includes aesthetic, food, adornment, ornament, service, spiritual and cultural, heuristic, scientific, educational, conservation planning and ecological values. These utilitarian values have two facets. The first is that they are there for us to enjoy aesthetically and be left alone. This is the Romantic-Transcendental Preservation Ethic (Callicot, 1990). This goes beyond just the insects themselves. It considers all their interactions and ramifications with other aspects of nature. It is an ethic that we adopt when we visit a nature reserve. The second utilitarian facet is that insects are there for sustainable use (Figure 1.1). This is the Resource Conservation Ethic. The harvesting of honey from honeybees is an example. But this ethic may apply to a wider, indirect set of services that insects supply, such as pollination and natural biological control. Where insects do not fit snugly into this ethic is when many actually do a disservice to our resources by nibbling, piercing and burrowing into plants, transmitting disease and killing animals. To entertain this ethic may indeed involve some control of insects.

In both the Romantic-Transcendental Preservation Ethic and Resource Conservation Ethic, humans are essentially separate from the rest of nature and

1.2 Building an Evolutionary-Ecological Land Ethic at the smaller spatial scale into Resource Conservation Ethic at the larger spatial scale. Here remant linkages of grassland are maintaining ecological and evolutionary processes, while the landscape as a whole is also being utilized to produce timber for export (KwaZulu-Natal, South Africa).

organisms have positive, negative or neutral value. In contrast, Leopold (1949) articulates in a subtle and charming way that other species have come about through the same ecological and evolutionary means as humans, and as such, deserve equal consideration. Humans nevertheless, reserve the right to use and manage nature as well as there being recognition of the intrinsic value of other species and the integrity of ecosystems (Figure 1.2). This is the Evolutionary-Ecological Land Ethic. Rolston (1994) goes a step further, and points out that culture has now emerged out of nature, which brings with it a responsibility for humans to nurture other organisms. Samways (1996d) then illustrates that culture has now become an evolutionary path and the human self-manipulating genome the driving force. Ideally, we now need to build into our new genome an environmental ethic.

1.2.2 *World in crisis*

The sharp increase in consumerism and human population growth over the last few decades has stimulated an acute awareness of the adverse impacts on the natural environment. A feeling has developed that not all is well in the world, and that wild nature, unsullied by humans, may even have ended (McKibben, 1990). There is also a growing awareness and accumulating evidence that our world is in crisis – but not necessarily doomed (Cincotta and Engleman,

2000). Out of these changes has developed a strong movement, that of deep ecology, which provides a sense of wisdom combined with a course for action (Naess, 1989). Pessimism is not allowed to prevail, and a sense of joy is the spirit behind the philosophy.

Deep ecology is not something vague as some have claimed. It is an ontology, which posits humanity as inseparable from nature, and with an emphasis on simplicity of lifestyle and on communication with all critics. Naess (1989) termed this approach ecophilosophy (shortened to ecosophy). It is the utilization of basic concepts from the science of ecology, such as complexity, diversity and symbiosis, to clarify the place of our species within nature through the process of working out a total view (Rothenberg, 1989). This is especially relevant to insect conservation, as the insect world is indeed complex and diverse, and it is one where symbioses in the widest sense are widespread. Also, it is at the core of the landscape approach to conservation, where focusing on individual species and interactions is insufficient to conserve the vastness of insect diversity. This emerging arena of ecophilosophy, ecopsychology or transpersonal ecology is likely to play a role in future conservation (Fox, 1993). Indeed, Johnson (1991) advances a potent argument on behalf of the morally significant interests of animals, plants, species and ecosystems. He notes that in a moral world, all living things, insects included, have a right to survival (Figure 1.3).

1.2.3 *Overcoming the impasse between utility and deep ecology theories*

Although deep ecology and even some schools of thought in landscape ecology (Naveh and Lieberman, 1990) include humans in the global ecological equation, it is nevertheless this very human factor that is threatening the planetary processes that in the past have led to the current, rich world-ecology. Although deep ecology purports a human omnipotence, the risk here is that a sense of place, and, in turn, places of wild nature, are left out. To ignore ecological differentials across the globe and to homogenize all would simply be sad. After all, it is the essence of conservation biology to conserve diversity, which, quite literally, is all the differences within nature and across the globe.

Norton (2000) argues persuasively that utility (instrumental value) and deep ecology (intrinsic value) theories are confrontational, and he then asks whether there is perhaps an alternative, shared value that humans may place on nature. The instrumental and intrinsic value theories share four questionable assumptions and obstacles: (1) a mutual exclusion of each other, (2) an entity, not process, orientation, (3) moral monism, and (4) placeless evaluation. To overcome these impasses, Norton (2000) suggests an alternative value system, which recognizes a continuum of ways that humans value nature. Such a spectrum would value processes rather than simply entities, is pluralistic and values biodiversity in place. Such a universal earth ethic values nature for the creativity of its

1.3 A road sign at Ndumo Game Reserve which emphasizes the ecophilosophy approach where all creatures have the right to survival, no matter how small or ecologically significant.

processes (Norton, 2000). This ethic is vital when we consider not only the sustainability of nature, whether for itself or for humans, but arguably and more importantly, it is crucial for maintaining the evolutionary potential of biodiversity, especially in extensive wild places (Samways, 1994).

The value of wild places is high, and such places are often the seat of interesting, curious and irreplaceable biodiversity. The problem with placing great emphasis on wild places is that reserves constitute less than 4% of the Earth's land surface (World Resources Institute, 1996). This emphasizes that much of nature is now within a stone's throw of humans, and the degree of anthropogenic modification varies from very little to very much. This spectrum has various degrees of ecological integrity. As such, a major goal of conservation is to conserve as much as possible of this remaining integrity, with due respect to the role of critical processes in maintaining that integrity (Hunter, 2000a). Indeed, even wild places are only likely to survive in the long run if recognized as wildland gardens that continue to be used with minimum of damage (Janzen, 1999).

Rolston (1994) has illustrated that there are various types or levels of values: natural and cultural, diversity and complexity, ecosystem integrity and health, wildlife, anthropocentric and natural intrinsic. All enter the essence of conservation biology, and all impinge on insect diversity conservation. It is this diversity of values, when maintained, that enrich the world, not just for us, but also for all the other organisms and all the processes that make this, as far as we know, a unique planet.

1.3 Insect utility

Although in practice insects are rarely harvested in the way of many other organisms, the principle of utility value still applies to them. The most significant feature of insects in terms of this utilitarian philosophy is to ensure continuance of their ecological services, so that ecological integrity and health are maintained (see Chapter 3). This is where we largely do not understand the consequences of our actions. To name one example, landscape fragmentation and attrition of landscape patches influence the insect assemblages such that the services they normally supply may no longer be possible (see Chapters 4 and 5).

In the agricultural context, it is not always possible to maintain ecological integrity, even though specific insects are being conserved and human intentions are good. Natural ecosystems adjacent to agricultural fields are often utilized for pools of natural enemies that invade the crop and control pests. On harvesting of the crop, the natural enemies then flood back to the surrounding natural ecosystem where they exert strong, albeit local, impact on natural hosts. This is a manifestation of the human demand for harnessing the interaction between host and parasitoid or predator. Biological control is one of the most sought-after services of insects, and one which is not without risks to ecological integrity (Figure 1.4).

Ecological services from insects include more than predation and parasitism. Technical details of these are addressed in Chapter 3. Among these is pollination of crops, both by wild insects and by captive honeybees. Encouragement of these pollinating insects can hardly be in excess, as the same insects can play a major role in maintaining indigenous plants, and hence in their conservation (Kwak *et al.*, 1996).

One area where insect overexploitation requires caution is in the case of colourful butterflies. Regulations need to be called into play, with many species on the *IUCN Red List of Threatened Species* (Hilton-Taylor, 2000). Local laws also

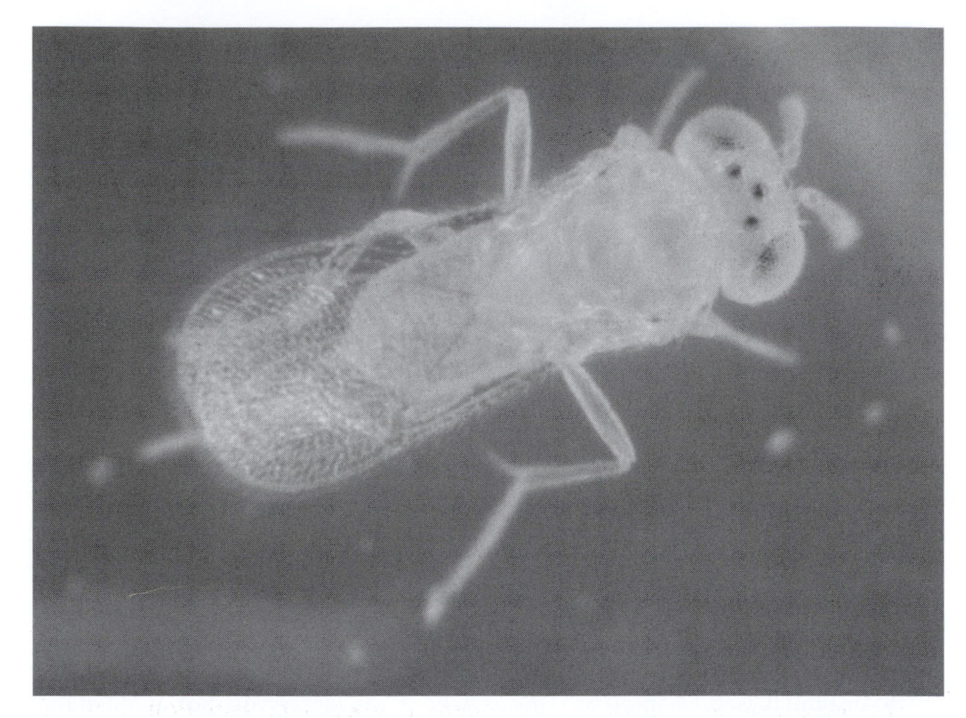

1.4 The parasitoid *Aphytis melinus* an important biological control agent against scale insects (Diaspididae). Biological control is one of the most sought after services from insects, but does carry risks (see Chapter 6). Although introduced specifically against Red scale (*Aonidiella aurantii*) in South Africa, *A. melinus* is now known to parasitize at least ten other species.

play a significant part, as does blanket protection of wild areas containing the habitats of these sought-after species. Insect farming can take pressure off the wild populations by providing reared specimens that are often in visibly better condition than wild-caught specimens.

Perhaps the utilitarian aspects of insects have been underexploited. While we are likely to see only limited progress in the direct harvesting of insects (simply because they are generally unpredictable, small and difficult to harvest) there may be some future for medical and novel silk products.

However, the heuristic value of insects in genetic research is undeniable, with *Drosophila melanogaster* virtually a household name. The future is likely to see particular insect genes, rather than the whole animal, having utilitarian value in many aspects of our lives.

Caring for the Earth (IUCN/UNEP/WWF, 1991), which is a world conservation strategy, implicitly addresses many facets of insect diversity conservation that underpin the well-being of humans. Insects and their activities are vital for conserving our life-support systems and for renewing our resources through services in addition to pollination, such as soil maintenance, population regulation, and in the food webs of terrestrial invertebrates. The value of insect utility therefore, is about conservation of insect diversity. This is part of the wider concept of sustainable use of the world's resources which involves a challenge, among others, which will require learning how to recognize and resolve divergent problems, which is to say a higher level of spiritual awareness (Orr, 2002).

1.4 Insect rights and species conservation

Individuals have rights so as to maintain and even improve their lives, and then they die. So then, do insects have rights? As this is such a difficult question to answer, it is best to put it in a converse sense. Do we have the right to assume that insects do *not* have rights? Bearing in mind the weight of their collective individuality, best we adopt the precautionary principle of keeping all the parts. This includes the moral option that, in fact, individual insects do have rights. From this standpoint, Lockwood (1987) proposes a minimum ethic: 'We ought to refrain from actions which may be unreasonably expected to kill or cause non-trivial pain in insects when avoiding these actions has no, or only trivial, costs to our own welfare.'

This may also be seen as a complement to species preservation, which is accentuated with increasing rarity of a species. The genotype and phenotype are naturally locked into a symbiosis. With great rarity and genetic bottlenecks, loss of individuals has the added responsibility of increasing risk to the species per se and its evolutionary potential.

It does not mean however, that the individual of an endangered species has any special rights over the individual of a widespread species. Both have nervous systems that demonstrate post-inhibitory rebound, both launch pheromones to attract a mate, and both will avoid a predator if they can. Both too, have an evolutionary offering to the future. Best we let individuals live (Figure 1.5).

But what of species? Is the fact that 99% of all species that have existed on earth are now extinct (Novacek and Wheeler, 1992) really a consideration? Perhaps not when we consider that many clades have evolutionarily advanced and that there is now more insect diversity on earth than ever before (Labandeira and Sepkoski, 1993). Yet no two species, or for that matter, subspecies or morphs (evolutionarily significant units, ESUs), are alike, and so all are special. This applies as much to the parasitoid as to the caterpillar host. Indeed, as polymorphisms are so rife in the insect world, it is essential to consider the caterpillar as well as its developmental polymorph, the butterfly. Reality has it, that we *have* to conserve both developmental morphs to preserve the species. The parasitoid and the butterfly polymorphism also remind us of the connectedness of all life, and that species value, in turn, is linked to the value of ecosystems and landscapes, the conservation of which are essential for maintaining all insect variety and interactions.

1.5 'The good, the bad and the ugly' – but there is not much between these three insects in terms of general biochemistry and physiology. All have a similar nervous system, but our perceptions of them vary considerably. (a) Basking malachite (*Chlorolestes apricans*), an attractive South African damselfly on the verge of extinction, (b) Citrus wax scale (*Ceroplastes brevicauda*), and other scale insect pests on a citrus twig that has been oversprayed with an insecticide (parathion), and (c) (see next page) the Leprous grasshopper (*Phymateus leprosus*), an African species that gives out an unpleasant protective foam when disturbed.

1.5 (*cont.*)

1.5 Spiritual conceptions

Jacobson (1990), with an educational perspective, has illustrated that conservation biology is an interdiscliplinary science and activity. It involves the basic biological sciences as well as the applied management sciences, such as wildlife management, forestry, range and forage management and fisheries. Impinging on conservation biology, besides the physical environment, is the implementational environment (planning, education, law, communication, public health, engineering and veterinary science) and the social environment (economics, political science, sociology, anthropology and philosophy). What is missing from this model, at least in explicit terms, is the role of spiritual beliefs. Yet conservation concerns everyone, as well as every organism. A conviction through spiritual involvement can play a major role in sustaining conservation action over and above the activities of scientists, managers and policy workers. This is particularly relevant in the case of insects, which are among 'the world's many creatures' and do not have the charisma of the large animals with which the western media are so absorbed.

Spiritual outlooks were brought to the fore in a major interfaith conference in Assisi, Italy in 1986. Various faiths made 'Declarations on Nature', sections of which directly relate to insect diversity as well as to other aspects of biodiversity, and are (Anonymous, 1986/7), in alphabetical order:

> **Buddhist** In the words of the Buddha himself: 'Because the cause was there, the consequences followed; because the cause is there, the effects will follow.' 'These few words present the inter-relationship

between cause (karma), and its effects . . . happiness and suffering do not simply come about by chance or irrelevant causes.' '. . . it (Buddhism) . . . attaches great importance towards wildlife and the protection of the environment on which every being in this world depends on survival . . .'

Christian '. . . man's dominion cannot be understood as licence to abuse, spoil, squander or destroy what God has made manifest his glory. That dominion cannot be anything else than a stewardship in symbiosis with all creatures . . .' 'Every human act of irresponsibility towards creatures is an abomination.'

Hindu 'Hinduism believes in the all encompassing sovereignty of the divine, manifesting itself in a graded scale of evolution. The human race, though at the top of the evolutionary pyramid at present, is not seen as something apart from the earth.' 'This leads . . . to a reverence for animal life. The Yajurveda lays down that "no person should kill animals helpful to all. Rather, by serving them, one should attain happiness."'

Jewish 'In the Kabbalistic teaching, as Adam named all of God's creatures, he helped define their essence. Adam swore to live in harmony with those whom he had named. Thus, at the very beginning of time, man accepted responsibility, before God, for all of creation.' '. . . when the whole world is in peril, when the environment is in danger of being poisoned, and various species, both plant and animal, are becoming extinct, it is our Jewish responsibility to put the defence of the whole of nature at the very centre of concern.'

Muslim 'Allah makes the waters flow upon the earth, upholds the heavens, makes the rainfall and keeps the boundaries between day and night.' 'Unity, trusteeship and accountability . . . the three central concepts of Islam, are also the pillars of the environmental ethics of Islam . . . It is these values which led Mohamed . . . to say, "Whoever plants a tree and diligently looks after it until it matures and bears fruit is rewarded."'

Throughout these declarations there is the common denominator that all in the world, including humans, are connected, and that protection of biodiversity and the environment is essential for a sustainable future. There is greater or lesser specific mention of organisms, although their role is implicit in the debate on interconnectedness. Such spiritual bases are now a fundamental underpinning for some major conservation donor agencies, such as the World Bank (Palmer and Finlay, 2003).

Like the philosophical approach to conservation biology, the religious one is also based on the writings of the intellectual forerunners. There is, of course, no guarantee that all followers will be strong adherents of a particular philosophy or religion. While the conservation of biodiversity needs positive philosophical, spiritual and active participation by all humans, this simply is not always going to happen. As Garner (2003) puts it 'Religion is not part of the problem; people are the problem'.

Nevertheless, philosophy gives guidance and draws attention to why we are doing what we are, and for whom. Religion then provides the spiritual underpinning. In turn, research explores the technical way forward, which is framed by policy makers, and implemented by managers.

1.6 Summary

Conservation activities require a philosophical base so as to reflect on why and for whom these activities are being undertaken. Insects, as they are so speciose, so numerous and so important in terrestrial ecosystems, are an important subject for environmental philosophy. A notable corollary however, is that not all insects are good for each other or for us.

There are various philosophical approaches, and among these are the utilitarian approaches of, on the one hand, the preservationist ethic (insects are there for us to enjoy) and the resource conservation ethic (insects provide useful goods and services for us). These philosophies set us apart from the rest of nature, and this has stimulated philosophies where humans and wild nature are considered together. One approach, of deep ecology, considers a total view and that all in the world is interconnected. More recent philosophies have emphasized that all organisms, including insects, have the right to survival. These philosophies, which portray omnipotence in nature, are being addressed with the added view that joy for nature and a sense of place are important. Furthermore, it is important to value nature at all hierarchical levels. It can be argued that insect individuals do have rights, but this is linked through the genotype-phenotype symbiosis, to species conservation. Polymorphisms, which are so rich in the insect world, are an additional consideration in this debate.

Declarations from some of the world's major religious faiths have the common denominator that all in the world, including humans, is connected. It is vital that natural ecological processes, of which insects are pivotal, must be sustained. Insect diversity conservation needs a philosophical and moral base so as to give reason to why it is being done. Religion spiritually underpins this, while research investigates the technical avenues available. Policy makers then provide the frame for these avenues, and managers implement them.

2 The special case of insects in conservation biology

Thus, as this class (insects) is prolific beyond computation, so are its varieties multiplied beyond the power of description. The attempt to enumerate all species of a moth would be fruitless; but to give a history of all would be utterly impractible: so various are the appetites, the manners, and the lives of this humble class of beings that every species requires its distinct history. An exact plan, therefore, of Nature's operations is this minute set of creatures, is not to be expected; and yet such a general picture may be given, as is sufficient to show the protection which 'Providence affords its smallest as well as largest productions . . .'

Oliver Goldsmith (1866).

2.1 Introduction

Insects have been hugely successful. There are possibly eight million species making up some four-fifths of all metazoans (Figure 2.1). The insect bauplan (their general design) has been a mouldable one, with flight being a hallmark.

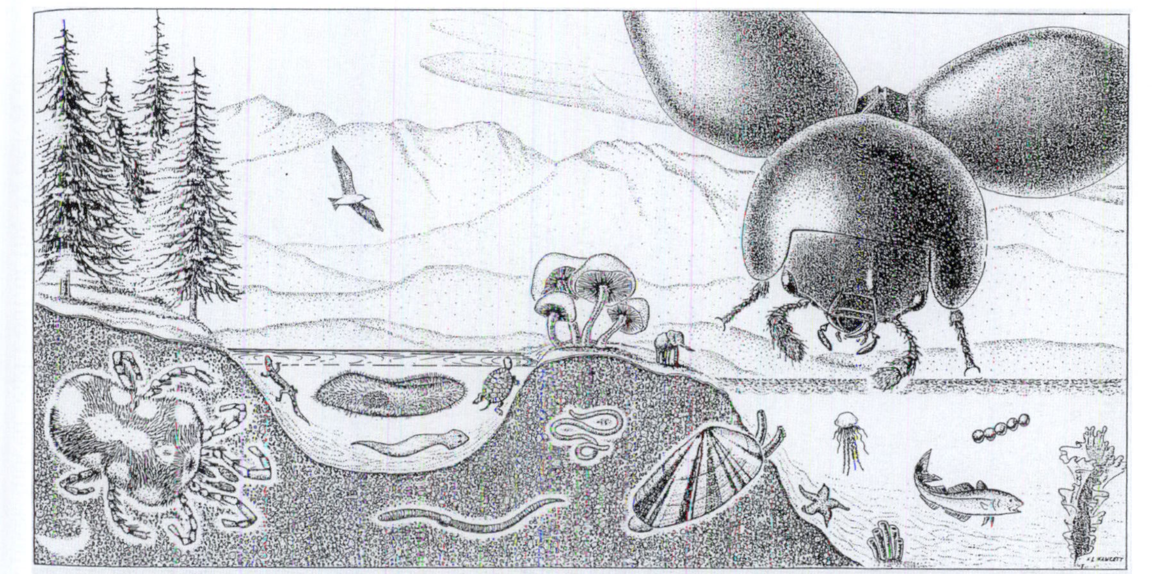

Size of individual organisms represents number of described species in major taxon.
Unit Area: ☐ = approximately 1000 described species.

Taxon	No. of Described Species	Taxon	No. of Described Species
1 Monera (Bactoria, Blue-green Algae)	4,760	11 Mollusca (Mollusks)	50,000
2 Fungi	46,983	12 Echinodermata (Starfish etc.)	6,100
3 Algae	26,900	13 Insecta	751,000
4 Planlae (Multicellular Plants)	248,428	14 Non-insect Arthropoda (Mites, Spiders, Crustaceans etc.)	123,161
5 Protozoa	30,800	15 Pisces (Fish)	19,056
6 Poritera (Sponges)	5,000	16 Amphibia (Amphibians)	4,184
7 Coelenterata (Jellyfish, Corals, Comb Jellies)	9,000	17 Reptilia (Reptiles)	6,300
8 Platyhelminthos (Flatworms)	12,200	18 Aves (Birds)	9,040
9 Nematoda (Roundworms)	12,000	19 Mammalia (mammals)	4,000
10 Annelida (Earthworms etc.)	12,000		

Illustration by Frances L. Fawcett. From O,D, Wheeler. 1990. Ann. Entomol. Soc. Am. 83:1031-1047.

2.1 The ‘species-scape’ of Wheeler (1990) in which the size of the organism represents the size of the described species in the overall taxonomic group (principally at class level). The beetle, for example, represents the insects; the elephant the mammals; the mite the non-insect arthropods; the pine trees the multicellular plants, etc. As so many insects are yet to be described, compared with many of the other groups, in real terms the beetle in the illustration should be much larger (from Wheeler, 1990).

Nevertheless, insects are structurally limited in the size to which they can grow. Thus insects generally remain small, yet often highly mobile. It is the tensile strength of the cuticle that has been the raw attribute for evolutionary sculpturing of a wide array of morphological modifications so vivid in the insect world. To take just the mouthparts: they can pierce, sponge, rasp and chew. So it is that insects have been able to inhabit a wide range of nooks, crevices and tunnels throughout all but the very coldest and most marine parts of the planet. Flight between these retreats enables searching for distant resources, dispersal to more salubrious settings and dispersal of gametes.

Insect genetic versatility has also been notable. Polymorphisms of all sorts are common among insects. This includes developmental polymorphism, where, from functional and conservation viewpoints, the caterpillar is a different animal from the butterfly. So both forms and both their habitats, must be equally conserved.

Although insects have been an immensely successful life-form that has taken so many diverse evolutionary paths, for many, their world is suddenly changing so much faster than possibly at any time before (with the exception of major meteor impacts). This is threatening for many of them, and they need our support and salvation, without which many will perish.

Let us now turn to these jewels of our planet, the insects, and review conservation of their diversity and how it might differ from that of other organisms.

2.2 Insect radiation

As the prow of the prirogue slices the water on its way to Nosy By, Madagascar, fast-moving sea skaters (*Halobates* sp.) skit across the surface. These are among the few insect mariners. The majority of insects are terrestrial, and/or aquatic. In both air and freshwater, they are among the most speciose and dominant of all organisms. This may not have always been the case, with insect families having steadily increased over the last 400 million years (Labandeira and Sepkoski, 1993) (Figure 2.2). Sometime during the Early Carboniferous, more than 325 million years ago, a massive insect radiation began, followed by peaks in fossil insect diversity during the Late Carboniferous and middle Permian. Labandeira and Sepkoski (1993) suspect a lower Triassic drop in insect diversity, reflecting the terminal Permian mass extinction, which was also the case with marine animals and terrestrial tetrapods. Indeed, eight out of the 27 orders of insects, making up 15 families, which were alive earlier, did not survive beyond the Permian.

About half of the orders that survived into the Recent underwent few to many diversifications through the Triassic and Jurassic. This shift in phylogenetic pattern of diversification is the most pronounced event during the history of

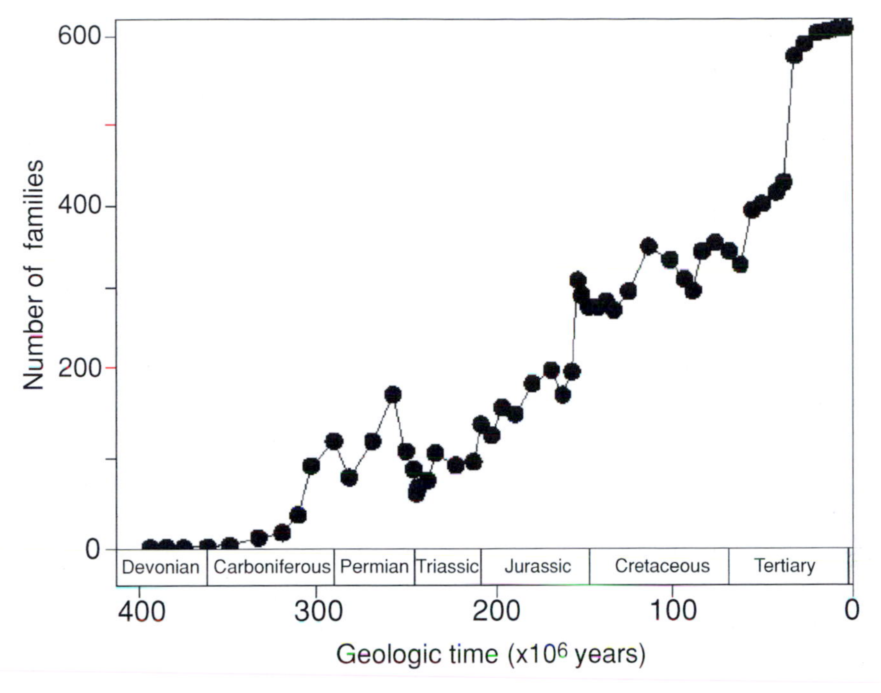

2.2 Family-level diversity of fossil insects through geologic time, plotted at the level of stratigraphic stages. D = Devonian, C = Carboniferous, P = Permian, Tr = Triassic, J = Jurassic, K = Cretaceous, T = Tertiary. (Reprinted with permission from Labandeira and Sepkoski (1993). Copyright 1993 AAAS.)

insects, and is the backdrop for the current huge variety that we see today (Figure 2.3).

At the family level, insects have shown little turnover throughout much of their recent history, with the Tertiary fauna being essentially the same as that today, and even the insect families of 100 million years ago were 84% the same. Labandeira and Sepkoski (1993) have cautioned that this analysis is at the family level, with insect families often consisting of many smaller taxa, the turnover of which, can carry the family through geologic time. Nevertheless, certain beetle (*Tetraphalerus*), crane fly (*Helius*) and leaf-mining moth (*Stigmella*) species appear to have existed over tens of millions of years.

Labandeira and Sepkoski (1993) then ask the question of whether the huge insect diversity seen today is attributable to the diversification of angiosperm plants (Strong *et al.*, 1984). However, using extrapolation based on constant rates of diversification, there is no necessity to invoke a huge angiosperm diversification as the specific reason for current insect familial diversity. The extant 980 or so insect families possibly represent saturation. Also, the radiation of insect families began more than 100 million years before angiosperms appeared in the

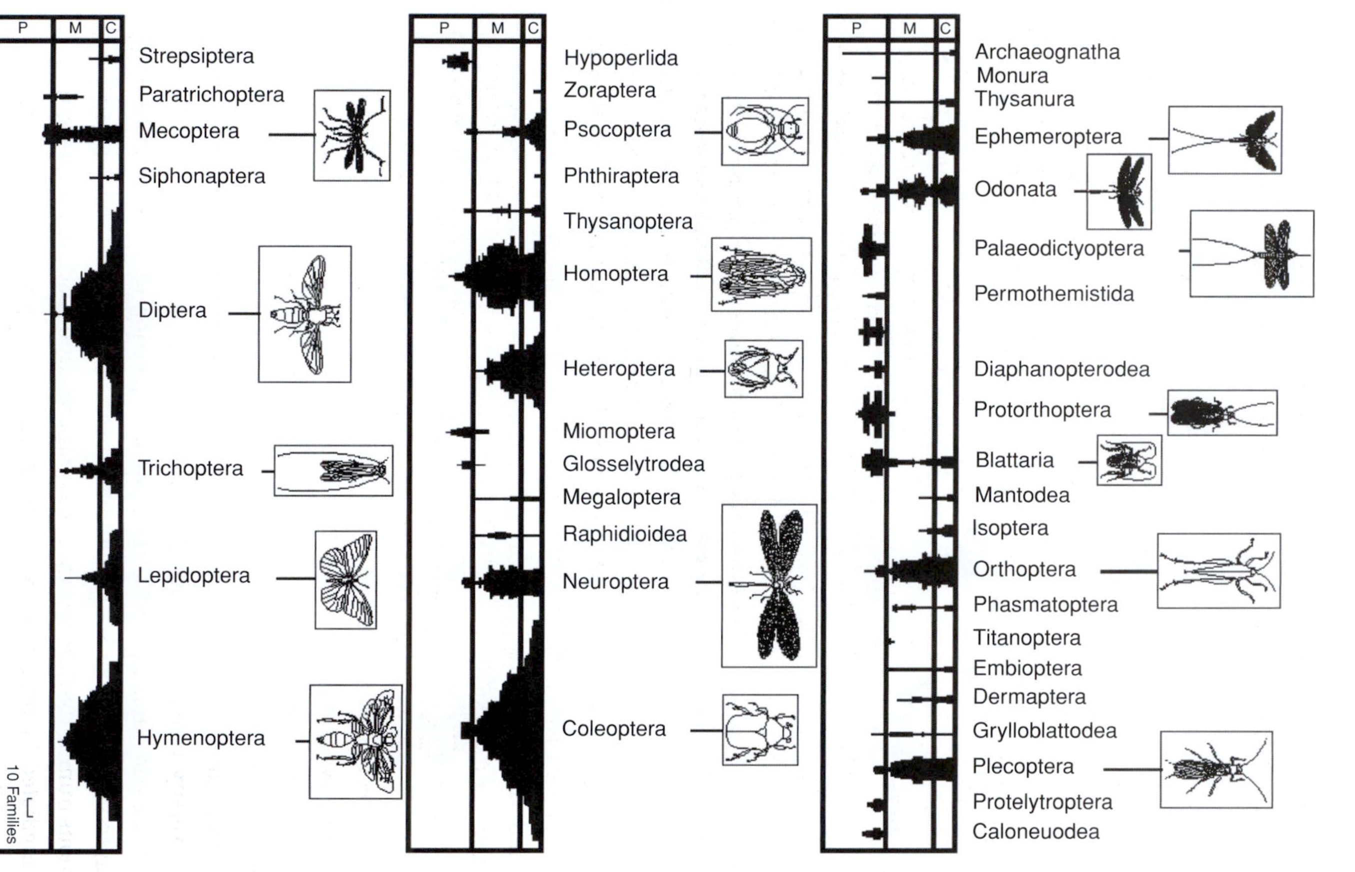

2.3 Spindle diagrams displaying diversities of fossil families within insect orders in stratigraphic stages of the Phanerozoic. P = Paleozoic, M = Mesozoic, C = Cenozoic. Boxed illustrations (not to scale) depict typical adult representatives of the more important orders. Angiosperm plants make their fossil appearance about two-thirds of the way up the band for the Mezozoic (that is, just above the 'M' in Mz). (Reprinted with permission from Labandeira and Sepkoski (1993). Copyright 1993 AAAS.)

fossil record. This is especially so for the highly speciose Coleoptera, Diptera and Hymenoptera, all of which apparently began their radiations in the Triassic and Jurassic, long before the ascendancy of the angiosperms.

A burning question however, is whether angiosperms have had a major effect on insect diversity below the level of family. This is a compelling point bearing in mind the close interrelationships between certain insect species and particular plants (Jolivet, 1998), which has become extremely fine tuned in the case of Yuccas and Yucca moths (Powell, 1992). An alternative explanation is that it was insects that triggered the radiation in plants in the mid-Cretaceous (Pellmyr, 1992). Only 10 out of the 30 extant orders of insects exploit the living tissues of higher plants. This relatively small proportion is surprising given the ready availability of green plant material. Life on plants therefore seems to have been a formidable evolutionary hurdle (Southwood, 1973), although once the hurdle was cleared, great opportunities existed and dramatic insect radiation occurred. The complexity of plant architecture and composition opened up opportunities for a range of insect exploiters. In turn, plant diversity increases in response to the pressures imposed by herbivores, especially insects (Strong *et al.*, 1984). Yet the flowers, in contrast to the vegetative parts, often provide a mutually beneficial interrelationship (Ehrlich and Raven, 1964), making plant diversification a process of adaptation to escape herbivorous insects and yet include pollinating ones. For many of the large holometabolous groups, especially the Lepidoptera, Diptera, Coleoptera and Hymenoptera, the developmental polymorphism of larva and adult often presents these two contrasting faces to plants.

In a comprehensive study of insect mouthparts, Labandeira and Sepkoski (1993) suggest that a number of insect feeding guilds diversified well before the appearance of angiosperms. By the Middle Jurassic, 65% (low estimate) to 88% (high estimate) of all modern insect mouthpart types were present, including those normally associated with flowering plants. This suggests, as does the radiation of insect families, that the special associations of specific insects with particular plant types and parts is a relatively modern phenomenon. It is a question of the taxonomic scale at which we are working as to whether insects and plants are to be considered as mutually evolutionary.

This effect of taxonomic scale has a major influence on our thinking relative to insect diversity conservation. Firstly, conservation prioritization and action based on higher-level taxonomic surrogates could be misleading as a means of conserving the huge species richness and all their subtle interactions. Secondly, the enormous number of insect–plant interrelations, as well as many other interrelations mediated by plants, is an evolutionarily finely tuned process, and, as Labandeira and Sepkoski (1993) point out, exploitative destruction of angiosperm communities could well reverse 245 million years of evolutionary success. This is underscored by the fact that it was the most specialized associations that were diverse and abundant during the latest Cretaceous and were the ones that

2.4 A giant dragonfly (*Meganeura* sp.) of the Carboniferous, with a modern-day sparrow for size comparison. (Illustration by courtesy of Valter Fogato.)

almost disappeared at the Cretaceous/Palaeogene boundary, while generalized associations regained their Cretaceous abundances (Labandeira *et al.*, 2002).

2.3 Bauplan, flight and insect conservation

The general design of insects has a profound influence on how we think about them in terms of their conservation. Flight is unquestionably one of their major features to consider. Indeed, giant dragonflies, such as *Meganeura monyi*, with a wingspan of 70 cm, were airborne some 300 million years ago (Figure 2.4). With airspace to themselves, and possibly elevated oxygen levels at the time (Dudley, 1998) (enabling highly efficient muscle action), these insects were the supreme flying predators. With pterodactyls and birds having not yet arisen, the Carboniferous airspace was dominated by flying insects.

Flight probably arose through a step by step improvement in wing form and action, as early stonefly-like insects continued to skim across the water's surface (Marden and Kramer, 1994). Insect diversification rate increased substantially with the origin of the Neoptera (insects with wing flexion) (Mayhew, 2002). Nevertheless, today, not all winged insects fly well. The threatened Apollo butterfly *Parnassius apollo*, although potentially a good flier (< 1840 m) is still constrained in its movement over host plant and nectar patches by segregation of adult and larval resources (Brommer and Fred, 1999). Nevertheless, flight enables some aquatic adult insects to reinhabit an optimal upstream habitat after the larvae have been swept downstream by unusually strong currents (Samways 1989a, 1996a).

Flight, besides being a mechanism to escape predators, is a means of exploiting food, shelter, mating and oviposition resources. Indeed, at any one time there are billions of insects on the wing, even at night and travelling over long distances (Riley *et al.*, 1995). Some of these movements are directional, and are migrations honed by natural selection (e.g. Monarch butterfly *Danaus plexippus* (Brower, 1977)). But many insect movements are short distance, but not necessarily random. The ladybird *Chilocorus nigritus* forages at the spatial level of habitat, prey patch and individual prey (Hattingh and Samways, 1995). Tree images attract the ladybirds first, then leaf shape; in turn, prey odour focuses their search on prey groups; then, a combination of visual and olfactory cues are used to detect individual prey.

Many other insects are at the mercy of wind currents, and they may even land on snowfields (Ashmole *et al.*, 1983), and colonize recently cooled lava flows (New and Thornton, 1988; Thornton, 1996). Some even travel on flotsam, arriving serendipitously on islands (Peck, 1994a,b). Small insects are known to regularly travel high up on wind currents (Berry and Taylor, 1968). Even first-instar moth larvae with silken threads can rise to 800 m on air turbulence and travel 19 km on wind currents (Taylor and Reling, 1986). On Anak Krakatau, Indonesia, around 20 arthropod individuals land per square metre per day (Thornton *et al.*, 1988), which equates to about 50 million individuals arriving each day on that volcanic speck.

2.4 Polymorphisms

Insects are well known for their great functional and morphological variation. This is the case within particular species, as well as between them. Indeed, ecomorphs are common in the insect world. Such morphs, or evolutionarily significant units (ESUs) (Ryder, 1986; Moritz, 1994; Vogler and De Salle, 1994) not only may have differing evolutionary potentials but also may have quite different conservation statuses. This is highlighted in the case of the Gypsy moth *Lymantria dispar*, which was extirpated from Britain around 1907 (Hambler and Speight, 1996). Yet, in June 1995, Britain saw a pest outbreak of an Asian variety of this species (Nettleton, 1996). Genetically, there must be something quite disparate here, which, by some extraordinary coincidence, makes the name *dispar* very appropriate. Moths are known to have ESUs (Legg *et al.*, 1996), and clearly, through subtle change in genetics, such an insect *species* can be extinct on the one hand yet be a pest on the other. This is why it is critical to recognize ESUs in insect diversity conservation as one of the crucial units of conservation, as a tiny genetic difference can lead to quite different ecomorphs (Samways, 1997b). This calls also for regular revision of insect species' threat category and the

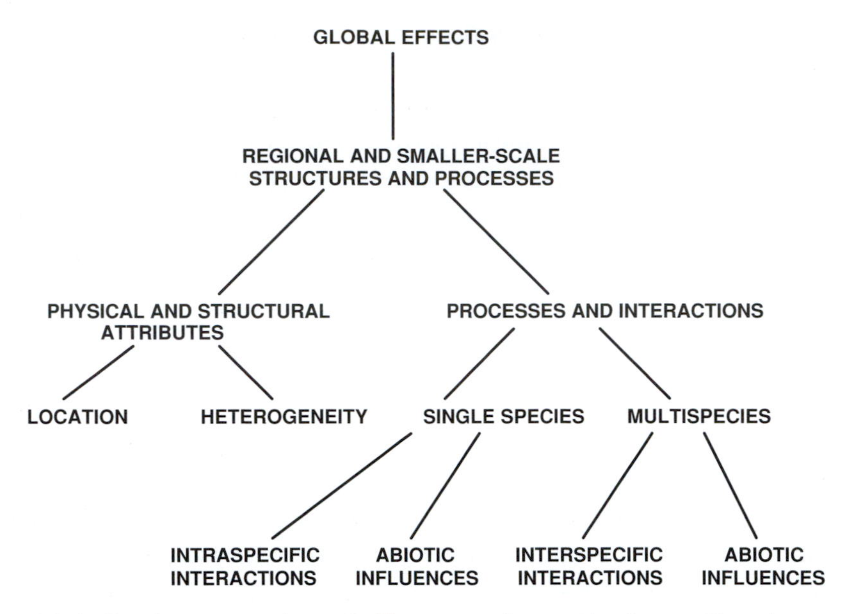

2.5 Global effects have a general overall effect upon regions and landscapes. These, in turn, may be divided into tangible, physical attributes such as geometry and architecture, and also into processes and interactions. Physical attributes can then be divided into location (where the physical structure is located) and heterogeneity (how the physical structures spatially change). Processes and interactions can then be viewed, and this is not the only way, under single-species situations ('fine-filter'), or, alternatively, under multispecies ones ('coarse-filter'). Further subdividing separates interactions between organisms on the one hand, and between organisms and their abiotic environment on the other. (From Samways, 1993a with kind permission of Intercept.)

clear need to emphasize what is regionally or nationally threatened as opposed to globally threatened. Hambler and Speight (1996) point out that the British ambrosia beetle *Platypus cylindrus* is classified as 'Rare' in the British *Red Data Book* yet it is a serious pest in parts of continental Europe and has become a pest of oaks even in Britain. Clearly, in insect conservation biology, we must recognize a dynamic and sometimes volatile genetic situation.

Other ecomorphs are those with alternative life-history attributes, such as apterous and alate aphids. Sexual morphism is also highly marked in insects, such as many scale insects and mealybugs. In addition to these are many other types of morphisms, some of which are listed in Samways (1993a). But above all, the most significant polymorphism feature that insect conservation biologists have to deal with, is developmental polymorphism. This is especially so for many species, where the larva is functionally a different animal from the adult. Any form of habitat conservation or management must consider survival of both these forms. Although this may seem intuitively obvious, there are few studies that really focus on restriction imposed on population survival through conditions having to be suitable for both larva and adult (including two types of feeding, predator avoidance etc.) in the same place at more or less the same time.

A further factor is that the immature and mature morphs interact mostly in different food webs, with caterpillars being significant herbivores and butterflies significant pollinators. Also, the caterpillar's and butterfly's natural enemies are mostly different. Any one caterpillar may have five species of parasitoids that are able to attack it, making up a food web that functionally may only overlap with that of the butterfly through bird predation.

2.5 Insect diversity and the landscape

Knowing that insects are major components of most terrestrial and aquatic ecosystems, does not help us much in focusing on conservation priorities. It does, however, indicate that conservation of the Earth's landscape heterogeneity and, indeed, conservation of the whole biosphere is a foundation for insect diversity conservation. All the while we need to be acutely aware of all spatial scales, and of compositional (how things are made up), structural (how things are arranged in space) and functional (the processes involved) aspects of biodiversity (Noss, 1990) (Figure 2.5).

As insects are small, often highly polymorphic organisms with hugely varying ways of life, we need to take cognisance of the 'sense of place' and be aware of these spatial scales, the subregional ones of which are given in Table 2.1.

Within a region, landscapes have attributes that are compositional and structural, as well as involving interactions and processes (Figure 2.5). Spatial scale is important in insect conservation biology because different behaviours of different morphs and different species each require specific scales to carry out their life functions (Chust *et al.*, 2003). But within a landscape are great complexities of composition, structure and processes. This has been emphasized by Kruess (2003), who showed that biological diversity and ecological functions within a plant–insect community are not only affected by local habitat factors but also by large-scale landscape characteristics. Furthermore, insect herbivores suffered most from parasitism in landscapes that were structurally rich and with a high proportion of large and undisturbed habitats (Figure 2.6).

So vast are these landscape complexities that we have to view the higher spatial scale (i.e. landscape level) as sufficiently large to accommodate most insects' activities yet small enough for practical management. This may be considered the 'black box' approach i.e. a whole range of composition and structural units plus a whole range of interaction types and strengths, as well as ecological processes (Figure 2.7), all of which have dynamics often with unpredictable outcomes. This is so through the principle of chaos (Capra, 1996), which leads to various and varying outcomes of huge complexity. For insects, and much other biodiversity, this 'coarse filter' landscape approach is a critical approach

Table 2.1. *Subregional framework for insect and biodiversity conservation research and management*

	Physical attributes		Biotic groupings, interactions and processes	
Scale and units of measurement	Location	Heterogeneity	Single-species (principally formational and involving size changes)	Multispecies (principally processes)
mm to cm (micro)	Microsite	Microscape	Intraspecific group, and their interactions/microhabitat	Intraspecific and intimate interspecific interactions/Guild/Community
cm to m (meso)	Biotope	Plantscape (Animalscape)	Subpopulation/Habitat	Community
m to km (meso)	(a) Landscape element (for fragmented landscape)	Heteroscape	Subpopulation/Population/ Metapopulation/Populations	Community/Communities/Ecosystem/ Ecosystems
	(b) Ecotone (for variegated landscape)	Multiscape	Population/Metapopulation/ Populations	Ecosystem/Ecosystems
km (macro)	Landscape	Landscape heterogeneity	Population/Metapopulation/ Populations	Ecosystems

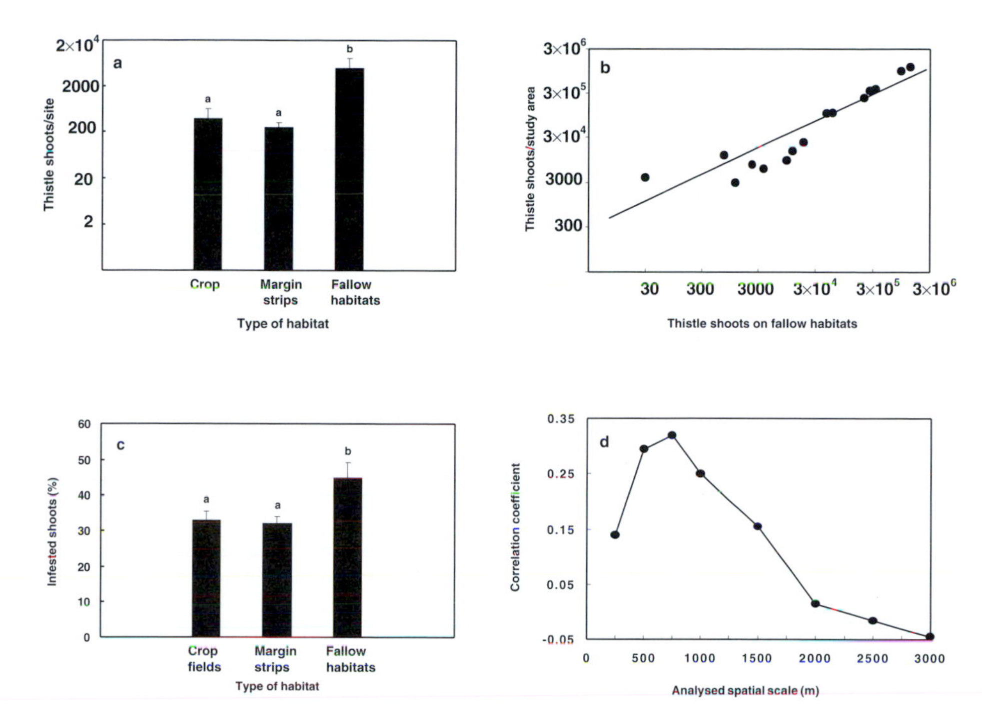

2.6 Effects of local habitat and large-scale landscape factors on species diversity and species interactions illustrated using Creeping thistle (*Cirsium arvense*). Above: Mean thistle abundance in fallow habitats was significantly higher than in crop fields and field margin strips (left) (a). Thus, total thistle abundance on landscape scale was mostly determined by thistle abundance in fallow habitats (right) (b). Below: However, the infestation rate of thistle shoots caused by the herbivorous agromyzid fly, *Melanagromyza geneoventris*, was significantly higher in fallow habitats than the other two habitats (left) (c). However, at the landscape scale, the infestation rate increased from 30% in areas with a low percentage of non-crop habitats to > 40% in areas with a high percentage of non-crop habitats. Moreover, this effect was scale-dependent because significant correlations were found for per cent non-crop area on a landscape scale of 500 and 750 m, whereas the correlation declines with further increase of the study areas (right) (c). (Redrawn from Kruess, 2003, with kind permission of Blackwell Publishing.)

for biodiversity conservation. This is not to decry the role of behaviour and the need for special, single-species, 'fine-filter' studies in special circumstances, especially when compiling the Red List (Hilton-Taylor, 2000). Indeed, the coarse-filter and fine-filter approaches are complementary. The disadvantage of the 'coarse-filter' approach is that it is blind to actually what and in which way the immensely complex contents are being conserved (Haslett, 2001). Nevertheless, conserving black boxes (i.e. whole landscapes with high connectivity, high ecological integrity and minimal human disturbance) is one answer in view of the magnitude of the biodiversity crisis and the shortage of time to conserve it. But it also requires critical consideration, which is the subject of Chapter 8.

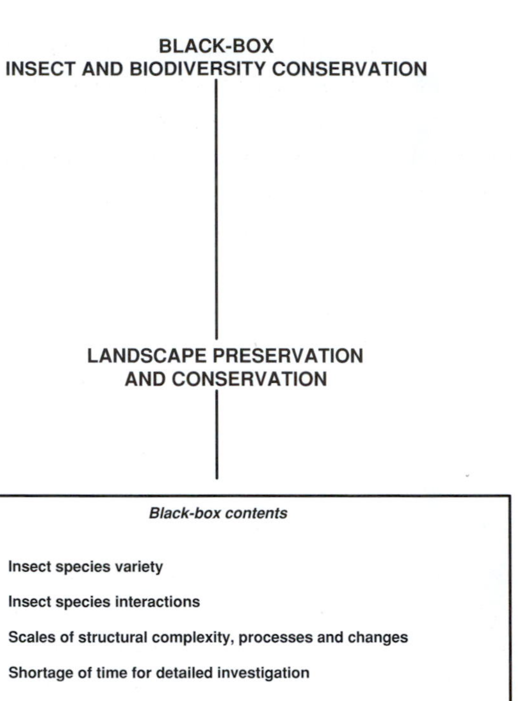

2.7 Black-box insect and biodiversity conservation. There are millions of insect species, of varied and varying abundances, and there are many times that number of interactions between insects and other biota. These interactions, in turn, are taking place at various spatial scales from centimeters to kilometers within the landscape. Within the time available for preservation of much of the world's biota, we cannot hope to define all the conservation parameters. Preservation (no management) and conservation (with management) of landscapes that are both unique or irreplaceable, and also typical, of regions are the only realistic approaches for maximal preservation of insect and other biota. Various approaches are being developed for doing this, and they are discussed in Chapter 8. (From Samways, 1993a, with kind permission of Intercept.)

2.6 Global insect species richness

The question of how many species there are on earth has fuelled some intense debates (Erwin 1982, 1988; May, 1990; Stork 1988, 1999; Gaston 1991; Hodkinson and Casson, 1991; Hammond, 1992; Ødegaard *et al.*, 2000). As the lexicographer Samuel Johnson pointed out we can argue only once we have defined what it is we are arguing. Here of course, it depends on what we define as species (Adis, 1990; May, 1990). The phylogenetic species concept of Nixon and Wheeler (1992), which is the smallest aggregation of populations (sexual) or linear (asexual) diagnosable by a unique combination of character states in comparable individuals (semaphoronts), is a useful starting point in insect diversity conservation. Nevertheless, there still are subspecies, evolutionarily significant units (ESUs) and polymorphisms, all of which also deserve conservation status. Species concepts based on implications of ancestry, while valuable in certain cases and for specific conservation questions, especially irreplaceability, are limiting for much practical insect diversity conservation because of the taxonomic insight and technical sophistication required. Invoking genetic concepts also poses problems in view of the inferential findings of Houck *et al.* (1991) that the semiparasitic mite *Proctolaelaps regalis* can transfer genes from one *Drosophila* species to another.

So any estimate of species numbers will be crude, not just because we are not quite sure when one species ends and another starts (Mallet, 1997), but also because our taxonomic knowledge of the insect world, and indeed invertebrates in general, is very sparse. Nevertheless, it is a healthy exercise for knowing the magnitude of what we aim to conserve and to estimate just how many species with which we share the world.

Erwin's (1982, 1988) estimates of 30–50 million species on Earth may be overly large (Ødegaard *et al.*, 2000), with the revised average figure being about 10 million (May, 1989) to 12.25 million (Hammond, 1992) species. Of these, insects may constitute 1.84–2.57 million (Hodkinson and Casson, 1991), 2.75–8.75 million (Gaston, 1992) or 5–10 million (Gaston, 1991) species. Bartlett *et al.* (1999) point out that estimates of number of species on Earth depend on understanding beta diversity, the turnover of species composition with distance. This involves also knowing whether species have widespread or localized distributions. Such studies are fraught with methodological problems of bona fide comparisons of geographical sites. Nevertheless, Bartlett *et al.*'s (1999) study suggests wasps may not have such restricted geographical ranges as Erwin's (1988) herbivorous beetles, which would certainly downsize world estimates of species numbers. Similar conclusions were reached by Thomas (1990) using heliconiine butterflies. Future efforts to refine biodiversity and insect diversity estimates should focus on site-region or region-region comparisons rather than on site-site comparisons of species beta diversity (Bartlett *et al.*, 1999). All such comparisons are also illuminating for spatial insect heterogeneity, and hence for prioritizing conservation actions.

2.7 Survival in prehistorical times

2.7.1 *Insect response to pre-agricultural impacts*

With the advance and retreat of the Quaternary (last 2.4 million years) ice sheets, so have insect populations similarly pulsed. There is a clear suggestion that there were not the widespread extinctions among insects as there were among the mammals (Elias, 1994; Coope, 1995; Ashworth, 1996). Furthermore, there is no evidence of widespread evolutionary change during this period (Ashworth, 1996) and constancy of species, and even species assemblages, appears to have been the norm for the last million years or so (Coope, 1995). This is a fascinating array of events that deserves closer scrutiny because it may indicate whether insects are equipped to survive the current, massive human impact.

Shackleton (1977) has shown that during the last three-quarters of a million years the Earth has been glaciated to a much greater extent than it is today, and the current state of affairs is a rare, warm post- or interglacial period. Also, the transitions between glacial and interglacial times were very fast indeed, with change from glacial to interglacial at the end of the Last Glaciation taking only 50 years (Dansgaard *et al.*, 1989).

During the transition from the last glacial to the current interglacial period, the climate was highly unstable. In Britain, about 13 000 years ago, the July temperatures rose from about 10 °C to 17 °C in less than a century (Coope and Brophy, 1972). Just a few centuries after this warming, a chill set in again, with sporadic drops, until 11 000 years ago, when again, glacial conditions prevailed, with summer temperatures no more than 10 °C and winter temperatures about – 15 °C. Then, about 10 000 years ago, there was a rapid thaw occurring over about a century. These climatic changes were global (Ashworth and Hoganson, 1993; Ponel, 1995) and during this time there were also global sea-level variations of more than 100 metres. Insect assemblages in Europe at a single location changed dramatically during these times, driven by vegetation changes in response to climatic ones (Ponel *et al.*, 2003).

Despite these huge climatic variations, some 2000 beetle species are known both from the Quaternary and today, with far less than 1% of the fossil beetles defying identification and possibly extinct. The fossil species, although formerly occurring in Britain, are today extant in more northerly latitudes (Figure 2.8). Coope (1995) points out that these species survived because they were sufficiently mobile, and space was available to receive them. But it must be remembered that there were late Tertiary extinctions, 2.47 million years ago, when the first major climatic oscillations began. This would have been a hurdle, with those species clearing it, capable of surviving the subsequent oscillations. The important point made by Coope (1995) is that the Quaternary survival strategy, at least in northern Europe, was to shift geographical ranges. This was coupled with the splitting and meeting of populations, with concurrent genetic mixing, and lack of speciation. This may not have been the case the world over. Eggleton *et al.* (1994) suggest that for termites in Africa, the particularly high level of Quaternary disturbance in this region has enabled high speciation.

2.7.2 *Early human impacts*

Although the prehistorical changes in plant and insect assemblages were driven primarily by climate, there is some evidence that humans have also played a role in changing the flora and insect fauna in these early times (Ponel

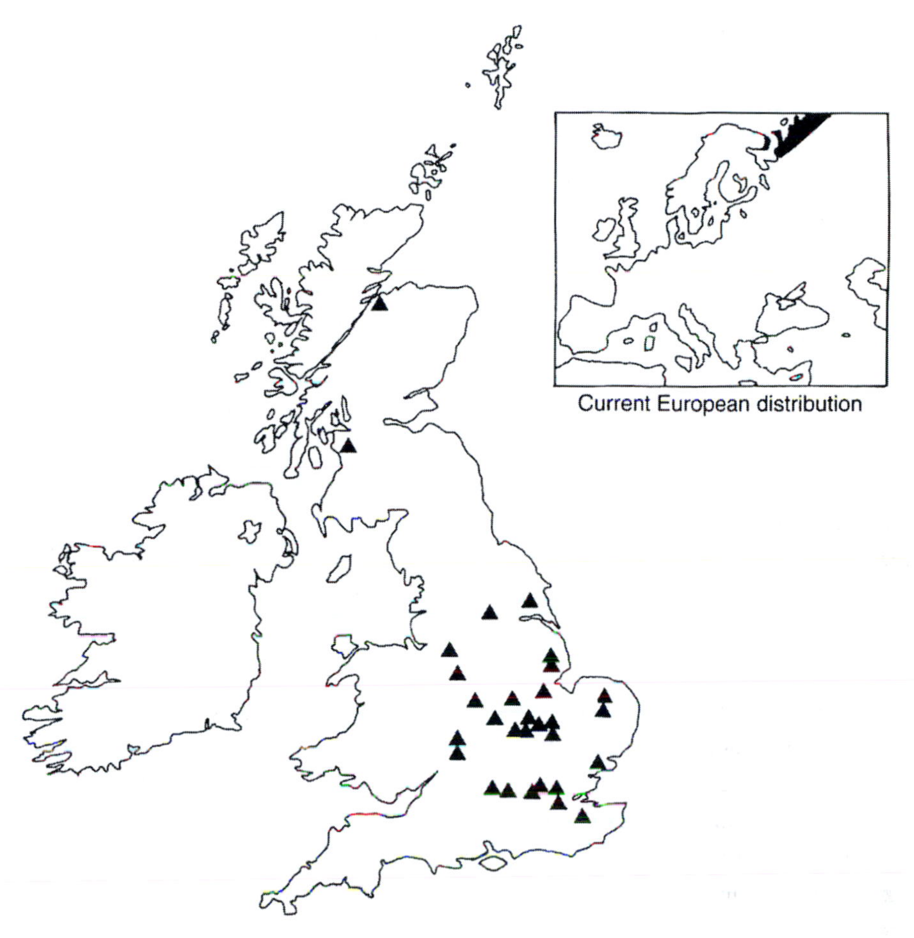

2.8 Fossil sites in the British Isles of the flightless beetle *Diacheila polita* which was common in Britain during the middle period of the Last Glaciation. It did not recolonize Britain at the close of the glacial period, and today is confined to much more northerly continental areas. Other, more mobile species that could fly, such as *D. arctica*, recolonized Britain during cold conditions. (From Coope, 1995. Reprinted by permission of Oxford University Press.)

et al., 1992). In Mediterranean France, the Neolithic human impact (4310 BP) has not only affected the local distribution of coleopteran fauna through a drastic change in hydrological regime, it has also induced major changes in the regional distribution of certain species through local extinctions (Andrieu-Ponel and Ponel, 1999). Neolithic landscape modification also impacted on insect faunas in Britain (Robinson, 1991; Dinnin, 1997), and in some cases, indirectly through increased sedimentation of rivers from soil erosion induced by forest clearance for agriculture (Osbourne, 1997). Girling (1982) lists 20 species of Neolithic saproxylic beetles that no longer occur in Britain. Not that the Neolithic forest clearances were all bad. With the cooling of the British climate

4500–3500 years, ago, the cleared landscape may have provided warm refugia for grassland Lepidoptera (Thomas, 1991).

In recent years, however, it has been the intensity as well as the rapidity of landscape change (Forman, 1995) that is causing concern for insect survival.

2.8 Current extinctions

There is a discrepancy in assessments of current extinction rates of insects. The dilemma lies in the difference between what we know and what we suspect. We categorically know of very few definite extinctions in recent years, yet, with loss of tropical forest in particular, it is suspected that 11 200 species of insects have become extinct since 1600, and that over the next 300 years a further 100 000–500 000 species of insect may become extinct (Mawdsley and Stork, 1995). The problem lies in not being quite sure when the last individual has died, and as Harrison and Stiassny (1999) have pointed out for fish, we must not be too hasty in declaring a particular species extinct, only to find out later that it is surviving, and in some cases, even thriving. Premature declaration of extinction runs the risk of forestalling further searches of remnant populations. Also, 'crying wolf' on extinctions can lose conservation credibility. Nevertheless, there is cause for concern, because McKinney (1999) suggests that the estimated number of insect species extinctions may be far too conservative, by as much as three orders of magnitude, with possibly a quarter of all species of insects under threat of imminent extinction. The level of this threat is re-inforced by McKee *et al.*'s (2003) model of threatened bird and mammal species, which suggests that the average nation is expected to have an increase in number of threatened species of 7% by 2020, and 14% by 2050.

Insects are particularly risky animals on which to declare extinction. Firstly, they are small and easily overlooked. Often the time window for the most apparent stage, usually the adult, is narrow. Populations also can be highly variable and shifting, particularly in a metapopulation context. Furthermore, while species extinction is the final result of population extinctions, it is probably not unusual that populations of generally abundant species occasionally and perhaps even readily become extinct. Indeed, an average of 24% of the populations of the Common cinnabar moth *Tyria jacobaea* become extinct every year, only to reappear in subsequent years (Dempster, 1989). This is a manifestation of sink populations being resupplied by source populations (Pulliam, 1988).

As extinction is forever, the listing of suspected extinct species cannot be taken lightly. The IUCN (2001) category 'Extinct in the Wild' is a formidable category, and as very rare and threatened insects are generally not easy to rear, this category, at least for insects, is virtually equivalent to category 'Extinct'. Harrison and Stiassny (1999) provide a useful key, developed for fish. Although complex, it has much merit for listing extinct or possibly extinct insects. Such a

2.9 Feared extinct and red-listed as Critically Endangered, the minute and unusual-coloured (purple) Cape bluet damselfly *Proischnura polychromatica* is now known to be extant but highly localized.

key should be followed before an insect Red List categorization involving extinction is stipulated. The fundamental tenet is that a species should not be declared 'Extinct in the Wild' until a full and thorough search has been undertaken. This is particularly important for insects, which are cryptic, and outlier populations may exist in unexplored locations (Figure 2.9).

This critical categorization of 'Extinct in the Wild' could also be performed on regional populations, which in some countries such as Britain are indeed considered as 'extinctions'. But as Thomas and Morris (1995) point out, such national extinctions are a warning knell for global extinction.

2.9 The taxonomic challenge

Insect diversity is great, and little known. Perhaps only 10% of species have scientific names, and among those that do, many require taxonomic revision. The situation is compounded by cryptic species (the adults of the hymenopteran parasitoids *Aphytis africanus* and *A. melinus*, for example, cannot be distinguished on morphology), evolutionarily significant units and polymorphisms. Such barely understood taxonomic variety has been described as the taxonomic impediment (New, 1984), and, more recently, the taxonomic challenge (Samways, 2002a).

There has been considerable debate concerning the value of taxonomy and systematics in biodiversity conservation (Stork and Samways, 1995; New, 1999). Much of the debate has been at cross-purposes with proponents of the necessity of overcoming the taxonomic challenge and those that see the value of morphospecies (recognizable, unidentified species) in biodiversity assessments. The problem is resolved when the conservation question is clearly posed (Oliver *et al.*, 1999). Depending on the conservation goal, identified species or, alternatively, morphospecies, may be the appropriate units of study. Much as we would like to, and should use named species, and their variation within (Austin, 1999), this is often not possible in some habitats in some areas, especially the tropics. So, how do we tackle the taxonomic challenge?

The loss of taxonomic expertise worldwide has generated much concern. As most insect groups are taxonomically relatively poorly known and expertise is thinly spread, some sort of prioritizing for particular groups to receive attention is essential. Even when that is done, there may be a real risk that estimated species richness values, for example from museum specimens, may be dangerously low and of limited use in conservation decision making (Petersen *et al.*, 2003). New (1999) points out that arguably the most urgent facilitating role for taxonomy in practical invertebrate conservation is to help increase the number of groups which can be appraised effectively. One way to achieve this would be to concentrate on lesser-known, but not poorly known, groups and make them into well-known groups. This is a choice process involving triage, where most attention is given to those taxonomic groups that would benefit most from the increased knowledge.

Interestingly, New (1999) emphasizes that facing the taxonomic challenge involves more than simply improving our descriptive knowledge of selected groups. It is also important to define, enumerate and describe the constituent elements of these groups in a way that is meaningful in practical conservation. This may involve, for example, the use of taxonomic keys that are readily comprehensible to non-specialists. McCafferty's (1981) *Aquatic Entomology: The Fishermen's and Ecologists' Illustrated Guide to Insects and their Relatives*, and Naskrecki's (2000) *Katydids of Costa Rica* (with CD) are superb examples (Figure 2.10). With the development of digital technology, the description process is being speeded up. *BioTrack*™, which is a development of The Key Centre for Biodiversity and Bioresources at Macquarie University, Australia, is an innovative biodiversity data management system that makes rapid and accurate identifications of specimens from digital images with only a small amount of taxonomic training. This directly addresses New's (1999) concern that taxonomic data must be worthy in the context of conservation management. *BioTrack*™ does just this, harmonizing taxonomic and ecological approaches to biodiversity conservation and management (Oliver *et al.*, 2000).

A final consideration is that these user-friendly taxonomic keys are available for use by non-specialists, especially parataxonomists (i.e. insect, or other taxa, identifiers without formal taxonomic training). Parataxonomic sorting of insect material can generate a huge amount of data, and when combined with

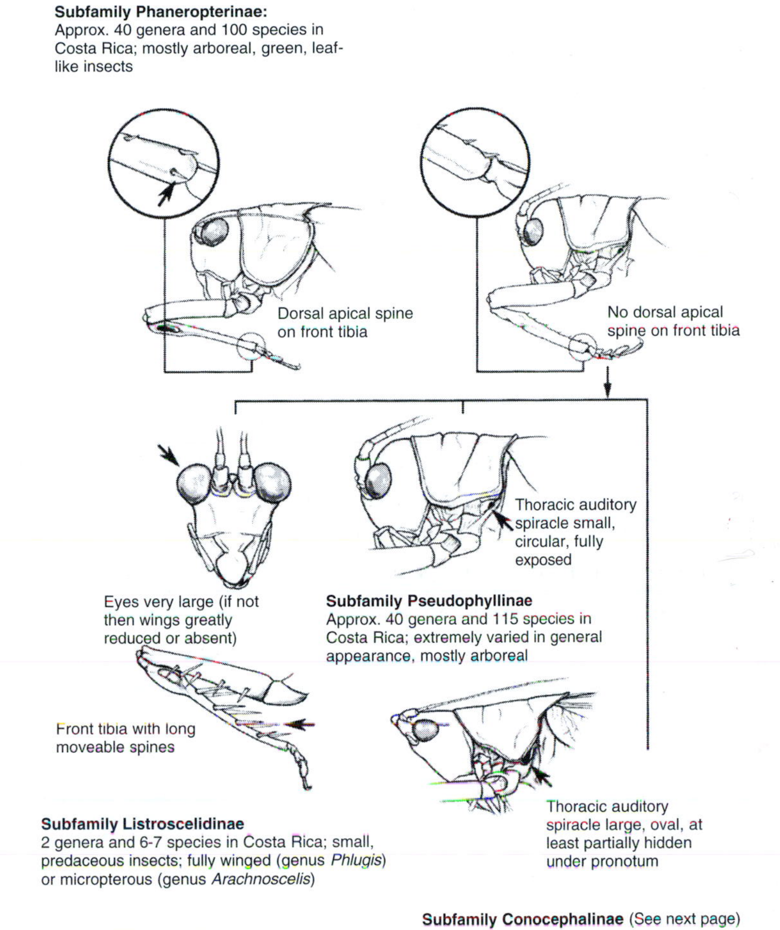

2.10 First-page extract from a user-friendly key that enables rapid keying out of taxa, in this case, of subfamilies of Costa Rican bush crickets (Tettigoniidae) (from Naskrecki, 2000).

supplementary activities such as caterpillar rearing, can produce a large amount of ecologically and conservation-useful information. There is however, a disadvantage to working with parataxonomists. According to Basset *et al.* (2000) 'the amount of field data amassed by them [parataxonomists] is forcing ecologists to become desk-bound, number-crunching writers of research papers instead of enjoying themselves in the forest'.

2.10 The perception challenge

It is the surroundings that make cockroaches dirty, and not cockroaches that make a situation unsanitary. In short, cockroaches are facultative carriers of bacterial pathogens, and the diversity of bacteria transported by them depends on where they live. Rat fleas, however, are carriers of the plague bacterium and mosquitoes of the trypanosome that causes malaria. Not surprisingly, people in general see insects as enemies rather than worthy conservation subjects. Butterflies, by virtue of their bright colours, delicate form, large size and association with flowers, are among the few insects that charm peoples' hearts (Kellert, 1986).

Yen and Butcher (1997) have pointed out that we need to overcome adverse attitudes towards invertebrates. At the Queensland Museum Reference Centre there are always twice as many queries concerning vertebrates than invertebrates, with most of those concerning the spectacular and the dangerous, especially spiders (Czechura, 1994). Indeed, *conservation* of invertebrates faces the 'tyranny of numbers' (Kellert, 1993). How can invertebrates be in trouble if we have so many of them? Currently, the mainstay is conservation of the aesthetic, harmless flagships, such as the Richmond birdwing butterfly *Ornithoptera richmondia* (Horwitz *et al.*, 1999) and the Wart-biter bush cricket *Decticus verrucivorus* (Pearce-Kelly *et al.*, 1998). Indeed, it is these very flagship, iconic species that are under threat, sometimes because of specialist habitat requirements, but also because they are sought after by collectors. In the case of crickets in China, the demand for 'fighting crickets' for gambling is estimated to involve up to 100 000 specimens per day during the peak period (Jin and Yen, 1998).

Smith (1999) suggests that the road to raising awareness is as 'perfected by our vertebrate cousins: stealth, persistence and subterfuge'. He also points out that stressing 'environment' or 'invertebrate' in a proposal is more likely to doom it to the filing cabinet.

New (2000a) provides an interesting slant on the perception challenge. Although many invertebrates are clearly 'webmasters' (Coleman and Hendrix, 2000) in ecosystems, many more are not, or we do not know whether they are. These unknown or unknowable organisms are what New (2000a) calls the 'meek inheritors'. They are probably best conserved by admitting our ignorance and seeking to cover them using an umbrella taxon or group whose values can be quantified more convincingly. (We revisit this topic in Chapter 8.) New (2000a) then emphasizes that the alternative, of promoting the 'meek inheritors' by suggesting positive, even extravagant, ecological values which we cannot prove, must inevitably lead to loss of credibility that we can ill afford.

2.11 Pest insects and population crashes

Pests are organisms that impact on our lives when and where we do not want them to. They do not know that they are pests, only we do. In terms of overall physiology and biochemistry, a pest insect is little different from a threatened one, and killing an aphid is morally little different from killing an Apollo butterfly. But small differences in DNA can have huge implications. For example, there is a large, brown, brachypterous variant (ssp. *monspeliensis*) of the familiar green European Wart-biter bush cricket *Decticus verrucivorus*, which was a pest in the late 1940s in southern France. The population then crashed and was thought extinct (Samways, 1989b) although there are reports that a small, remnant population might still exist. Such population surges and crashes do not seem to be a feature of the more familiar *D. v. verrucivorus*, which appears to remain at considerably more constant levels (Cherrill and Brown, 1990; Pearce-Kelly *et al*., 1998).

Among higher taxa there are also huge differences in pest versus threatened status, especially among Orthoptera species (Samways and Lockwood, 1998). A first glance might suggest a 'typical threatened orthopteran species' to be a narrowly distributed, specialist-feeding, relatively local bush cricket or cave cricket. In contrast, a 'typical pest orthopteran species' might be a widely distributed, polyphagous, highly mobile grasshopper or locust. However, the Rocky Mountain grasshopper (*Melanoplus spretus*) was so abundant in the western United States and Canada prior to 1880, that it caused the wheels of locomotives to slip (Swinton, 1880). Yet by the early 1900s it was extinct (Lockwood and DeBrey, 1990). The population changes that occurred in this pest were not so much due to changes in genetics as changes in the landscape. A similar situation appears to be happening today in the Moroccan locust, *Dociostaurus maroccanus*, where deforestation and overgrazing stimulate its increase, whereas converting grasslands to croplands suppress it (Latchininsky 1998). The point here is that abundant, even pestiferous species, can be at risk, once their habitat has been modified. Abundance at one time, and notoriety as a pest, is not necessarily a guarantee against extirpation or even extinction. Insects can be passenger pigeons too.

2.12 Summary

Insects have been enormously successful in terms of species richness and abundance. The main platform for this success has been the evolution of flowering plants. Insect conservation, particularly of specialist species, therefore goes hand in hand with plant community conservation. Destruction of plant communities could therefore reverse, and is indeed reversing, the evolutionary success of insects. Nevertheless, many insects are superb dispersers, often turning up in remote places, surprisingly rapidly. This suggests that many non-specialists, at least, will survive the future.

Although we so often speak of insect 'species', we must at all times be acutely aware that many insect species are a complex of polymorphisms. Among these is developmental polymorphism, where the larva is a functionally different animal from the adult, and conservation of the habitat means conservation of conditions for long-term survival of both these forms. In turn, this means conservation of compositional, structural and functional aspects of the landscape. Such a coarse filter approach (i.e conservation of landscapes) is the only realistically all-embracing approach to future insect diversity conservation, especially bearing in mind that less than 1% of the 10 million or so species have scientific names. However, this does not preclude the fine-fitter, single-species, approach when resources are available for such fine focus.

Although in the past, many species shifted geographical ranges in response to climate change, the current anthropogenic impact on landscapes is causing mass extinctions. This mass extinction is mostly to unnamed, 'Centinelan' species, which most humans do not know about, or perhaps even care about. Indeed, some pest insect species can be prone to extinction. With landscape transformation, an abundant species can become a rarity in a matter of a few years.

3 Insects and the conservation of ecosystem processes

... But in an African forest not a fallen branch is seen. One is struck at first at a certain clean look about the great forests of the interior, a novel and unaccountable cleanness, as if the forest-bed was carefully swept and dusted daily by unseen elves. And so, indeed, it is. Scavengers of a hundred kinds remove decaying animal matter – from the carcase of a fallen elephant to the broken wing of a gnat – eating it, or carrying it out of sight, and burying it on the deodorising earth.

Henry Drummond (1889)

3.1 Introduction

Insects, being good dispersers and exploiters of virtually all types of organic matter, can be found almost everywhere. That horrendous term 'bug splatter' on the front of moving vehicles, bears witness to the general abundance and omnipotence of insects and all those little lives lost. Insects do not occur in permanently frozen areas, nor at any depth in the oceans. Some, however, come close to these extremes. In North America, under snowfields, bittacid scorpion flies continue to hunt.

The ecological grandeur of insects is in their ability as a group to transfer vast amounts of energy. As such, they are determinants of community structure and shapers of habitats. Some, like termites, are such notable movers of physical materials, that they are known as ecological engineers.

One of the most beneficial attributes of insects is that many are pollinators. Not that insects always have the upper hand, as many are food items for vertebrates, as well as for each other. Conservation of insects, therefore, goes hand in hand with conservation of plants, vertebrates and other invertebrates. But insects can also be harmful to other organisms, and some are also vectors of plant or animal disease. These facts lead to the inevitable conclusion that understanding conservation of insect diversity is largely about appreciation of ecosystem processes.

Insect assemblages and communities are shaped by the compositional and structural land mosaic. In turn, the floral and faunal features of the landscape can be shaped by insects. This interaction between insects and the landscape may not always be linear, with weather and climate changing the direction and magnitude of the interactions. Conserving insect diversity is therefore about conservation of the integrity of ecological systems.

3.2 Insects as keystone organisms

A keystone species could be considered as one whose impact on its community or ecosystem is large and disproportionately large relative to its abundance (Power *et al.*, 1996). This concept, however, has been criticized as it threatens to erode the utility of the keystone concept (Hunter, 2000b). Paine's (1969) original idea was that the species composition and physical appearance of an ecosystem are greatly modified by the activities of a single indigenous species high in the food web. Such a keystone species influences community structure and ecological integrity, with persistence through time. Mills *et al.* (1993) have pointed out that the term keystone has since been applied to a plethora of species, at different levels in food webs, and with very different effects, both qualitative and quantitative, in their communities.

In terms of conservation, it is not so much the keystone species per se that is significant but its keystone role (Mills *et al.*, 1993). Following on from this, these authors advocate the study of interaction strengths and subsequent application of the results into management plans and policy decisions, instead of using the keystone/non-keystone dualism. Such an approach recognizes the complexity, as well as temporal and spatial variability of interactions. This complexity and the different effects in different places at different times has a familiar ring to practicing insect biocontrol workers. It is well known that any one insect host may well have 10 or more natural enemies (Miller, 1993). Some of these are generalist, and others specialist or host-specific. In turn, many of these natural enemies are in competition with each other, and with other species on other hosts. These interactions shift and change in strength with weather, season, climate and elevation, which, in turn, increases temporal and spatial complexity enormously.

Nevertheless, very strong competition sometimes occurs among keystone species. Risch and Carroll (1982), for example, observed an increase in abundance in 24 other insect species and a decline in three species when certain species of ant were excluded from agricultural fields. The dynamics of severe competition was emphasized among other ant species, where asymmetrical competition became so severe that there was complete amensalism and local extinction of nests of one species by another (Samways, 1983a).

Although interaction strengths is a scientifically more rigorous concept than keystone species, De Maynadier and Hunter (1994) argue that we must not too hastily discard the term 'keystone', as it can effectively rally public understanding and protection. The point is, so long as we use it and understand its ramifications and implications, then we do indeed have powerful imagery with which to work in practising conservation.

3.3 Insect ecosystem engineers and soil modifiers

Ants are well known to influence certain terrestrial ecosystem processes, at least in the tropics (Folgarait, 1998) and in arid areas (Whitford, 2000). They are able to do this because some species are major modifiers and controllers of the physical state of abiotic and biotic materials (Samways, 1983a). In this way, they may be regarded as ecosystem engineers (Jones *et al.*, 1997). The Funnel ant *Aphaenogaster longiceps* in Australia, for example, single-handedly is responsible for moving some 80% of the soil that is moved to the surface by soil fauna (Humphreys, 1994).

Another major engineering taxon is the termites (Whitford, 2000). Nests of macrotermitines in West Africa can cover as much as 9% of the land area and have a volume of 300 m^3/ha (Abbadie *et al.*, 1992). Such mounds have a higher organic carbon and nitrogen content than the surrounding soil. Termites also play a significant role in global carbon fluxes. Global gas production by termites in tropical forests represents 1.5% of carbon dioxide and 15% of all methane produced from all sources (Bignell *et al.*, 1997).

These insect ecosystem engineers can locally influence structural, compositional and functional biodiversity. West African termites, by modifying water dynamics and organic matter status, increase local tree diversity (Abbadie *et al.*, 1992), and some ants allow unique plants to exist (Folgarait, 1998). Besides these effects on vegetation diversity, engineers with distinct and/or large structures provide homes for many symbiotic organisms (Wilson, 1971).

The concept of ecosystem engineers is far from fully explored and there may be far more less explicit engineers than is fully realized. Grasshoppers in the arid region of South Africa produce frass which is finely divided and provides nutrients to plants far faster than that, say, of sheep (Milton and Dean, 1996), which in turn influences plant diversity (Stock and Allsopp, 1992). Although plant diversity inevitably increases insect diversity, there is not necessarily a direct relationship. Again, in the southern part of South Africa, it does not seem to be simply the plant variety that has generated the insect diversity, but that the insects have speciated through allopatric isolation, as did the plants themselves (Cottrell, 1985). The point being that although we generally think of ecosystem engineers working at the recent, ecological temporal scale, it is likely that they have had, and continue to have, a differential influence on the evolution of other organisms, depending on how dependent those organisms are on the controlling role played by the engineers. A very strong controlling role has thus generated the symbionts.

3.4 Insects as food for other animals

Conservation of invertebrates is intimately tied in with conservation of ecosystems. This is because many species are major components in those ecosystems (Coleman and Hendrix, 2000), although many others are not (New, 2000). Insects are major prey for many vertebrates, and of course for many invertebrates, including other insects. Insects provide a large food resource. Pimentel (1975) estimates that in the USA, their fresh biomass is about 450 kg/ha, about 30 times that of humans in the same country. In the Brazilian rainforest, ants and termites make up more than one-quarter of the faunal biomass, and ants alone have four times the biomass of all land vertebrates (Wilson, 1991). In terms of shear abundance, the record holders are probably the Collembola, which can occur at densities of between 10^4 and 10^5 per m^2 (Hopkin, 1998). This biomass is inevitably a major foodbase for many dependent faunal elements. Even in freshwater, the role of insects is pivotal, with the fly-fishing industry, to name one, being built on the functional role of insects as food.

Certain alien predators are having a detrimental threatening impact on certain insects, such as poeciliid fish on endemic damselflies in Hawaii (Englund, 1999). It is not certain the extent to which many red-listed amphibia, reptiles, birds and mammals depend on specific insect prey items for their survival. Certainly, some are generalist predators, with the Seychelles endemic skinks *Mabuya sechellensis* and *M. wrightii* depending on local arthropods, while interestingly, the highly threatened Seychelles Magpie Robin *Copsychus sechellarum* has the introduced cockroach *Pycnoscelus indicus* as a main food item (LeMaitre, 2002). In Britain, agricultural change in recent decades has caused a decline in birds, which appears in part to be due to the decline in quality and quantity of their invertebrate food source (Benton *et al.*, 2002).

3.5 Insect dispersal

3.5.1 *Differential mobilities*

Many insects are remarkably mobile. This appears to be part of the natural dynamics of many insect species as a survival strategy for dealing with natural, local extinctions. Perhaps this diffusive dispersal is more common than generally realized, especially considering the discovery that there is long-distance gene flow in an apparently sedentary butterfly (Peterson, 1996). Certainly for an assemblage of British butterflies it was not dispersal ability limiting their overall abundance but lack of suitable habitat (Wood and Pullin, 2002).

Not all insects are so vagile. There are what Adsersen (1995) has described for plants, 'fugitive species'. These are species that instead of radiating and diversifying, originally dispersed and then remained localized. For fugitive Seychelles damselflies, they can be surprisingly tolerant of droughts, even maintaining territories when water is no longer present (Samways, 2003b). Similarly, populations of *Maculinea alcon* butterflies in Denmark are more isolated than counts of flying adults or eggs on foodplants indicate (Gadeberg and Boomsma, 1997). Nevertheless, in the long-term, some apparently low mobility species can disperse long distances and colonize remarkably remote oceanic islands (Figure 3.1) (Peck, 1994a,b; Samways and Osborn, 1998; Trewick, 2000).

3.5.2 *Tracking resources*

As the subtitle of Drake and Gatehouse's (1995) book on insect migration indicates, movement is about tracking resources through space and time. The critical point is that knowledge of a species' dispersal ability, and inhibitions to that dispersal, are fundamental to conservation of a species. To name one example, Appelt and Poethke (1997) illustrated that populations of the grasshopper *Oedipoda caerulescens* undergo local extinction when there are stray fluctuations in environmental conditions. This leads to the concept of metapopulations, the framework of which explicitly recognizes and provides a conceptual tool for dealing with the interactions of within- (e.g. birth, death, competition) and among-population processes (e.g. dispersal, gene flow, colonization and extinction) (Thrall *et al.*, 2000). The within- and among-population interactions are well-illustrated in *Prokelisia* spp. planthoppers, where there is a negative relationship between dispersal capability (i.e. genetically determined per cent macroptery) and habitat persistence (Denno *et al.*, 1996).

Whereas not all species whose populations have undergone fragmentation fit the definition of a metapopulation, the metapopulation paradigm (Harrison and Hastings, 1996) (see Figure 10.6) is nevertheless a useful management tool. As Thrall *et al.* (2000) point out, a metapopulation perspective ensures a process-oriented, scale-appropriate approach to conservation that focuses attention on among-population processes that are critical for persistence of many natural systems.

3.5.3 *Myriad of mobilities*

The problem is that we do not know what each insect species, each evolutionarily significant unit and each insect polymorphism needs under all

3.1 The aptly named Globe skimmer *Pantala flavescens*, the only dragonfly to have colonized the world's most remote island, Easter Island (Rapa Nui). After colonizing the island, its morphology and behaviour have changed. It has undergone natural selection and become adapted to surviving on this small speck of land, where to wander away from the island would mean almost certain death. On the island, it flies in small circles close to the ground, whereas on continents it flies high and wanders (see Samways and Osborn, 1998).

environmental conditions. This is illustrated by the various movement patterns in a butterfly assemblage at any one point. Each species, although seemingly simply flying by, is reacting sensitively to the various landscape vegetational structures (Wood and Samways, 1991). This serves to illustrate that conservation of insect diversity encompasses a vast complexity of interactions that in themselves vary over space and time. Against this background, the landscape may be considered as a continuously varying differential filter (Ingham and Samways, 1996), and try as we may, cannot always be managed to provide optimal conditions for all species all of the time. This argues strongly for the conservation of larger spatial scales (i.e. landscapes and larger) such that all aspects of an insect's behaviour, and all types of insects are supported. It also means a whole

3.2 Specialized pollinator systems are under threat. Here, a Convolvulus hawkmoth *Agrius convolvuli* inserts its long proboscis to the bottom of a long corolla to obtain nectar.

range of interaction types, interaction strengths, ecological processes, and biotic units are conserved, without us knowing the details. This returns us again to the familiar coarse filter approach (Hunter, 2000a), so critical for long-term insect diversity conservation, although nevertheless requiring critical evaluation, the subject of Chapters 8 and 10.

3.6 Insect pollinators

Flowers show a range of specialization to pollinators (Johnson and Steiner, 2000) (Figure 3.2). Some flowers like the Madagascan orchid *Angraecum sesquipedale*, which has a 30 cm long tubular nectary, is pollinated by a moth *Xanthopan morgani praedicta* with a correspondingly long proboscis. Another Madagascan orchid *Angraecum 'longicalar'* has been discovered, with an even longer nectar spur, 36–41 cm long, which presumably has a matching pollinator (Schatz, 1992).

Although certain flowers do have specialized pollinator systems, there are strong differences among plant families. Indeed, some plant specializations for pollinators may not be a determinant for an evolutionary 'trend' (Ollerton, 1996).

There is no doubt that certain pollinator systems are threatened and probably more so in the case of the most specialized systems (Bond, 1994), and for plants in small, isolated landscape remnants. But the main issue is that whereas plants occupy virtually every point on the continuum from extreme specialization to extreme generalization, in realistic terms, we know very little of the ecological dependency of plants on pollinators, both for seed production and for population viability (Johnson and Steiner, 2000). In particular, besides broad-scale community studies, we need more studies on pollinator effectiveness (Schemske and Horvitz, 1984), so as to determine which are the important pollinators and where the weak links lie in terms of ecological integrity of the system. This is particularly so as the break down of mutualisms can have long time delays. This arises from compensatory mechanisms such as clonality, longevity and self-pollination, which enable the plant species to survive only temporarily, albeit over many years (Bond, 1994). It is now timely to identify keystone species that sustain particular pollinator systems (Corbet, 2000).

Dependency on pollinators relates to crop plants as well as to wild plants (Bonaszak, 1992; Allen-Wardell *et al.*, 1998). Human-transformed landscapes are likely to have complex, serious and largely unpredictable consequences for wide-scale, long-term conservation of both insects and plants (Kearns *et al.*, 1998; Kremen and Ricketts, 2000). This serves to underscore again that conservation of insect diversity is not an isolated activity but is intimately linked with that of plants, although each group is not necessarily an exact surrogate for the other.

3.7 Insect herbivores

The world total biomass of plants to animals has the ratio 99.999 : 0.001, whereas the total number of higher plant species to animal species may be 0.026 : 99.974 (Samways, 1993b). In other words, there are few but bulky plants compared with many and varied, but not bulky, animals. And four-fifths of the animals are probably insects. This suggests that many insects are feeding on few plants. Yet it is not easy to feed on plants, where environmental conditions are harsh. Also, much of the plant material is non-nutritious and protected by noxious substances. Nevertheless, the throughput of energy of insects feeding on plants can be enormous, with some grasshopper species transferring 5–10 times the amount of energy as a bird or mammal species in the area. Even in the African savanna, where megaherbivores visually dominate the landscape (or used to), the grasshopper assemblage can ingest 16% of the grass cover (Gandar, 1982).

The insect–plant interaction is highly significant for maintaining biocycles. The Brown locust *Locustana pardalina* in the Karoo of South Africa produces 2.26 million tonnes of frass during an outbreak, which represents 14 700 tonnes of nitrogen (Samways, 2000a). The point is that the Karoo is an arid ecosystem, poor in organics and nitrogen. Large quantities of nutrients can become available only after rains and principally in the upper 30 cm of soil. As insect frass is fine grained, it means that locusts and termites convert plants to a nutrient-rich

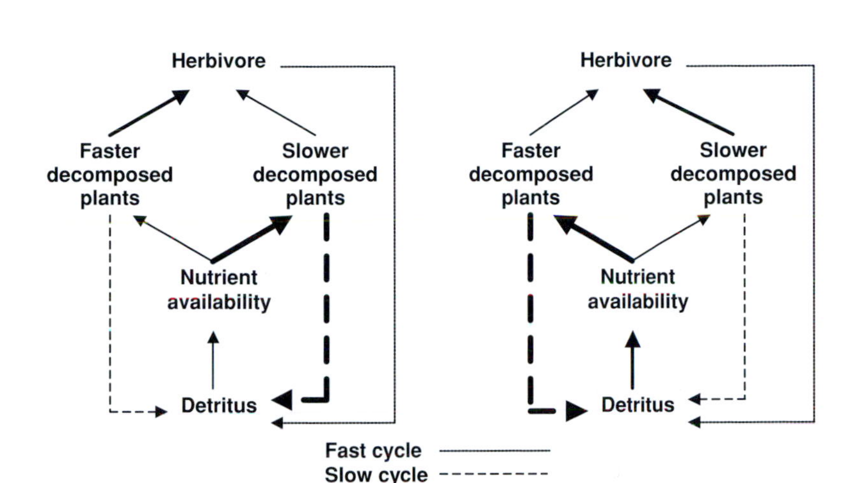

3.3 The conditions for herbivores to decrease nutrient cycling and primary productivity (a) and to increase nutrient cycling and primary productivity (b) are shown. The solid lines reflect the trophic transfers within the ecosystem. The broken lines reflect the pathways by which nutrients are recycled in the ecosystem. Line thickness reflects the relative magnitude of the trophic transfer or recycling pathway. (Redrawn from Belovsky, 2000, with kind permission of Kluwer Academic Publishers.)

resource that is immediately available to the plants (Milton, 1995). Similarly, in the USA, it is the high abundance of grasshoppers that enhances grassland productivity (Belovsky, 2000) (Figure 3.3). This contrasts with the forest situation, where plant primary production is stimulated more by low to medium levels than by high intensities of phytophagy (Schowalter, 2000). These seemingly contradictory results are probably related to differing conditions and the extent to which nutrients are resupplied through the fast cycle of frass and herbivore corpses as opposed to the slow cycle of litter decomposition (Belovsky, 2000).

While it is well known that plant species and assemblages influence insect diversity, less well known is the fact that insects can determine the secondary succession of certain plant assemblages (Brown *et al.*, 1987). Such interactions between herbivores, plants and ecosystem processes are now known to involve complex interactions with feedback loops (Schowalter, 2000), which vary according to the time span under consideration (Anderson, 2000). Furthermore, these interactions involve a host of biochemical and physiological reactions in addition to the ecological ones (Dyer, 2000) (Figure 3.4). Let us not forget too, that these interactions are also taking place on the roots as well as on the aerial parts.

The interaction between insects and plants must also consider seeds. Insects, especially ants, distribute and bury seeds, an essential survival strategy for many plants. This may even involve the production of special attractant seed structures

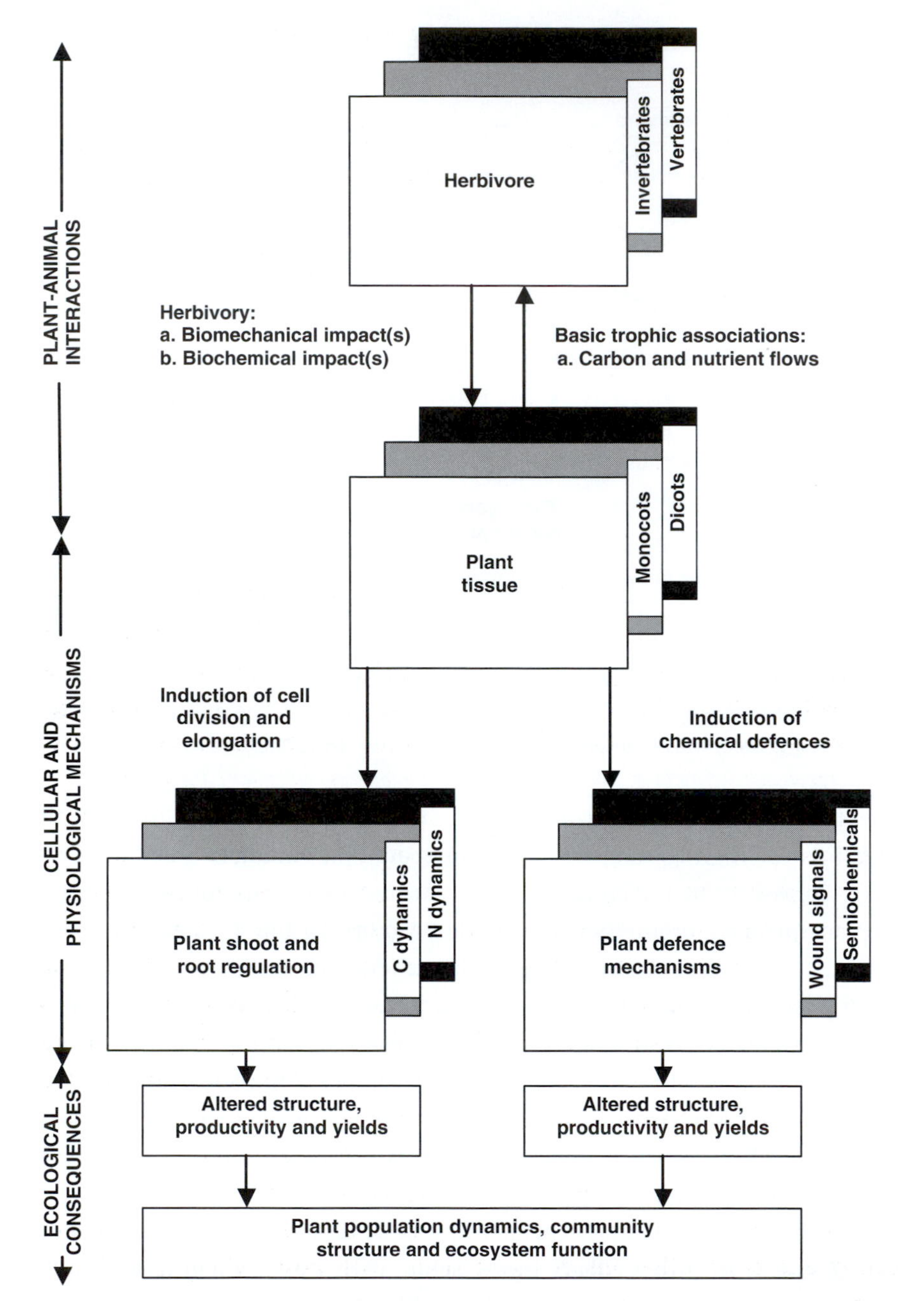

3.4 This theoretical framework places scale of association into plant–animal associations via the need to study differing concepts to answer differing questions that one can pose at each level. The plant–herbivore association suggests two lineages involving induction impacts: one, considered more traditional in terms of numbers of reported studies, involves negative feedback induction of chemical defences which can alter the population structure of the community, and the second involves positive feedback through a signal from the herbivore to the plant in which the plant reorganizes its growth and development within the community, which one measures in terms of productivity. (Redrawn from Dyer, 2000, with kind permission of CABI Publishing.)

called elaiosomes (Jolivet, 1998). In contrast, seeds also suffer from the ravages of insects feeding on them, which may have a major effect on seedling recruitment (Janzen, 1978; Wright, 1994).

In the plant-rich Cape Floristic Region, it is host-plant characteristics, such as infructescence structure, that determine insect borer diversity (Wright and Samways, 1999). This re-emphasizes that plants can determine insect assemblage diversity, with host-plant effects generally more profound in the case of endophagous insects than exophagous insects (Price *et al.*, 1995), particularly in the case of galling insects (Wright and Samways, 1998). The upshot of this in terms of insect diversity conservation is illustrated in Hawaii where five moth species have become extinct as a result of plant extinctions (Gagné and Howarth, 1985).

3.8 Insect parasitoids and predators

Great insect diversity has been generated in part because so many insects are parasitoids, parasites (Askew, 1971) or predators (New, 1991) as well as hosts. Arguably, feeding on prey is an easier way of life than feeding on plants, bearing in mind plants' abundance of indigestible material, toxic compounds and tough cuticle. Parasitism is such an opportunistic way of life that many species have become adapted to it, with the hosts inevitably becoming increasingly sophisticated at avoiding attack (Bernays, 1998). Nevertheless, it is not unusual for one host species to support many parasitoid species (Memmott and Godfray, 1993). But similarly, parasitoids are opportunistic, with most being surprisingly host non-specific (Prinsloo and Samways, 2001). This inevitably leads to complex food webs where hide-and-seek behaviour determines both species richness and abundance (Berryman, 1996) and this is determined by the strength of refuges occupied by hosts (Hawkins, 1993). Inverting this argument, given the current biodiversity crisis, parasitoids are more prone to extinction than their herbivorous hosts leading to a reduction in herbivore mortality. This is likely to have cascade effects with possible destabilization of some communities, as parasitoids can have roles as keystone species (LaSalle, 1993; Combes, 1996).

Some predators have a similar effect, and certain ladybirds can have major regulating effects on other insect populations, with host switching and optimization on resources being an integral part of the interaction (Erichsen *et al.*, 1991, Majerus, 1994). This behaviour can be so strong that it is the availability of hosts that may dictate predator diversity across the landscape. This, in turn, overrides the effects of landscape transformation (Magagula and Samways, 2001), which is normally considered as a major concern for the conservation of insect diversity. But there are varied and complex behaviours involved (Hattingh and Samways, 1995) which vary substantially between individuals. While some individual dragonflies, for example, remain at their natal pond, others disperse

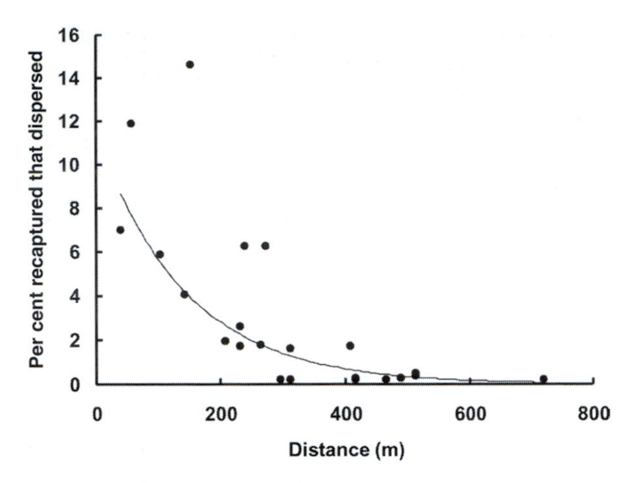

3.5 Decreasing distance dispersed by individuals of seven species of British Odonata, with fitted exponential regression. The important point is that some individuals disperse relatively far, even across a transformed landscape, which is important for maintaining founder populations and genetic viability. (Redrawn from Conrad *et al.*, 1999.)

long distances (Figure 3.5) (Conrad *et al.*, 1999). Such dispersal can result in founder populations as opportune conditions arise. This occurred on the formerly waterless island of Cousine, Seychelles. The establishment of dragonfly diversity was that which was predicted by island biogeography theory (Samways, 1998a) (see Figure 10.8).

Parasitoids can inflict high mortality on insect hosts. Theoretically, particularly in a constant environment, there can be rapid co-evolution of host and parasitoids. But as hosts often have several parasitoids, and environments are not constant, co-evolution is probably rare (Lapchin, 2002). Rarity and low apparency then becomes an important survival option. This in turn is likely to be an important driver of insect diversity through allopatric speciation.

A final consideration in the parasite–host equation, as regards insects, is the loss of insect parasites such as lice and fleas, when their vertebrate host becomes extinct, particularly when there is tight coupling of the interaction (Stork and Lyal, 1993). However, caution is always needed as we know very little of the taxonomy of some of these ectoparasites which may be more generalist than at first may seem the case.

3.9 Insects and disease

The generation or degeneration of biological diversity may in part depend on the role certain insects play as vectors of disease organisms. The point is that disease organisms are parasites that play a major role in the biodiversity of free-living species by interfering in processes as diverse as competition, migration, speciation and stability (Combes, 1996). The importance of pathogens is underlined by the fact that sex itself came about as a disease-dodging strategy through reshuffling of the genetic code, despite the fact that it is costly energetically and physiologically (Bush *et al.*, 2001).

While host–parasite co-evolution may reach some sort of compromise for both host and parasite, and can also generate insect population cycles (Briggs and Godfray, 1996; Dwyer *et al.*, 2000), the impacts of disease organisms are likely to increase with environmental stresses such as fragmentation, pollution, invasive aliens and global climate change (Holmes, 1996). Loss of habitat, combined with hunting and perhaps avian malaria transmitted by mosquitoes, has seen synergistic impacts on the Hawaiian bird fauna (Pimm, 1996). Insects are well known to harbour many diseases, as illustrated by concerns expressed when biological control agents are introduced against weeds (Kluge and Caldwell, 1992). Yet many hosts are remarkably scarce. This begs the question of whether scarcity itself, among other things, has been a disease-avoiding strategy. The trade off would be the Allee effect (Stephens and Sutherland, 1999), where there is inverse density dependence at low population densities.

The relationship between insects, disease and diversity has many other ramifications. Among them would be the impact that insects vectoring disease have on plant communities. It is not likely to be simply a one-way process of insect-plus-vector impacting on plants, but may also involve a feedback loop. If the pathogen is too potent, it is likely to stress the plant too much, which would then succumb to various other impacts including competition from other hosts. This would then deprive the insect of its host plant. Little is known of the converse situation involving impact of plant pathogens on the insect vector. Interestingly, while the bacterial disease 'greening' has a debilitating effect, sometimes mortally, on *Citrus* spp., the psyllid vector *Trioza erytraea* (see Figure 6.8), despite carrying a huge parasite load in its salivary glands, appears to be little affected by this particular bacterium.

Insects, disease and diversity is an under-explored field. However, if the role of insects as vectors in agricultural disease is a pointer, this is an area we need to explore more fully. Infectious microorganisms that undergo vertical transmission from parent to offspring, such as *Buchnera* endosymbionts of aphids, may have evolved from an earlier situation of poor co-adaptation and associated virulence towards becoming benign and then finally mutualistic (Moran and Telang, 1998). The fascinating situation between the bacterium *Wolbachia pipientis* and its insect host has taken an alternative evolutionary route. *Wolbachia* invades and survives in an arthropod host by manipulating the reproductive success of that host, with the result that females that vertically transmit the bacterium to the next generation have a reproductive advantage over their uninfected counterparts (Bourtzis and O'Neill, 1998). In such a situation, it may be essential to maintain the insect–*Wolbachia* relationship, as it is an enabling process for competitive survival.

3.10 Ecosystem diversity and insect diversity

Terrestrial and aquatic compositional ecosystem diversity is largely about insect diversity. Contrastingly, in structural diversity, insects would feature weakly in favour of plants, but nevertheless, as we have seen in earlier sections of this chapter, insects are major drivers of ecosystem processes. They are thus important components of functional diversity. But our knowledge of the relationship between organismal functional diversity and ecosystem functional diversity is only beginning to be understood. Among plants at least, functional diversity as measured by the value and range of species traits, rather than species numbers alone, strongly determines ecosystem functioning (Diaz and Cabido, 2001). Furthermore, species diversity is often an inadequate surrogate for functional diversity.

But does this help us answer the question that if we conserve ecosystem diversity we automatically conserve insect diversity? This question begets a multitude of subquestions because of the multitude of species, polymorphisms, interactions and mobilities. These in turn are affected by physical features and complexities of the ecosystem (Haslett, 1994), and by seasonal changes of weather that influences the quality of the microhabitats (Ferreira and Silva, 2001). This means that the landscape is subject to changes and fluxes, in a way that the variegated landscape (Ingham and Samways, 1996) could be considered a sort of ecological tissue. This being the case, it calls into question concepts such as umbrellas (Haslett, 2001), surrogates (Prance, 1994) and black-box landscape conservation (Samways, 1993a) as possibly not holding all the answers to insect diversity conservation. We will return to these questions, but meanwhile managers are calling for positive guidelines (Whitten *et al.*, 2001) for conserving the immense complexity that we know as insect diversity. And we are going to need simple yardsticks if we, as insect diversity conservationists, are to make a credible case.

3.11 Insects and the naturally changing landscape

The old aphorism that there is only one certainty, and that is 'change is inevitable', is an added complexity and synergism to the already immensely complex task of insect diversity conservation. But insects are pastmasters and -mistresses at the game of change. The reassembly of specialist fig wasp pollinators on Krakatau after the volcano blew its top in 1883 illustrates this well (Thornton, 1996). It is almost as if insects ebb and flow through the ecological tissue as conditions change, whether daily, seasonally or over geological time. The question that now arises is whether biodiversity is about to undergo a major discontinuity, where change is about to be non-linear and sudden, with catastrophic consequences (Myers, 1996a). It is well known that insect population surges and local extinctions are not uncommon, and that their populations have often shifted their geographical ranges to past climatic events. This suggests that a reserve network might be too rigid for insects in the ecological tissue. Yet human-induced loss of insect populations, which finally leads to loss of species, is becoming more common and widespread. It appears not so disastrous as it is for vertebrates when we simply peruse the *IUCN Red List of Threatened Species*. But perhaps we are deluding ourselves, with Centinelan extinctions (extinctions of which we have no actual records) being far more rampant than we thought (Mawdsley and Stork, 1995). Without doubt, the world is becoming increasingly species-poor yet more homogeneous in its insect fauna.

As insects ramify virtually all terrestrial ecosystem processes, as a consequence, they are almost certainly on the brink of a range of major discontinuities (Samways 1996b). A discontinuity is a catastrophic shift in community composition. With such dramatic changes, we are obliged to invoke some sort of landscape triage (Samways, 1999b) because much as we would like to invoke the precautionary principle (maintaining all the parts and processes because we do not know their function), we will have to be selective in what we conserve, simply because human and financial resources are limited. The really important point in insect diversity conservation becomes: How much irreplaceable and evolutionarily potent insect diversity will be maintained across the transformed land mosaics? Reserves have a role, but it is the overall context of the ecological matrix and its capacity to retain the flux of insects and other biodiversity that will dictate the final outcome. The insects that survive this transformation will be those that have been honed by past natural changes and are adaptively equipped for the current changes. Those that are not so equipped will need our constant attention. Are we prepared to do that, forever, for a little, six-legged creature?

3.12 Significance of ecological connectance

Important as insects are in compositional and functional biodiversity, they are not the whole story. We need to view them as part of the bigger ecological picture. As insects function at different trophic levels, any consideration of insect diversity conservation must consider food webs and their stability. Williams and Martinez (2000) have developed a 'constant-connectance hypothesis' which states that there exists a particular balance in food-web complexity, so that connectance remains fairly constant at about 10% no matter how many species are added to the web. Williams and Martinez (2000) built a niche model that predicts the niche any species will adopt in terms of its connectedness to other organisms in the system. But it is likely that there are also basic physical laws involved, including that of 'small worlds', which is a network with many nodes that have a few links as well as a few nodes with many links, following a power law distribution.

In terms of food webs, the highly connected nodes represent keystone species, and removal of only 5–10% of these highly connected species can lead to radical ecosystem change (Sole and Goodwin, 2000). For insect diversity conservation, this means that where there are high levels of connectance between species in an ecosystem, an adverse impact could reverberate across many species, causing a cascade of secondary extinctions as closely connected associates are affected. Furthermore, loss of just a few important predators of grazers can have a disproportionately large effect on ecosystem diversity, sometimes with long, and perhaps, unrecognized time delays (Duffy, 2003). Evidence is accumulating that changes in biodiversity can be both a cause and a consequence of changes in productivity and stability. This bi-directionality creates feedback loops, as well as indirect effects, that influence the complex responses of communities to biodiversity losses. Food webs mediate these interactions, with consumers modifying, dampening and even reversing the directionality of biodiversity–productivity–stability linkages (Worm and Duffy, 2003).

This emphasizes yet again the value of conserving whole landscapes with all levels and types of connectance intact. Such connectance should also include the rare 'meek inheritors', because evidence from plant communities suggests that the collective effect of rare species increases community resistance to invasive aliens and minimizes alien invader impact on communities (Lyons and Schwartz, 2001) (Figure 3.6). Nevertheless, we are still likely to see future catastrophic regime shifts in some ecosystems, where accumulation of pressures reach a point when an ecosystem changes from one state to another and then remains more or less in the new state (Scheffer and Carpenter, 2003). In turn, ecosystems could, at least theoretically, be present in different states across the region, and even move back and forth between states over the long term.

3.13 Summary

Insects are important players in terrestrial ecosystem processes, both as herbivores and as parasitoids or predators. Some species, particularly among ants and termites, are so numerous that they alter the soil to such an extent that they may be termed ecosystem engineers. Insects can also be on the receiving end of predation, and hugely important as food in maintaining many vertebrate populations.

3.6 Easter Island (Rapa Nui) is an example of an ecosystem that has been through a catastrophic regime shift. Once forested, it is now grassland on fertile, volcanic soil. A millennium of tree removal, especially for transport of statues (maui) from the quarries to the sites of erection, resulted in extinction of many species, including the endemic forest palm. The island is rich in species, but mostly aliens, and is now set on a new course of ecological succession.

Insects are variously mobile, flight being their hallmark among terrestrial invertebrates. This mobility is about tracking resources in space and time, which is being thwarted for many species as a result of landscape fragmentation. However, the situation is very complex, because of the huge numbers of species and their various biologies. This means that the natural, and particularly the modified landscape, is a huge differential filter where certain species become isolated in habitat patches, but others less so.

Among the insect interactions most affected is that of pollination, which has become a global concern for plant and insect alike. This is likely to have a cascade effect because as the pollinator/plant mutualisms decline, plant communities will change, which is then likely to impinge on a range of other insect herbivores. This in turn, will affect ecosystem processes, including nutrient return to the soil, as well as loss of the seed bank. As most insect herbivores have a host of parasitoids and predators there are also likely to be large changes in food web structures. These stresses might be further aggravated by the impact of pathogens.

In sum, the maintainenance of insect diversity is a pivotal part of the maintenance of ecosystem form and function, with diverse ecosystems having diverse insect communities. The ecological changes that are currently taking place are so complex and have such far-reaching ramifications that we are likely to see major ecological discontinuities and radical ecological shifts to new community states in the future. Coupled with this, is likely to be an increased homogenization and impoverishment of the world insect fauna associated with changes in ecosystem processes. Maintenance of insect diversity is thus part and parcel of maintenance of ecosystem integrity in any given geographical area.

II Insects and the changing world

Verily it has seemed, looking back on the past, as if our insect hunters of the future were doomed to a sport composed of an influx of Colorado Beetles, White Butterflies, Hessian Flies and Woody Oak-galls; and that the flowery wood clearings and purple heaths of our forefathers, with their basking and fluttering fauna, were all to be heartlessly swept away in the present era of steam and telegraphy. It is quite certain that our butterflies, especially the Orange Fritillaries, are fast disappearing.

A. H. Swinton (*circa* 1880)

Humans are modifying the world from moss greens and serendipidity pinks into one of monotonous greys and browns. The soft edges of plant and insect nature are being hardened into steely points where the interactive processes that are the hallmark of this planet are being simplified and impoverished. The human appetite for consumption of the biosphere has become insatiable. This modified world is hard and unbending, requiring even more energy input for maintenance. Humans are rapidly losing touch with the intricacies of nature, its variety, its dynamics and its exciting unpredictability. We are forsaking natural, evolved complexity for predictable blandness. By ignoring our inherent connection with nature, the planet will reach a point of impoverishment; sadly, overriding its unimaginably long bona fide. We need to know the type and magnitude of these human-induced changes taking place, so as to have wisdom to redress our plundering of the planet.

I, THE MANTIS

In vain I turn to look. . . .
only the most erased memory of that which was.
The path ahead is torn,
its holes scattered with tattered remains.
Remnants of trees reach, with black bony fingers,
in futile pleading agony
to a sky turned blind and deaf
by waves of human impacts. . . .
And the silent tick tock of the clock
are only the echoes of
the song of the Cricket
that has gone
Forgotten
Silent
Forever
Dead . . .

4 Degradation and fragmentation of ecosystems

In the current cycle one last natural selection has begun to operate of which man is the delegated agent, and the demesnes of oak and heather, home of the David and Saxon, are fast ceding their primitive mysteries before the steady march of building, agriculture, horticulture, floriculture, and domestication of species; handmaidens of an era of civilization, the creative wonders of whose potent wand can scarcely compensate an entomologist for the loss of his breezy heath and sylvan shade where the Fritillaries once sunned, or the lonely classic fen-land where the Large Copper Butterfly, *Chrysophanes Hippothoë*, variety *Dispar*, once flew.

A. H. Swinton (*circa* 1880)

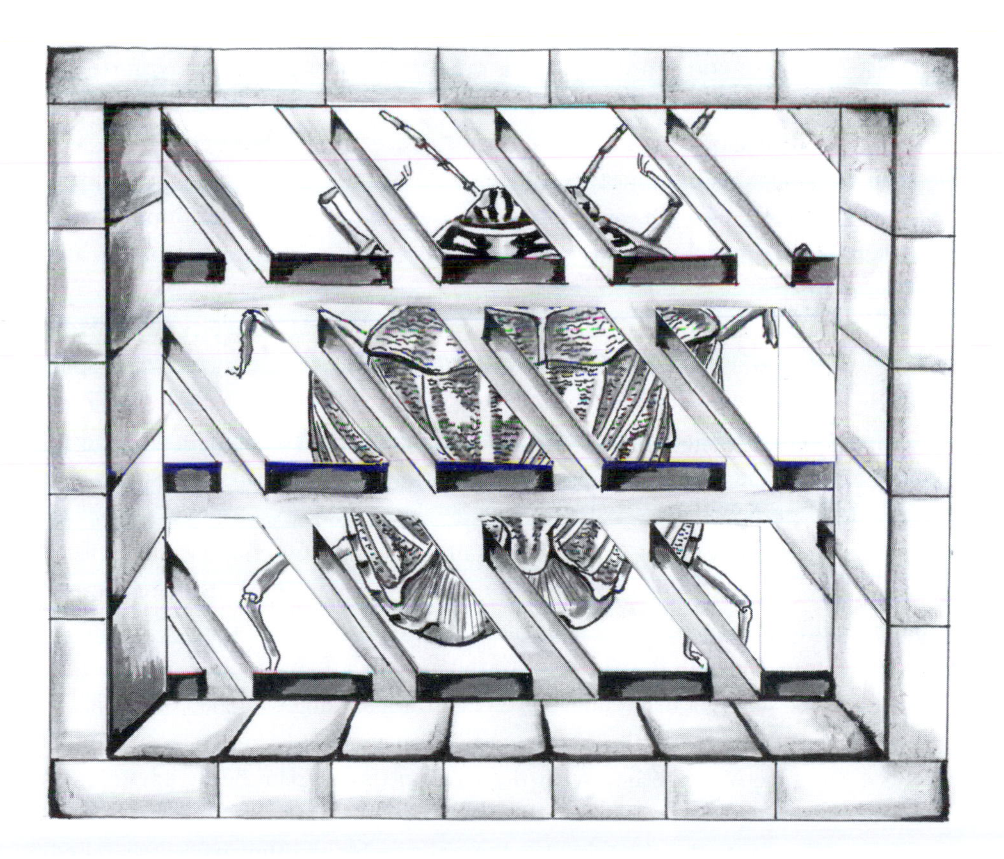

4.1 Introduction

A combination of human population increase and elevated consumerism is changing the character of the biosphere. Contamination, mostly from byproducts of manufacturing and from agricultural chemicals, are the major substances impacting on biological diversity. Pesticides are among these substances and require special focus with regard to insect diversity conservation.

In addition to contamination, landscapes worldwide are being fragmented. In turn, the fragments are gradually being made smaller through a process termed attrition. Among the causal factors are more extensive and intensive agroforestry systems.

Urbanization has become a major force on insect diversity. Quite simply, wildlands, and even agricultural land, are lost when buildings are erected. This is not to say that urbanization is unchallenged by insect diversity conservation. One of the current major opportunities is to ecologically landscape urban areas so as to maintain biological diversity.

Of general concern worldwide is the loss of wilderness. Some would say that there is no wilderness left, with no place being unsullied by human impact, immediate or distal. Unquestionably, for insect diversity conservation, the loss of forests, especially tropical ones, has been and continues to be devastating. Insect species extinctions are occurring on a daily basis as tropical forests are removed. Yet for insects this is not the only biome under severe pressure. Grasslands are also rich in insect life, and they too are being degraded or lost at a phenomenal rate. High domestic animal stocking rates are among the impacts on grasslands and savanna, and encourage desertification. Such simplification in the composition and structure of ecosystems changes insect assemblages and reduces diversity.

Deterioration and loss of aquatic systems is of great concern worldwide, mainly as the quality of human life also deteriorates. Insects have been particularly hard hit from these adverse changes. Many insect species have suffered geographic range retraction and even extinction as water systems have deteriorated. In fact, changes in insect diversity are often one of the first signs of water quality deterioration.

Overcollecting may not at first seem to fall easily into this chapter. But a closer look sees overzealous removal of insect specimens as targetted deterioration of ecological integrity. This can be especially important in those areas with small-range endemics that are highly sought after by collectors.

Many of the impacts listed above are essentially changes to the composition, structure and function of the land mosaic. Some of these impacts are multiplicative when acting together, with the resultant synergism being particularly harmful.

Insect responses to the changing land mosaic are being intensively researched, and are addressed in more detail in the next chapter. Other synergistic effects, such as arrival of new organisms and genes, are also important (Chapter 6), and all these impacts are overlain by the blanket effect of global climate change (Chapter 7).

4.2 Environmental contamination

4.2.1 *Pollution*

Contamination of ecosystems can come about from agriculture, industry and urbanization (Freedman, 1989). The combination of increased human population and increased consumption of resources and energy has, as measured by Gross Domestic Product, increased by 460% over the last century (Maddison, 1995), with current figures likely to increase by 240% by the year 2050 (National Research Council, 1999).

Among systems most affected in terms of changing insect diversity, as a result of environmental contamination, are riverine systems. Some species, such as Tobias' caddis-fly *Hydropsyche tobiasi*, may even be extinct as a result of industrial and urban contamination of the River Rhine (Wells *et al.*, 1983). Indeed, it is well known that aquatic insect assemblages change in response to water pollutants. This has led to the development of water-monitoring programmes based on the relative abundances of certain taxa (Resh and Jackson, 1993). However, the aim of such programmes is not usually to monitor named endemic species but rather to reflect pollution levels. In other words, they do not measure biodiversity at the species level but rather they indicate environmental stress on the system. Thus, they monitor ecological health rather than finer aspects of ecological integrity.

Air pollution has frequently been suggested as a cause for the decline of some butterfly species. But there is little evidence as to whether this is in fact so. It is not clear what exactly is the causal mechanism between levels of air pollution and butterfly decline. If the pollutants affect the larva via ingestion, or any stage, via direct deposition, one would expect most species of Lepidoptera to be affected (Corke, 1999). But since larvae usually select the youngest foodplant leaves, direct ingestion of pollutant deposits should be minimal. Similarly, adult Lepidoptera feed mainly from nectar, which also rarely contains airborne deposits. In contrast, honeydew and sap feeders are much more exposed, and circumstantial evidence strongly suggests that they are indeed vulnerable (Corke, 1999). Interestingly, the most famous industrial melanic of all, the Peppered moth *Biston betularia*, which has survived in smoke-polluted habitats for many decades, as have other industrial melanic moths, is a species that does not feed in the adult stage.

Nevertheless, evidence so far indicates that insect diversity can be remarkably tolerant of air pollution. Although Russian noctuid moths were heterogeneous in their response to pollution from a smelter, neither species richness nor diversity were affected by the pollution (Kozlov *et al.*, 1996). In a further Russian smelter study, studies on a geometrid moth *Epirrita autumnata* produced the surprising result that parasitism rates of this moth were not associated with pollution, indicating that parasitoids were no more sensitive to pollutants than their herbivorous host (Ruohomäki *et al.*, 1996). In contrast, although the larvae of the butterfly *Parnassius apollo* in Finland can excrete metals, this appears to be insufficient to enable it to tolerate high levels of this pollution on its host plant. Relaxation of this heavy metal pollution has enabled this butterfly to widen its geographical range (Nieminen *et al.*, 2001).

4.2.2 *Synergistic effects*

Still very little is known of the effects of pollutants on insect diversity. Although in some cases, pollution impacts may be relatively benign, as in the case of detergents on dragonflies on the island of Mayotte (Samways, 2003a), there may be in reality three other factors to consider. Firstly, pollution may have a long-term effect that is not detected in short-term studies. Secondly, pollution can pulse depending on intensity and frequency of emission activities, and may go undetected unless the monitoring is taking place at just the right time and in the right place. Thirdly, pollution is very often synergistic with other impacts, especially fragmentation and threats from invasive aliens, making it difficult at times to pinpoint the pollution threat and to make recommendations for sound conservation management. Pollution effects can also be synergistic with global climate change (see Chapter 7). Both gaseous pollutants and increased CO_2 concentrations are likely to alter the amount of insect damage to trees. In addition to direct harm to trees by pollutants, damage is often increased through larger populations of herbivores, which cannot be controlled by predators and parasitoids (Docherty *et al.*, 1997).

4.2.3 *Lack of evidence for negative effects*

Not all aspects of pollution are negative for insect diversity. Brändle *et al.* (2000) looked at plant and insect diversity along a pollution gradient in Germany. Their results relate to previous emissions from a smelter that increased the pH-values of the soil as a result of particulate deposition. The prior emissions increased both plant and hemipteran herbivore species richness close to source. In particular, the proportion of specialized herbivores increased with closeness to the smelter, which favours the 'specialization hypothesis' rather than the 'consumer rarity hypothesis' which purports more plant productivity and hence more herbivore individuals which, in turn, would lead to more species. Interestingly, the predatory bugs did not follow the pollution/plant/herbivore gradient, possibly because of (1) a combination of factors including energetic loss between trophic levels (so damping gradients) (Huston, 1994); (2) predators not being host specific (so leading to niche limitation and increased interspecific competition) (Dolling, 1991); and (3) larger home ranges of predators (so damping measurable differences across the landscape) (Brändle *et al.*, 2000), as is the case with ladybirds (Magagula and Samways, 2001).

4.2.4 *Long-term effects*

We cannot leave the topic of environmental contamination without considering possible long-term effects, especially as some metal pollutants have extremely long residence times in soils. Although some insect species are able to survive because they are adapted to cope with a wide range of metal concentrations in their diet (e.g. polyphages), others are vulnerable to poisoning since the levels of metals in their food are normally within limits (e.g. sap suckers). When essential metabolic metals are present in high concentrations due to pollution (along with those not required, such as cadmium, lead and mercury), they disrupt normal biochemistry. Although certain sensitive insects may die from acute or chronic poisoning, others die from deficiency of an essential element through antagonism from a non-essential metal in the diet (Hopkin, 1995). Yet species that are tolerant to pollution, as some of the European examples above, may respond to the subsequent lack of competition and much higher population densities than in an uncontaminated area. These changes at the community level may lead to disruption of ecological processes such as plant litter decomposition (Hopkin, 1995). Similarly, the gaseous pollutants SO_2 and NO_2 increase the performance of herbivorous insects, while the situation with O_3 is more complex, with a range of possible responses (Brown, 1995).

4.3 Pesticides

4.3.1 *Threats posed by pesticides*

Pesticides, especially insecticides and acaricides, would by their very name, appear to be the antithesis of insect diversity conservation (Pimentel, 1991), especially as 5 million tonnes are used annually. But what is the evidence? There has been much speculation but little concrete evidence has been forthcoming. Indeed, there is apparently no verified case of an insect species going extinct primarily from insecticide usage.

Most insecticides are used in the agricultural sector, while those used in urban pest control pose little threat to insect diversity conservation (Samways, 1996c). The problem with pesticides lies mostly in their impact on food chains through bioaccumulation (Moore, 1987). The important point is that it is not generally how poisonous per se a compound is, but rather the persistence of its toxic impact. This is why certain organochlorines, which have been used widely for mosquito and leaf-cutter ant suppression, are so environmentally threatening. Those environmental threats coupled with human health hazards and high costs, are reasons why pesticide usage is being reduced where possible (Pimentel, 1995).

In comparison with biological control, pesticides are generally much more spatially and temporarily explicit. They are sprayed in a particular area and last a particular length of time. This is why insect diversity has proportionately been little affected by insecticides compared with landscape fragmentation and habitat loss.

4.3.2 *Impacts on insect populations*

Longley and Sotherton (1997) have reviewed the effects of pesticides upon butterflies inhabiting arable farmland. Factors determining a species' exposure and susceptibility to particular compounds range from chemical properties of the insecticidal or herbicidal compound, intrinsic susceptibility of the species, exposure of the butterfly-related plants to drifting pesticides, and species-dependent ecological factors determining their within-boundary behaviour and dispersal. This is also why conservation headlands, which are the outer 6 m-wide edges of cereal fields, and which receive reduced and selective pesticide inputs, are a feasible option for maintaining hedgerow insects as well as other organisms (Dover, 1991, 2001), including insectivorous birds (de Snoo, 1999).

One of the problems with pesticides, and principally insecticides, relative to insect diversity conservation, is that insect natural enemies are often differentially killed, and/or their suppressing effect reduced, relative to the host. A manifestation of this is familiar to insect pest managers as secondary pest resurgence. Often, the natural enemies, being more mobile and less cryptic than the pest, are more exposed. The difference in susceptibility between natural enemy and host may also be magnified by the pest being partially chemically resistant. Even outside the crop environment, there may be similar effects. Collembolans, for example, are not affected by DDT and often increase in numbers in treated soils after depletion of their acarine predators, which are susceptible to it (Curry, 1994).

Ivermectin is used to control nematodes and parasitic arthropods in cattle, and has been speculated to be a risk to dung beetles. Although there is indeed an initial depression of dung beetle diversity, after 2 months, populations return to normal (Scholtz and Krüger, 1995) (Figure 4.1). However, some results from Irish pastures have indicated that the use of chemical fertilizers and veterinary drugs such as ivermectin alongside removal of herbaceous field boundaries can be detrimental to dung beetle diversity (Hutton and Giller, 2003).

In the case of grasshoppers treated with deltamethrin, one day after treatment, numbers were significantly reduced. There was no local loss of species, although population levels, especially of flightless bushhoppers, were reduced, even after summer rain (Stewart, 1998). Nevertheless, caution is required. Not only may insecticides be adversely synergistic with other impacts but, as demonstrated by Cilgi and Jepson (1995), they can have subtle effects such as reduced fitness on larval and adult butterflies when deltamethrin is applied at only 1/640th of the field dose rate.

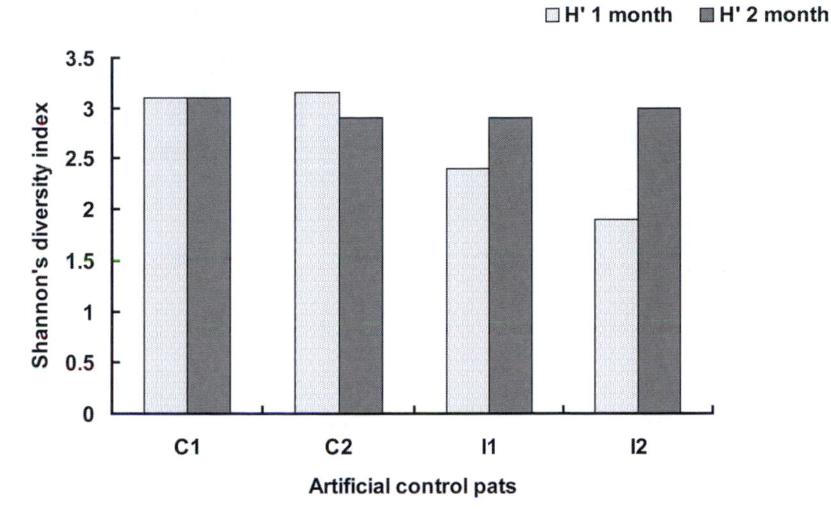

4.1 Comparison of Shannon's diversity index (H′) of dung beetles in artificial control pats 1 month and 2 months after treatment of cattle with a single standard injection of the insecticide/acaricide ivermectin; control paddocks (C1 and C2) and paddocks with treated cattle (I1 and I2). Recovery is virtually complete after 2 months. (Redrawn from Scholtz and Krüger, 1995, with kind permission of Elsevier.)

4.3.3 *Other impacts*

It is not only insecticides that have an impact on insect diversity. The herbicides atrazine and pentachlorophenol can reduce soil collembolan populations by 80%, as well as reducing staphylinid beetles and spiders (Curry, 1994). Such effects are likely to change community structure, albeit temporarily and locally (Ellsbury *et al.*, 1998).

The question that now arises is that with a projected increase in pesticide usage of 270% compared with present levels by the year 2050 (Tilman *et al.*, 2001), the environmental impacts need to be considered in more detail, especially as these impacts are synergistic with other impacts, from increased fertilizer input to landscape fragmentation and invasive alien organisms.

4.4 Agriculture and afforestation

4.4.1 *Scale of the challenge*

Demand for food by a wealthier and 50% larger global population will be a major driver of global environmental impacts. Should past dependencies on agriculture continue, 10^9 ha of natural ecosystems would be converted to agriculture by 2050, with a 2.4- to 2.7-fold increase in nitrogen- and phosphorus-driven eutrophication of terrestrial, freshwater and near-shore marine ecosystems (Tilman *et al.*, 2001). If the global human population stabilizes at 8.5–20 billion individuals, the next 50 years may be the final episode of rapid

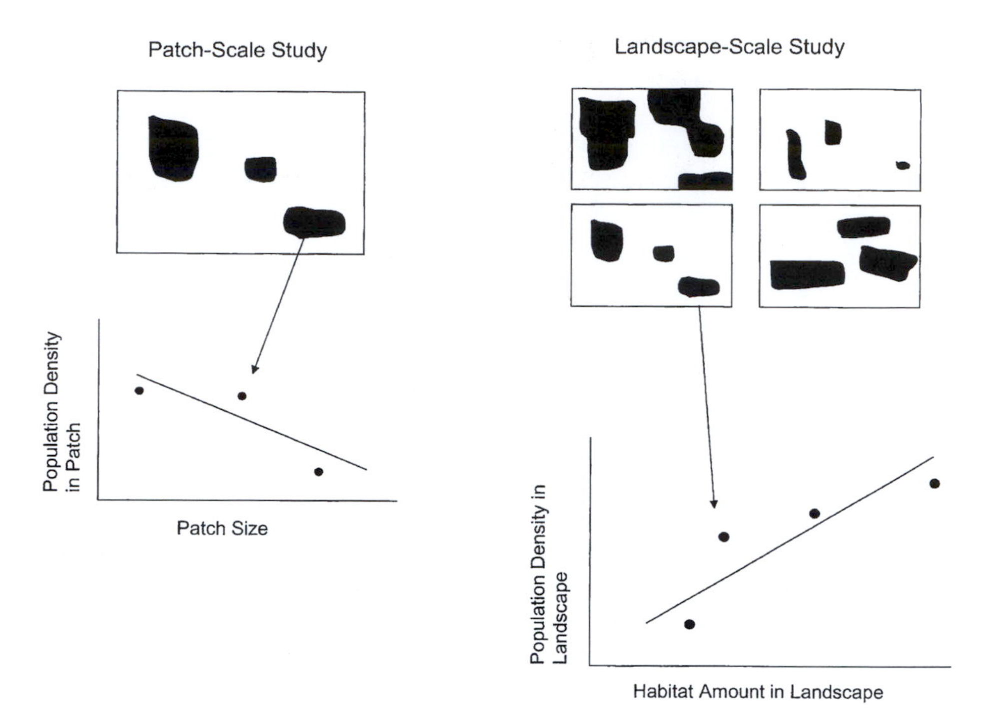

4.2 (Left) Patch-scale study. Each observation represents the information from a single patch. Only one landscape is studied, so sample size for landscape-scale study inferences is one. (Right) Landscape-scale study. Each observation represents the information from a single landscape. Multiple landscapes, with different structures, are studied. Here, sample size for landscape-scale inferences is four. (From Fahrig, 2003.)

global agricultural expansion. These impacts are likely to be devastating for many insect species, with synergistic effects of pollution, pesticides, fragmentation, invasive aliens and global warming taking an immediate toll, followed in time by the effects of ecological relaxation as populations are gradually lost from the small remaining natural fragments.

The results of Léon-Cortes *et al.* (2000) suggest that some common species may decline at the same rates as ecological specialists. In the case of smaller indigenous patches, there is likely to be greater threat from invasive alien plants as well as from proportionately less interior habitat. This decline in patch quality may matter as much as patch size for some of these species (Dennis and Eales, 1997) (see Chapter 10). The debate as to the relative importance of patch quality relative to patch size is an intense one. Fahrig (2003) provides some valuable insights emphasizing that empirical studies of habitat fragmention are often difficult to interpret because (a) many researchers measure fragmentation at the patch scale, not the landscape scale, and (b) most researchers measure fragmentation in ways that do not distinguish between habitat loss and habitat fragmentation per se, i.e. the breaking apart of habitat after controlling for habitat loss (Figures 4.2 and 4.3). Evidence suggests that habitat loss has large, consistently negative effects on biodiversity, whereas this might not be the case with fragmentation.

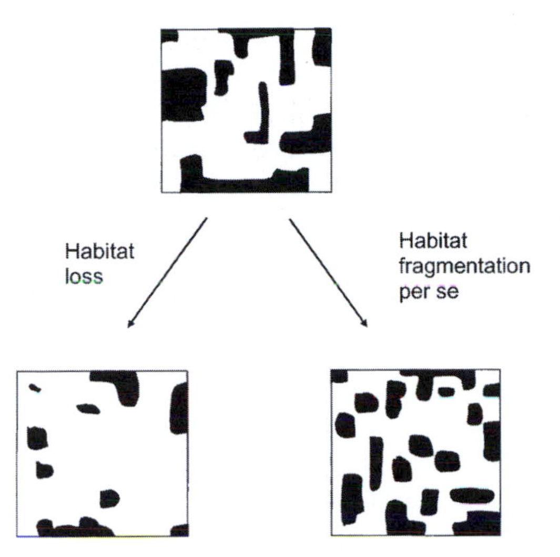

4.3 Both habitat loss and habitat fragmentation per se (independent of habitat loss) result in smaller patches. Therefore, patch size itself is ambiguous as a measure of either habitat amount or habitat fragmentation per se. Note also that habitat fragmention per se leads to reduced patch isolation. (From Fahrig, 2003.)

However, the situation is complex and depends very much on the species concerned and the spatial and historical context of habitat loss and landscape fragmentation.

4.4.2 *Some comparative issues*

Duelli *et al.* (1990) suggest that not all species will suffer declines as the landscape mosaic changes. In Switzerland, 'hard-edge' species tend to be specialists for undisturbed perennial habitats while 'soft-edge' species with a diffuse distribution are mainly associated with annual crops. It seems that in cultivated areas, a mosaic landscape of small-sized crop fields and semi-natural habits maximizes arthropod diversity and decreases the probability for overall extinction, even of rare species. This apparent contradiction between these results and some others may be because the Swiss landscape has been highly anthropogenically disturbed over many centuries, even millennia in places, meaning that the original palaeobiodiversity was ameliorated a long time ago. Nevertheless, the highly modified European landscape is still losing species. Among carabid beetles, large-bodied, habitat specialists are declining the most, seemingly because of their lower reproductive output and lower powers of dispersal (Kotze and O'Hara, 2003). Size, per se, is not the important factor, with weak dispersers and habitat specialists (which can be small), being the most prone (Kotze *et al.*, 2003).

Fragmentation and its synergistic consequences may be more severe in the tropics where, as Soulé (1989) puts it 'the demographic winter will be more severe and longer'. In Costa Rica, forest fragments have more moth species than surrounding agricultural habitats with many of these species utilizing both natural and transformed habitats, such that the forest fragments have halos of high species diversity 1.0–1.4 km from the forest edge (Ricketts *et al.*, 2001). Conversely, alien pine-tree patches have a 30–50 m halo of reduced indigenous grasshopper populations (Samways and Moore, 1991). Similarly, alien pines can reduce macroinvertebrate diversity in streams formerly subject to litter input from *Nothofagus* (Albariño and Balseiro, 2002).

Generalizations however, are likely to be difficult, especially as agricultural habitats differ greatly in vegetal and litter structural and compositional diversity, so resulting in differential effects on different insect species (Samways *et al.*, 1996). Afforested plots in Cameroon with the tree *Terminalia ivorensis* encouraged a rich butterfly assemblage, although they did not provide habitat for some of the indigenous forest species (Stork *et al.*, 2003). Overall interpretation of research results must also be done carefully. Insect species richness declined dramatically in an afforested African grassland landscape, but then it also did in natural forest, indicating that it is essential to compare like with like in terms of habitat characteristics in transformed versus untransformed patches (Kinvig and Samways, 2000). Sun coffee plantations, for example, are less suitable for foraging army ants than shade coffee plantations (Roberts *et al.*, 2000).

Yet some agricultural features enhance insect survival. Indeed, it is difficult to generalize because some insect species decrease in abundance with land-use intensity while others increase, as does species richness, even among taxonomically close groups (Klein *et al.*, 2002) (Figure 4.4).

The damselfly *Lestes barbarus* uses hedges for maturation (Hill *et al.*, 1999), while structural features of the habitat mosaic such as trees, can alter dispersal distance of scarabaeid beetles (Conradi *et al.*, 1999).

Using data for butterfly species in the fragmented European landscape, Thomas (2000) showed that species of intermediate mobility have declined most, followed by those of low mobility, whereas high-mobility species have survived well (Figure 4.5). Compared with the more sedentary species, those of intermediate mobility require relatively large areas where they can breed at slightly lower local densities. Intermediate mobility species have probably fared badly through a combination of metapopulation (extinction and colonization) dynamics and the mortality of migrating individuals which fail to find new habitats in fragmented landscapes. Habitat fragmentation is likely to result in the non-random extinction of populations and species characterized by different levels of dispersal, although the details are likely to depend on the taxa, habitats and regions considered.

4.5 Urbanization and impact of structures

4.5.1 *Cities and insect diversity*

Urbanization is a high-impact activity characterized by high human densities (> 620 individuals per km^2) and an urban–rural gradient (McDonnell and

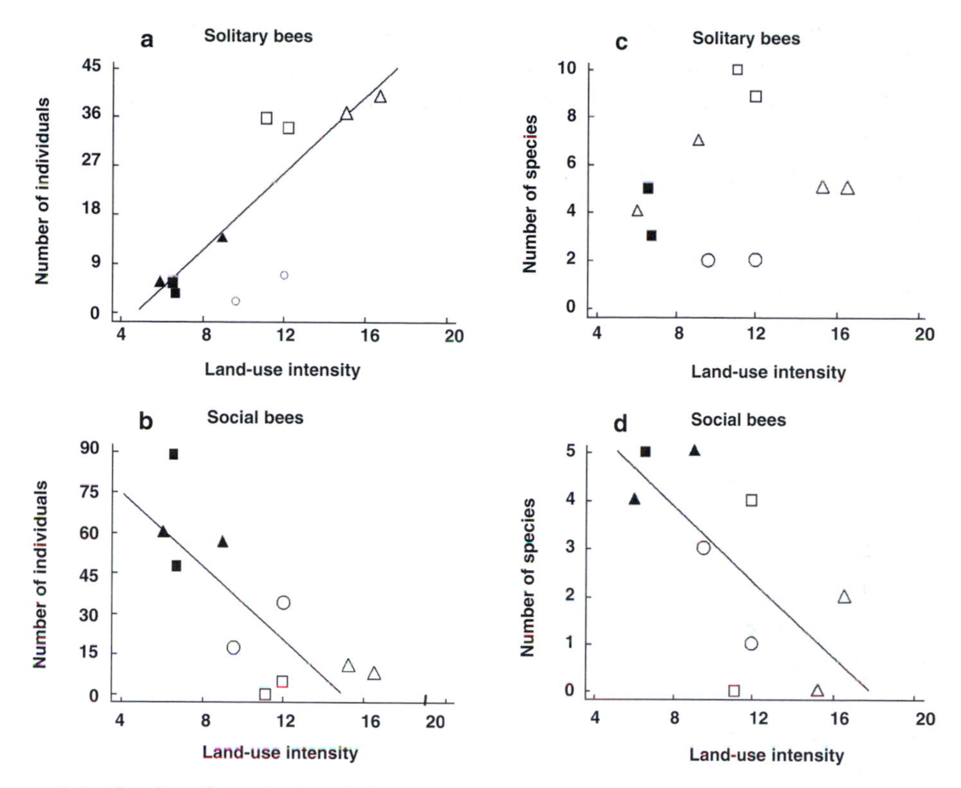

4.4 Land-use intensity in relation to the abundance and species richness of solitary and social bees. Social bees decline in abundance and species richness with increasing land use intensity, while solitary bees do not. (The different symbols represent different land-use types from low-intensity, near-natural forest through to intensively managed agroforestry.) (Redrawn from Klein *et al.*, 2002, with kind permission of Blackwell Publishing.)

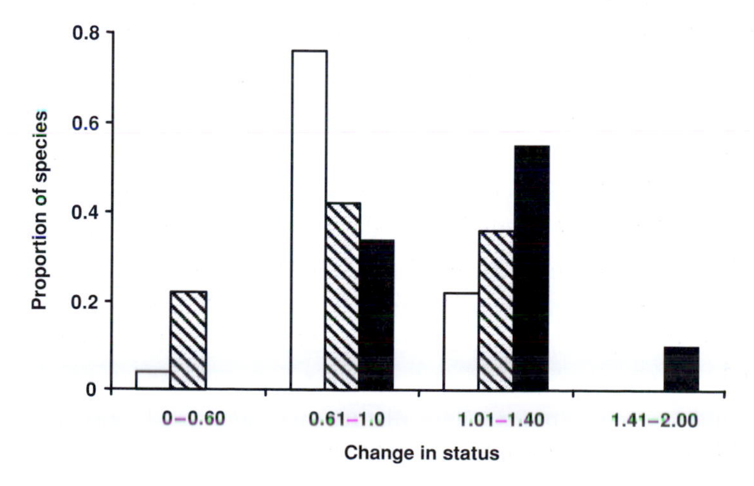

4.5 Rate of butterfly population change in the United Kingdom county of Dorset, between 1970–1984 and 1980–1994, as a function of butterfly mobility: open bars, sedentary species; hatched bars, intermediate species; black bars, mobile species. (From Thomas, 2000, and data from Thomas and Webb, 1984, and Thomas *et al.*, 1998a.)

Rickett, 1990; Pickett *et al.*, 2001). Urban environments are generally warmer than surrounding areas (the 'heat island effect') as a result of rapid heating of urban, hard surfaces.

Generally there is a decline in insect diversity as the city centre is approached (Davis, 1978). However, city parks and gardens, which are effectively 'green nodes', can be rich in species (Bradley and Mere, 1966; Owen, 1991), particularly some butterflies, which can persist in small patches where larval and adult foodplants are available, and vagrancy re-supplies propagules (Hardy and Dennis, 1999). Nevertheless, at least in Australia, maintenance of threatened specialist butterfly species as opposed to more generalist and widespread species may demand rather different approaches for practical conservation (New and Sands, 2002). This is because most species capable of persisting in urban areas depend on their adults adapting to modified habitats, and their immature stages utilizing cultivated alien or indigenous food plants. Alien weeds and inappropriate fire regimes are the greatest threats to these butterflies marooned in patches of remnant bushland.

Using carabid beetles in three northern hemisphere cities, Niemelä *et al.* (2002) emphasized that an urban–rural gradient is modified by local factors and their interactions. Nevertheless, overall beetle species richness decreases, small-sized species increase in dominance and opportunistic species gain dominance with increasing urbanization; the result is no significant changes in total abundance and richness, both in carabids and spiders (Alaruikka *et al.*, 2002).

The local factors that influence insect diversity can be immensely complex, synergistic and variable over time. This means that the mosaic of human structures and disturbances act as a differential filter, allowing certain species to penetrate the urban centre, so long as resources are there to sustain them. In the case of gall-inhabiting Lepidoptera, these species are the robust generalists with regionally wide distributions, broad climatic tolerances and apparently good dispersal abilities (McGeoch and Chown, 1997). Kozlov (1996) further emphasizes that in urban environments, among microlepidoptera at least, extinction of stenotopic species and those with limited dispersal is practically irreversible in the city centre. This underscores the importance of maintaining structurally complex green nodes (Clark and Samways, 1997; Whitmore *et al.*, 2002) within urban environments, and also maintaining specific microclimatic conditions when certain species of conservation significance are the focus. For urban dragonflies, increased habitat heterogeneity begets increased species richness (Samways and Steytler, 1996). Maintenance of urban insect diversity may involve some intermediate disturbance (Blair and Launer, 1997), as was illustrated when a pipeline was laid through an urban butterfly reserve (specifically for *Aloeides dentatis dentatis*) in South Africa, ironically making conditions more suitable for the butterfly (Deutschländer and Bredenkamp, 1999).

Understanding conservation of insect diversity in urban environments also means understanding insect functional types (McIntyre *et al.*, 2001) and trophic relationships. Some surprises have surfaced. Monophagous herbivores, for example, were found to be less affected by urbanization than polyphagous herbivores (Denys and Schmidt, 1998). This of course depends on which plants are available for particular insects. If the foodplant is available, certain insects can thrive in an urban environment, and this may not necessarily be because natural enemies have been excluded (Ruszczyk, 1996). Nevertheless, natural enemies can be affected, with parasitoids, especially rare ones, being more strongly affected by isolation than predators (Denys and Schmidt, 1998).

4.5.2 *Structures and insect behaviour*

Structures, whether vegetational or constructional, affect the behaviour and movement of insect species across the landscape (Wood and Samways, 1991). Even within species, and even similar but differently constructed structures, can make a difference. Bridges over a German river, for example, had a differential impact on the flight behaviour of different individuals of the damselfly *Calopteryx splendens*, with the widest and lowest bridge inhibiting the movement of 70% of the individuals (Schutte *et al.*, 1997).

Limestone quarries in the Czech Republic are beneficial for xerophilous butterflies as they create considerable habitat heterogeneity, from earliest-succession barrens to later-succession scrub (Beneš *et al.*, 2003). Importantly, these quarries increase in value when natural xerophilous sites in the area are also preserved and adequately managed.

4.5.3 *Impact of roads*

Roads are well known to influence or limit insect movement, including grasshoppers (Weidemann *et al.*, 1996), ground beetles (Mader, 1984), and bumblebees (Bhattacharya *et al.*, 2003), although it depends on the size and type of road, and traffic density. Small dirt tracks may not inhibit tettigoniid movement (Samways, 1989b) and may even encourage ant nesting because conditions are dry and warm (Samways, 1983b). In the case of busy roads, traffic may cause mortality of many insects, and thus supply food for scavenging ants, so enhancing their numbers (Samways *et al.*, 1997). Also, small-scale disturbances caused by vehicle activity can be of value in producing locally abundant forage resources in less intensively managed British grasslands (Carvell, 2002). Similarly, British trackways bounded on both sides by hedgerows ('green lanes') had more flowers and more bumblebees than field margins (Croxton *et al.*, 2002).

Road impacts are not to be underestimated, with 4 784 351 ha of land in the USA alone being given to tar and tyre with enormous ecological consequences (Trombulak and Frissell, 2000). The impacts often extend well beyond the actual road itself, changing stream courses (Jones *et al.*, 2000), affecting wetlands

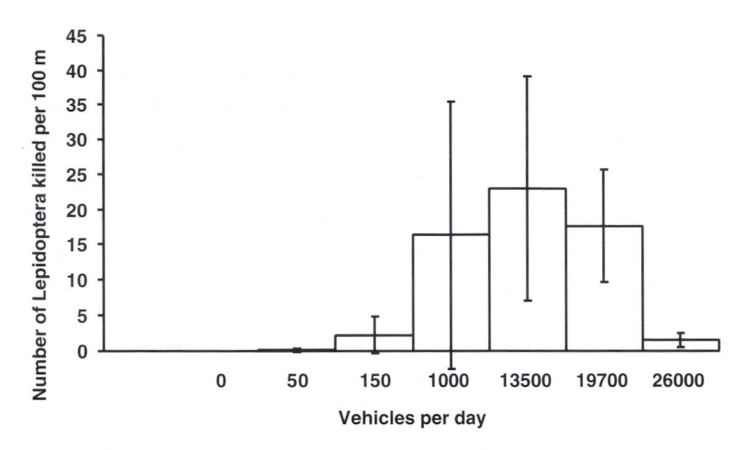

4.6 Average (± 1 S.D.) number of individuals of dead Lepidoptera sampled along 100 m of roadway at various traffic rates in Illinois, USA. (Redrawn from McKenna *et al.*, 2001.)

(Findlay and Bourdages, 2000), soil fauna (Haskell, 2000) and encouraging plant invasions (Parendes and Jones, 2000). The corridor adjacent to the road where ecological effects extend out from the road is termed the 'road-effect zone' (Forman and Deblinger, 2000). This may be very narrow (4 m) for eurytopic scavenging ants (Samways *et al.*, 1997) but wider (20 m) for more stenotopic ant species (Keals and Majer, 1991). This is not to say that road verges of 4 m or even 20 m are sufficient for all insect species, some of which require much larger areas, either because they are prone to road-kill or their habitat has been adversely affected by the road, i.e. they can survive only beyond the road-effect zone.

Although road verges can be highly beneficial for some species, they nevertheless suffer mortality from traffic. Between 0.6% and 7% of butterfly and burnet populations are killed by traffic on British main roads (Munguira and Thomas, 1992). Although this figure is insignificant compared with natural mortality factors, road traffic mortality may be very high in some areas (Samways, 1994). In Illinois, USA, McKenna *et al.* (2001) estimated that the number of Lepidoptera killed along roads in the state was likely to be more than 20 million individuals, with the number of Monarch butterflies *Danaus plexippus* killed exceeding 500 000 individuals. The level of mortality was proportional to traffic volume, with a peak at 13 500 vehicles per day, but dropping at very high volumes (Figure 4.6).

4.5.4 *Canalization*

In terms of threat, there is one further major structural consideration in insect diversity conservation, which involves hydrological engineering. Canalization (i.e. channelization) has had an adverse impact on aquatic insects in industrialized countries, particularly dragonflies in Germany (Ott, 1995). Regulated rivers in Germany also reduce the number of floodplain gravel bars from 60 to 3.5 per kilometre. This inevitably affects the abundance of ground beetles and rare beetles, as well as other arthropods, particularly in the detritivore guild (Smit *et al.*, 1997). It is likely, however, that there are many subtle general changes in insect diversity in urban areas, because changes in hydrology associated with urbanization create 'hydrologic drought' by lowering water tables, which in turn alters soil, vegetation, and pollutant removal functions (Groffman *et al.*, 2003).

4.5.5 *Reservoirs*

Although small, weed-fringed, farm dams with constant water-levels encourage a range of South African dragonfly species, these are all geographically widespread, habitat generalists (Samways, 1989c), with the rare and endemic species mostly inhabiting clear, unspoilt streams and natural wetlands (Samways, 2002c). In another study, increases in local abundance of Odonata species were two-fold, with concurrent loss of one lotic species, when a dam was constructed (Steytler and Samways, 1995). Such increases occur in other taxa, with certain *Simulium* fly species increasing in the rapids below impoundments (DeMoor, 1994). As well as impeding flow, impoundments also regulate it, such that natural drought/flood cycles and ecological succession are forestalled, depriving species that need a particular seral stage or appropriate habitat (Homes *et al.*, 1999).

4.5.6 *Urbanization and insect extinction*

The impacts of urbanization on insect diversity are complex and often synergistic. Connor *et al.* (2002) point out that, for the San Francisco Bay area, the clash between high insect diversity and high human density led to the first recorded extinction, the Satyr butterfly *Cercyonis sthenele sthenele*, as well as other species, in the USA. In this area, habitat loss and invasive species, such as the Argentine ant *Linepithema humile*, and pathogens causing mortality of oaks and pines, are continuing to ameliorate insect diversity.

4.6 Deforestation and logging

4.6.1 *Scale of the impact*

The tree canopy is the ultimate domain for insect diversity (Watt *et al.*, 1997), although this is not to overlook the richness of the forest floor. Besides millions of species living in the canopy, there are countless billions of interactions taking place. Just stand and listen to the number of tettigoniids alone singing from the tropical forest canopy. Along with coral reefs, the forest canopy

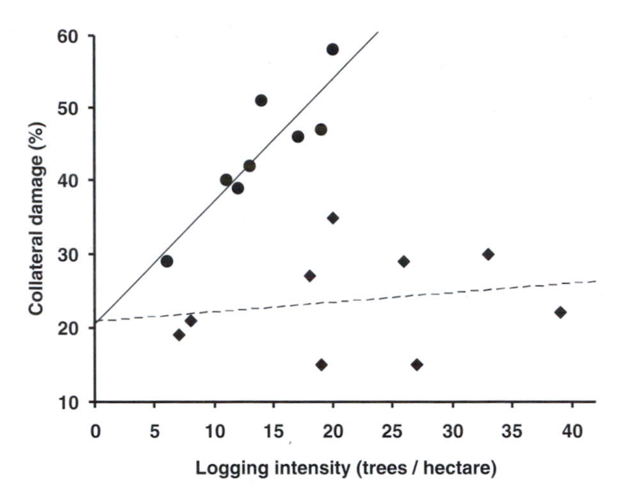

4.7 Unregulated commercial logging can cause heavy forest damage. The relationship between harvest intensity and collateral damage in tropical forests subjected to unregulated (South-East Asia; closed circles) and regulated (north Queensland, Australia; closed diamonds) logging is illustrated. Broken and unbroken lines were fitted by linear regressions. Collateral damage is defined as the percentage of non-harvested trees ($\leq$10 cm diameter) that were inadvertently killed during logging. (Redrawn from Laurance, 2000; data compiled from Crome *et al.*, 1992, with permission from Elsevier.)

is fabulously rich and complex in biodiversity. Yet all is not well with this biotic jewel.

By far the biggest land-use change on the planet, in the shortest time, has been conversion of 16 million square kilometers of forest in 50 years (World Commission of Forests and Sustainable Development, 1999). Tropical forests are being lost at least at 130 000 square kilometers per year, although we must bear in mind that not all is of ancient origin (Mayle *et al.*, 2000). Besides being home to half of the world's species and most of the insects, these forests regulate movement of water across the land, modulate local and regional climates through transpiration, and they play a major role in determining current atmospheric concentration of CO_2 through their high above- and below-ground productivity and huge standing crop (Malhi and Grace, 2000).

The root of the problem is that logging is usually the fastest way for corporations and investors to make lots of money quickly in the tropics (Laurance, 2000) (Figure 4.7), although in reality, the proximate causes are diverse and vary from region to region (Geist and Lambin, 2002). And the stakes are high. In developing nations, forest tracts currently allocated for logging are at least 8–10 times larger than the limited areas set aside for parks and reserves (Johns, 1997). Even reserves are not safe, with accidental deforestation from forest fires becoming greater than deliberate conversion in some areas (Cochrane and Schulze, 1999), threatening unburned patches (Siegert *et al.*, 2001).

So, what is to be done? Whitmore (1999) points out the difference between deforestation (complete removal of trees, often with concurrent burning) and timber extraction. While deforestation dramatically changes structural, compositional and functional biodiversity (with concurrent changes even in soil-inhabiting termites (Eggleton *et al.*, 1996)), timber extraction, in contrast, simulates natural gap-forming processes. This is so long as only a few trees per hectare are removed carefully, and damage to the forest floor is minimal. This low-impact logging, encapsulating good practice, is becoming increasingly the norm (Whitmore, 1999). Nevertheless, much research is required relative to management practices.

4.6.2 *Tropical forest logging*

Forest clearance in Cameroon, and conversion to farm fallow, caused a 50% drop in butterfly and termite species richness (Watt *et al.*, 1997). Clearance and conversion to forest plantation in comparison with complete clearance caused a reduction in butterflies and leaf-litter ants of 15% and of 40–70% in arboreal beetles and termites. These figures are likely to be conservative as more species will probably disappear over time after logging. Furthermore, type of management plays an important role, with partial manual plantation plots having more insect diversity than plantation plots established after complete clearance. Similarly, in Indonesia, termite species and abundance decreased in proportion to intensity of land use. Primary forest had 34 termite species, while cassava gardens had only one (Jones *et al.*, 2003).

Liberation thinning, which is a management method in favour of potential crop trees, seems to most affect the more specialized West African nymphalid butterfly species with smaller geographic ranges, thus risking loss of regional diversity (Fermon *et al.*, 2000). Similarly, selective logging of tropical forests in Indonesia significantly decreases butterfly diversity, at least during the first 5 years (Hill *et al.*, 1995). Furthermore, there is distinct spatial heterogeneity in vegetation structure within these logged forests, which in turn leads to heterogeneity in butterfly abundance corresponding to availability of suitable forest (Hill, 1999). Some of this may be due to natural population dynamics, such as colony foundation and extinction in termites (Eggleton *et al.*, 1996), which underlies the effect of disturbance.

Nevertheless, there is a significant loss of diversity and taxonomic quality with increasing levels of forest disturbance, with some taxa such as moths being more sensitive to these changes than other taxa, such as beetles (Holloway *et al.*, 1992). Indeed, changes in one taxon following disturbance do not necessarily correlate with those of another (Lawton *et al.*, 1998). Certain taxa, such as arboreal dung beetles in Borneo (Davis and Sutton, 1998) and ants in Ghana (Belshaw and Bolton, 1993) can even survive in agricultural areas after removal of primary forest. Such findings however, must not undermine efforts to conserve tracts of primary forest, which may be the last refuge for some species (Fermon *et al.*, 2001). This is emphasized by Castaño-Meneses and Palacios-Vargas' (2003) findings that Mexican tropical deciduous forest disturbance results in a decrease in ant density and diversity with a resultant change in the energy recycling in the ecosystem.

4.6.3 *Disturbance and maintenance of late successional stages*

The significance of having forest heterogeneity in cooler forests is similar to that in tropical forests, with certain insect species preferring a particular level of disturbance (Spagarino *et al.*, 2001). This underscores the importance of rotational or selective logging, which benefits particular species (Kaila *et al.*, 1997), as well as the importance of maintaining patches of virgin forest with sufficient extent of interior conditions to circumvent gradual loss of species through ecological relaxation (Lövei and Cartellieri, 2000). Indeed, virgin patches may be critical for maintaining local stenotopic species (Trumbo and Bloch, 2000), which are often large-sized, reluctant-to-disperse endemics (Michaels and McQuillan, 1995). This serves to emphasize the importance of maintaining virgin, old-growth forest nodes which are exempt from any form of rotational management. Such patches (the larger, the better (Horner-Devine *et al.*, 2003)) and patch groups need to be represented across a region (Niemelä, 1997). Such old-growth forests also represent habitat predictability, essential for many insect species (Nilsson and Baranowski, 1997). Such late successional patches may have special habitats such as soil conditions (Eggleton *et al.*, 1995), litter depth (Lomolino and Creighton, 1996), fungi (Økland, 1996; Jonsell *et al.*, 1999) or simply large logs (Økland *et al.*, 1996; Grove and Stork, 1999; Kelly and Samways, 2003) which particular insect specialists need. In addition, special habitats, such as tree-holes, where particular ecological interactions take place (Fincke *et al.*, 1997), or unique sites, such as Monarch butterfly roosts (Brower *et al.*, 2002), must also be considered. In summary, the precautionary principle of keeping all the parts, especially all the parts of old forests, is essential for the survival of current insect diversity.

Natural spatial heterogeneity of butterflies in primary tropical forest may be obscured by logging (Willott *et al.*, 2000). Such spatial considerations are important because, at least in Borneo, low-intensity logging in forest in close proximity to primary forest does not necessarily reduce the species richness or abundance of butterflies, although assemblage composition is changed. This is possibly because moderate levels of disturbance may increase butterfly diversity and also because the primary forest was not in its final successional stage, owing to earlier disturbance, possibly from drought and fire about 100 years ago. This may be why these results are apparently conflicting with others where there was a reduction in lepidopteran diversity in logged forest (Willott *et al.*, 2000) (Figure 4.8).

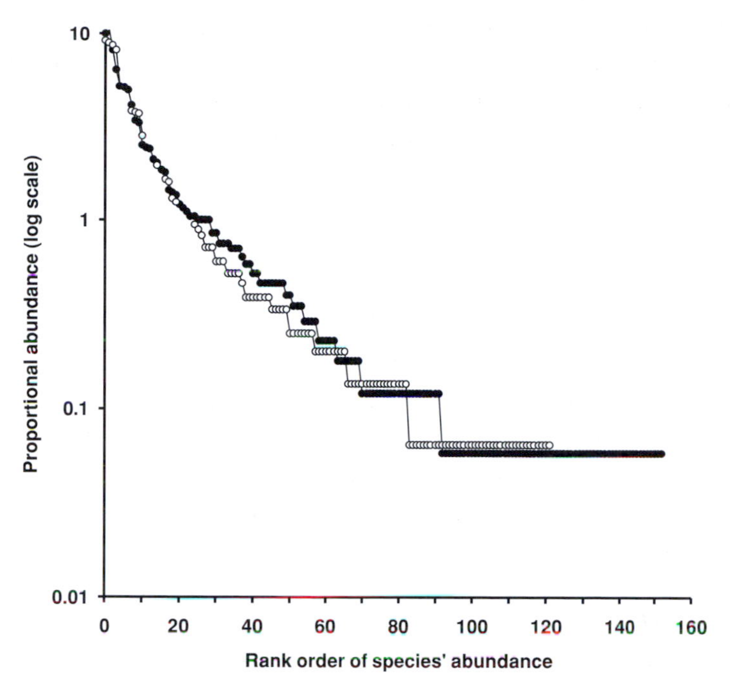

4.8 Bornean butterfly species ranked according to their abundance in primary (open circles) and logged (closed circles) forest. (Redrawn from Willott *et al.*, 2000, with kind permission of Blackwell Publishing.)

It is essential that stenotopic insect species with small geographical ranges are considered in any large-scale assessment, as these species may be confined to particular closed-canopy habitats tolerant of only very minimal disturbance (Spitzer *et al.*, 1997). Conversely, some species are lost locally when there is no disturbance (Brown, 1997). A further point is that different taxa respond differently to disturbance. In Bolivian tropical forests, ants and cockroaches were more abundant in undisturbed areas, whereas grasshoppers and lepidopteran larvae were more abundant in burned and partially logged areas (Fredericksen and Fredericksen, 2002), this being a reflection of life styles and preference for particular amounts of insolation. While much focus has been on moist tropical forests, some dryer forests can also be rich in insect and other arthropod species, with Australian eucalypt forests being much more species-rich than previously thought (Majer *et al.*, 2000).

Saproxylic insects are a rich and varied dominant functional group which depend on dead wood and the old trees that generate it (Grove, 2002). These insects are sensitive to forest management, with managed or secondary forests generally having a depauperate fauna. Forest hygiene, where old and dead wood is removed, coupled with saproxylic insects' often weak powers of dispersal and long-lived larval stages, has severely threatened this group of insects, with many species in Western Europe now regionally extinct.

4.7 Transformation of grasslands, savanna and Mediterranean-type ecosystems

4.7.1 *Grasslands*

Grasslands in the northern hemisphere alone once covered some 600 million ha, with very few natural remnants left today. Interestingly, across the northern hemisphere there is remarkable similarity in the grasshopper fauna (Lockwood and Sergeev, 2000). Such grasslands are not homogeneous, with the shrub and desert zones deserving highest conservation priority as they have the highest number of rare species. In some respects, this situation is magnified in the southern hemisphere. Having had no extensive Pleistocene ice sheets, yet long periods of relative geological stability coupled with an erosional, topographically complex range of landscapes, the south has an even richer endemic insect fauna (Samways, 1995), with grasshoppers, for example, 47% endemic in South Africa and 90% endemic in Australia. Human-induced fire (Greenslade, 1993; Burchard, 1998), direct habitat destruction such as removal of thick bush (Scholtz and Chown, 1993), afforestation, agricultural development, as well as hunting, pastoralism and livestock rearing have all had major impacts on insects of the southern non-forest ecosystems (Samways, 1995).

Overgrazing has also been extensive in other areas (Samways and Sergeev, 1997), resulting in depression of populations of some species but also providing opportunities for outbreaks of others (Samways and Lockwood, 1998; Lockwood *et al.*, 2000). In North America, tallgrass prairie once covered 68 million ha but has now been reduced to scattered remnants. Besides the effects of loss of area per se, the prairie fragments are subject to invasion of agricultural pests from the surrounding fields. Corn-rootworm beetles *Diabrotica* spp. damage the flowerheads of adjacent prairie composites, reducing seed set (McKone *et al.*, 2001).

Human impacts on these open, non-forest areas have probably been intense for long periods of time (Flannery, 1994) and what we are seeing today in many areas, from the Russian steppes (Bei-Bienko, 1970) to the Australian wilderness (Greenslade, 1993), is a post-disturbance fauna, and one that has been very under-researched (Scholtz and Chown, 1993). One of the problems with determining the causal factor of human impact on insect diversity is that the various forms of disturbance are synergistic. Fire, overgrazing, impact of invasive aliens all interact, and can aggravate natural drought/flood cycles. Burning of natural grasslands, for example, amplifies the effect of natural cold-air drainage on grasshopper abundance and diversity (Samways, 1990).

Increasing levels of domestic livestock impoverish insect diversity, but at least in Africa, this simulates the effect of indigenous hoofed mammal trampling and grazing (Rivers-Moore and Samways, 1996), which becomes amplified at waterholes that are congregation points (Samways and Kreuzinger, 2001). Even tourist impact can be additive upon natural impacts, with the Greek moth *Panaxia quadripunctaria* suffering human trampling (Petanidou *et al.*, 1991).

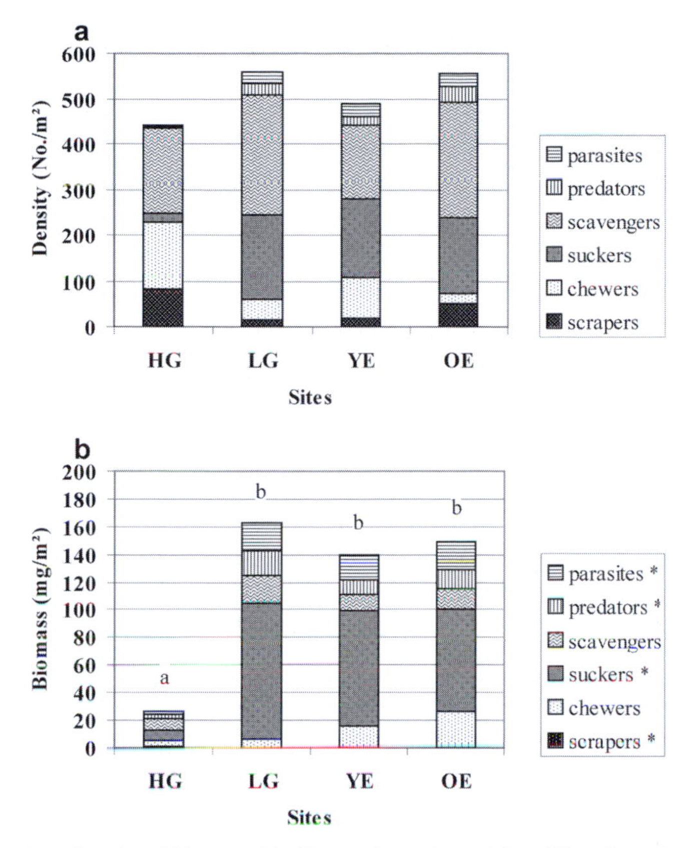

4.9 Density (a) and biomass (b) of insects in each trophic guild, collected at four sites under different grazing regimes and age of exclusion from cattle, at a montane grassland in Central Argentina. (HG) Heavy grazing; (LG) low grazing; (YE) young enclosure (7 years old); (OE) old exclosure (19 years old). Different letters above bars denote significant differences in biomass (total and for (*) marked guilds) among sites. (From Cagnolo *et al.*, 2002, with kind permission of Kluwer Academic Publishers.)

Grassland deterioration and consequent loss of insect populations and extinction of species have illustrated the vulnerability of the grassland fauna, many species of which have a substantial impact on soil fertility and plant growth (Curry, 1994). Grasshoppers in the arid Karoo, for example, convert plants to nutrient-rich frass, and do so much faster than sheep (Milton and Dean, 1996). Also in the Karoo, monkey beetles (Scarabaeidae, Hopliini), which are pollinators, are influenced by level of grazing, with a shift away from perennial and bulb pollinator guilds towards those favouring weedy annuals in overgrazed areas (Colville *et al.*, 2002).

Cagnolo *et al.* (2002) showed that in the grasslands of montane Argentina, abundance, richness, diversity and biomass of insect assemblages were minimal in the most intensively cattle-grazed area (Figure 4.9). In addition, besides changes in taxonomic composition, intensively grazed areas had fewer secondary consumers, with chewers replacing suckers as the most abundant herbivore group. Nevertheless, there are also many functionally insignificant species in grassland and some of these are threatened and restricted to 'island' reserves. Especially threatened are some Orthoptera species (Rentz, 1993; Samways, 1997a) and Lepidoptera species (New, 1993). In reality, it may not necessarily be that grasshoppers and butterflies of open habitats are any more threatened than other taxa in the same or different habitats. It may be that we are simply seeing declines and losses in these conspicuous insects.

It is among the huge species richness of the tree canopy (Watt *et al.*, 1997) where the greatest population and species losses are actually taking place (Mawdsley and Stork, 1995). The important point, however, is not which ecosystem is worse off, but are the threats similar, and are there principles that can ameliorate the threats whatever the ecosystem? This point is addressed in Part III.

4.7.2 *Mediterranean-type ecosystems*

Another low-canopy ecosystem that deserves special mention is the fynbos (Cape Floristic Region), where many narrow endemic plant–insect species interactions occur (Wright, 1993). This, and other Mediterranean-type ecosystems (MTEs), are under severe pressure, probably proportionately more than in any other system (Hannah *et al.*, 1995). This inevitably means that many insect species in these systems are threatened (Samways, 1998b). The salient point here is that this anthropogenic pressure must be viewed against the fact that many insects in MTEs are narrow endemics (e.g. Wright and Samways, 1998). As such, they are 'pre-adapted' to surviving in a small area, and their survival becomes an all-or-none affair. If the human impact is intense and squarely overlays their focal population, their demise is highly likely. If, however, the impacts leave them in an undisturbed fragment, their chances of survival are high, so long as there is not too much of an adverse context effect from the surrounding disturbance matrix. It is almost as if we have 'unlucky' versus 'lucky' endemics. The giant flightless cockroach *Aptera fusca* and the Conspicuous malachite damselfly *Chlorolestes conspicuus* have a 'lucky' home on top of Table Mountain despite being surrounded by the city of Cape Town (Figure 4.10).

In the MTE of California, the response of arthropods to habitat fragmentation is complex, and depends very much on the taxon in question (Bolger *et al.*, 2000). Nevertheless, fragment area and edge effects were generally so significant, along with the impact of the alien Argentine ant *Linepithema humile*, that in all likelihood, trophic relationships within the community are changing.

4.8 Deterioration and loss of aquatic systems

4.8.1 *Canalization and synergistic impacts*

Besides canalization and impoundments (Section 4.5), there are other changes to aquatic systems that threaten insect diversity. A first consideration

4.10 The giant flightless cockroach *Aptera fusca* (a) and the Conspicuous malachite damselfly *Chlorolestes conspicuus* (b) are narrow-range, Western Cape endemics that find refuge on top of Table Mountain, yet they are surrounded by the city of Cape Town (c) (see next page).

4.10 (*cont.*)

is that stream invertebrate communities are structured to some extent by the type and intensity of disturbance (Power *et al.*, 1988; Resh *et al.*, 1988). But how resilient are stream faunas? The answer depends on the intensity and frequency of the disturbance, and relative sensitivity of the responding taxa and ecological relationships. In contrast, when disturbances are spatially confined and relatively short-lived, recovery of the invertebrate aquatic community can be very rapid, with more complex communities at the small scale (< 1 m^2) being the most resilient (Death, 1996). Even at the landscape level, certain aquatic insect populations can recover remarkably quickly, with populations returning to former levels within a year after severe floods (Samways, 1989a).

The threats to aquatic systems are often synergistic with loss of habitat (including damming, canalization, water diversion and draining), isolation of source habitats, pollution and threats from alien organisms all contributing, as seen by the threats to British aquatic insects (Shirt, 1987). One of the reasons threats to aquatic insects can be so severe is that water bodies are relatively small, with lakes and marshes generally only small patches in the extensive terrestrial matrix (Angelibert and Giani, 2003). Arrival of new propagules becomes increasingly small as water bodies are lost and populations in source areas decline. Indeed, loss of wetlands worldwide is of considerable concern. In Finland, for example, the decrease in butterfly species has been greatest in those species living in bogs and fens (Saarinen *et al.*, 2003).

An additional factor is pollution which can also be synergistic with other impacts. A pollutant can easily disperse through a relatively small body of water, as well as be carried downstream. Recolonization can be restricted, as the lateral dispersal of adult insects of headwater streams appears to be very limited (Griffith *et al.*, 1998).

4.8.2 *Dragonflies as an example*

Work on dragonfly conservation in recent years has enabled a focus on what really are the threats rather than supposed threats. For dragonflies at least, perhaps because they are predators, it does not matter very much whether the riverine canopy is composed of alien or indigenous vegetation (Samways, 2003a,b) as long as the right proportions of sunlight versus shade are present (Steytler and Samways, 1995). This does not however, mean that alien plants are not harmful. The problem comes when the trees are invasive and convert grassland banks to alien forested banks. Such dense-canopy aliens include *Acacia mearnsii* and *A. longifolia* which are threatening at least 12 highly localized sun-loving odonate species (Samways and Taylor, 2004). Almost perversely, the interpretation behind this is seen when plantation trees, such as pines, are introduced along a grassland stream. What actually happens is that conditions now mimic forest stream conditions so disfavouring the grassland species while favouring forest species (Kinvig and Samways, 2000).

The dragonfly fauna also depends on the type of stream and whether it is constant in flow and level, or not. Large, savanna streams subject to the vagaries of El Niño, for example, generally have an opportunistic habitat-tolerant fauna, components of which, nevertheless, can be washed away when conditions are particularly harsh (Samways, 2003c). In more climatically stable areas, where streams have a more constant, perennial flow, endemic species packing can be high, with threats such as overextraction of water, alien plant invasion and alien fish predation all impacting on the dragonflies (Samways, 1995). However, close-focus on actual cause is essential. Englund and Polhemus (2001) point out that introduced rainbow trout *Oncorhynchus mykiss* have posed little threat to the endemic Hawaiian damselflies. This contrasts with alien poeciliid fish, which are clearly destructive (Englund, 1999).

4.8.3 *Aquatic ecotones and marshland*

Aquatic habitats, which also include wetlands, worldwide are of major biodiversity significance. The important point also being that many of these waterbodies are unique, from the tiny and vulnerable clear streams in the caves of Table Mountain, Cape Town, with an unusual endemic fauna (Sharratt *et al.*, 2000), to the enormous Okavango system of Botswana. Yet it is not simply a wetland on the one hand or a dryland on the other, rather a series of aquatic ecotones (Samways and Stewart, 1997). Pullin (1999) focuses on these dynamic, marginal areas, and points out that these habitats have long been threatened by drainage for agricultural use and by loss of water to surrounding land. On top of this, is increasingly severe drying out caused by anthropogenic climate change. But the problems extend beyond simply drying up, to those associated with rapid changes in water levels. Excessively long submergence increases mortality of sawflies (Lejeune *et al.*, 1955) and butterflies (Webb and Pullin, 1998; Joy and Pullin 1999). The extinction of the British Large copper butterfly *Lycaena dispar dispar* and decline of the Heath fritillary *Mellicta athalia* are likely to herald changes in insect diversity in wetlands throughout the world, and that localized, habitat specialists need to be watched very carefully as early responders of possible permanent change.

4.9 Pressure on special systems

4.9.1 *Oceanic islands*

Many oceanic islands have been subject to intense human disturbance (Quammen, 1996). Besides direct loss of indigenous habitat, especially on lowlands, there are many introduced species on islands, from ants to *Albizia* trees. Such introduced species may be additive upon the island system or they may eliminate or suppress indigenous species through competition, predation or disease. Global climate change is apparently already aggravating the situation (Chown *et al.*, 2002). The overall prognosis for some of these islands thus remains grim, with the current challenge being simply to maintain existing biodiversity (Clout and Lowe, 2000).

Insect diversity on islands is often skewed, with some taxa not being represented. This is the result of the sweepstake effect, where only certain taxa successfully land and colonize an island. Many of those that naturally invaded new islands or were marooned on islands that became separated from the mainland developed into island endemics. This is because the island is separated from neighbouring suitable habitat by a hostile environment, the sea. Counterintuitive as it may seem, many island endemics are not more unusual than their mainland counterparts. Besides adaptive radiation, where species have evolved sympatrically into different niches and where they have acquired evolutionary stability, there is also fugitive radiation. This, according to Adsersen (1995), is the appearance of 'weak' species, which are very local and have to evolve further to avoid extinction. Many insects appear to fall into this latter category and maintain remarkably small populations (Samways, 2003a,b), which presumably are highly susceptible to adverse changes that have not previously been encountered. This emphasizes the risks of synergistic effects of global warming and invasive aliens, which are impacting severely on some island faunas.

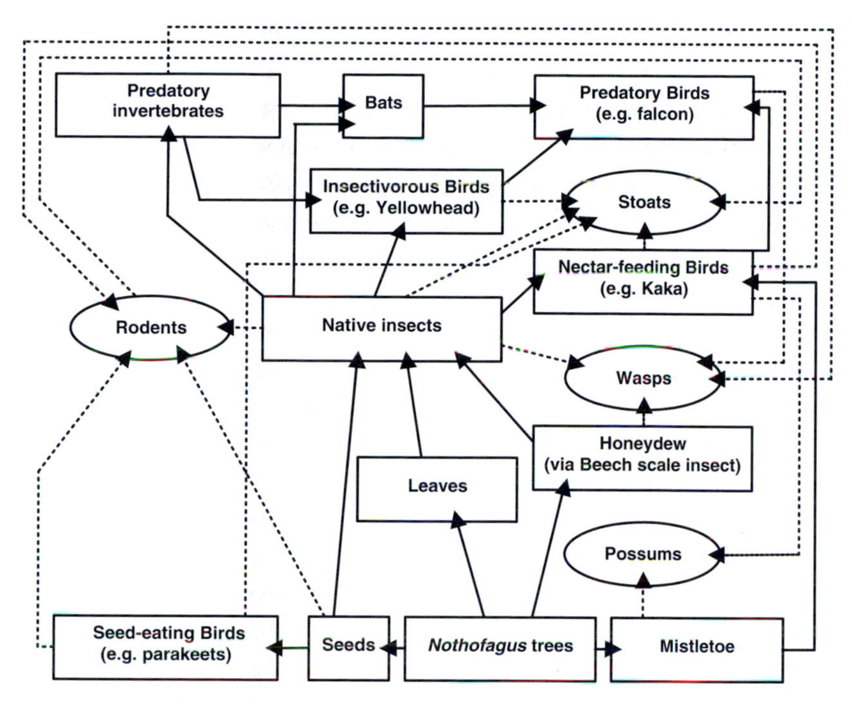

4.11 Simplified food-web of lowland *Nothofagus* forest in the northern South Island of New Zealand, illustrating the impacts of invasive animals (circled). Direction of energy flow is shown by arrows, with solid lines connecting indigenous elements and dotted lines showing predation on indigenous biota by the invasive animals. (From Clout, 1999, with kind permission of Kluwer Academic Publishers.)

These impacts are virtually meteoric, with many of the invertebrate introductions on the relatively unvisited Gough island having occurred in the last 50 years (Jones *et al.*, 2002). On islands that are visited frequently, the invasion frequency of insects is extremely high. Hawaii accumulates 20–30 new insect species per year (Beardsley, 1991) and Guam accumulates 12–15 new species (Schreiner and Nafus, 1986). Evidence is now accumulating that invasive insects, along with other agents of change are affecting certain island food webs, such as *Nothofagus* woodland in New Zealand (Clout, 1999) (Figure 4.11).

4.9.2 Caves

Special environments inevitably are islands surrounded by less favorable habitat, and need not necessarily be oceanic islands. Many other habitat islands exist, which are pockets of special and/or threatened insect diversity (Stanley and Weinstein, 1996). Among these are caves. Their obligate cavernicolous inhabitants tend to show a high degree of very localized endemism (Barr and Holsinger, 1985). The ancestors of these species may have taken refuge in the humid cave habitats during periods of inclement surface conditions, so as well as being rare and having well-developed troglomorphy, they also have specialized habitat requirements (Howarth, 1987). Inevitably, in the face of anthropogenic pressure, some of these faunal components are threatened (Howarth, 1981; Culver *et al.*, 2000), mostly because a relatively small disturbance, at least in terms of surface standards, can have major repercussions for stenotopic troglobites and stygobites (Balleto and Casale, 1991; Sharratt *et al.*, 2000). Some of this disturbance may be allochthonous, with, for example, the cave cricket *Speleiacris tabulae* and its co-inhabitants being dependent on bat guano, and hence on survival of bats that forage in the surrounding urban environment. This emphasizes that cave arthropod conservation depends largely on an integrated, whole-ecosystem approach (Harrison, 1964; Culver *et al.*, 2000).

There are many other 'special environments' or special localities, and it is one of the aims of prioritization to discover and assess threats to geographically unique or habitat-unique localities. These may then become Sites of Special Scientific Interest or equivalent, and generally require a total protectionist approach. Although the 'island' species have survived genetic bottlenecks, and may not necessarily exhibit metapopulation dynamics, they are likely to be very vulnerable to instantaneous anthropogenic impacts, many of which are synergistic. What we do not know yet is how cave insects and other faunal components will survive the new surface changes, especially global warming. Will their insulated home be enough to pull them through?

4.10 Overcollecting

Certain insects have human appeal, as food items, aesthetic collectables or as scientific curiosities. In short, certain insects have utilization value (Morris *et al.*, 1991). But when does consumption outstrip supply? This usually happens as a ramification of fragmentation, or more precisely, in terms of Forman's (1995) model, dissection of the landscape. As paths and roads penetrate natural habitat, this then encourages increasing human traffic and hence exploitation of resources, including insects. This has led, for example, to removal by tourists of 100 000 *Panaxia quadripunctaria* moths per generation in the Valley of Butterflies, Rhodes, Greece (Petanidou *et al.*, 1991).

Like so many aspects of conservation, overcollecting must be put in perspective and on a rational, non-emotive level. For butterflies at least, which include the most collected of all insects, New (1997) points out that the adverse effects of collecting are probably far less than that of habitat change and that simple bans on collecting play only a minor role, if any, in conservation. However, it is essential that collecting be monitored carefully because certain species with small total populations, and which may also be slow breeders, may be susceptible to overcollecting. Nevertheless, we must be sensitive to the fact that for certain species overcollecting has caused extinction. The British Large copper butterfly *Lycaena dispar dispar* appears to have been collected out of existence by 1848 (Duffey, 1968). For the 33 species of butterfly listed under the United States Endangered Species Act, 30% are threatened from overcollecting. This human harvesting is a specialized form of predation and as such can result in localized overexploitation and extirpation. This can be extended to harvesting of wild insects. With regards to indigenous African silk moths, it is essential to establish levels of sustainable utilization (Veldtman *et al.*, 2002).

4.11 Summary

Human population growth over the last century has resulted in an increase in consumption of resources by 460%. There is little concrete evidence that industrial pollution (as opposed to the physical footprint of urbanization) has had any major effect on terrestrial insect diversity. Impact on some aquatic systems, however, has been severe. Identifying pollution as a key factor eroding insect diversity is a complex issue and so we may be underestimating its effect. Impact of pollution is a matter of toxicant concentration, with only high levels having a major effect in some cases. Most concern is for the long-term, especially long-lived contaminants, especially in the soil.

Despite the huge consumption of pesticides worldwide, there is no evidence that pesticides have been singularly responsible for any insect extinction. This is because non-persistent pesticides are generally applied only over a limited area over a relatively short time period. Persistent pesticides are more insidious, upsetting predator–prey relationships. These effects are often synergistic with other impacts, including the increased use of fertilizers and herbicides.

Agricultural fragmentation of the landscape has many ramifications, with remnants of natural vegetation in agricultural areas often home to considerable insect diversity. This is especially so for the architecturally more complex agricultural landscapes. Similarly, diverse and complex green nodes in cities can be remarkably rich in insects. The converse situation is perforation of the natural landscape with agricultural or urban patches, as well as road corridors. These impacts have effects that go beyond the patch edge. Some insects benefit from these modified patches, while others do not. The upshot is that the landscape is best viewed as a differential filter, letting some insects through (physically and genetically) but not others.

As the tree canopy is so important for insect diversity, the current rates of deforestation, principally in the tropics, are devastating insect diversity. The remnant reserves are often too small and too vulnerable to disturbance to guarantee long-term survival of many species. The point is that large remnants of intact, virgin areas of forest are often critical for maintaining large-sized, ecologically sensitive, narrow-range endemics. The precautionary principle, of 'keeping all the parts', especially all the important parts of old forests, is pivotal for maintaining current levels of insect diversity. Evidence is pointing to a future dominated by weedy, ecologically tolerant species with tramp invaders among them.

Grasslands and Mediterranean-type ecosystems have also suffered insect loss, principally from the synergistic effects of fire, overgrazing and impact of invasive aliens, which can aggravate natural drought/flood cycles. Of concern is that the decreased arthropod diversity in these systems is reducing soil fertility.

Aquatic communities are remarkably tolerant to natural flood/drought cycles, having been honed over millennia. Human impacts, however, present a new multifaceted force on aquatic insects. Canalization streamlines water flow, while invasive alien plants change the character of banks and water alike. In turn, cattle trample natural riparian vegetation and agricultural run-off contaminates the water. Wetlands, which are the soaks and cleaning agents of the hydrological landscape, are under siege. Wetlands are 'special environments', like caves and islands, whose insect diversity is under enormous pressure. Added to this, are risks of overcollecting, which is a form of specialized predation by humans of mostly showy species. The bottom line is that these various impacts operate together, and we may not be detecting which is most harmful or able to determine exactly what the long-term impacts will be.

5 Responses by insects to the changing land mosaic

In the days of Moses Harris' book, and in the good time before, the Marsh Fritillary flew near Kingsbury and adorned the Wormwood Scrubs; now it is even scarce at a distance from London.

A. H. Swinton (*circa* 1880)

5.1 Introduction

Arguably, the greatest threat to maintenance of insect diversity is loss of habitat through landscape fragmentation and attrition. Insects react to these changes at multiple temporal and spatial scales. At the smallest scale, individual insects show measurable behavioural responses to changing land mosaics. At the next largest scale, changes in population dynamics come into play, which may involve local extinction of populations. Metapopulation dynamics feature strongly at this scale.

At a still larger scale, insect assemblages show responses to the altered land mosaic, although the responses may vary from one assemblage to another. Functional groups (e.g. herbivores, predators) may also show differential responses to the changing land mosaic. Similarly, interactions, both between insects themselves and with other organisms, vary depending on the type and intensity of change in the land mosaic.

At progressively larger spatial and temporal scales, some generalized predictions may be possible, such as extinction first of habitat specialists as the intensity and frequency of disturbance increases. However, with so many variables, plus the impact of stochastic events and synergisms, generalized predictions over large areas and long times becomes increasingly uncertain. Meanwhile, genetic changes also manifest. While this may involve direct adaptations to survive the changed conditions, in other cases there may be genetic impoverishment. Such reductions in genetic variety (genetic bottlenecks) can arise when the disturbance is long, intense and frequent.

The aim of this chapter is to review insect responses (behavioural, ecological, evolutionary) to both short- and long-term changes in the land mosaic, at small and large spatial scales. It is important to identify which types, aspects and intensity of land mosaic change impose genuine threats to insect diversity, as a background for appropriate management.

5.2 Behavioural responses

5.2.1 *Species differences*

Forman's (1995) model of increasing anthropogenic impact on landscapes begins with dissection of the landscape, leading to its perforation, fragmentation and finally attrition (Figure 5.1). The landscape may then be anthropogenically transformed into a mosaic (Wiens, 1995). These landscape-scale, spatial models become a powerful hypothetico-deductive foundation for testing effects of agriculture and tree farming on insect diversity. However, although a mechanistic view may be taken, one must ask 'mechanistic for whom'? While a landscape has many measurable features, these may not directly relate to the whole community of organisms. One species may respond to our measured patch edge, for example, differently from the next species. Even whole assemblages may respond differently, which has important implications for the use of indicators.

Cynically, one could choose an insect indicator group to illustrate what one wants (up to a point), and, as Andersen (1999) has pithily put it, 'my bioindicator or yours'. As we add species to our picture, so the edge becomes fuzzier and less and less visually obvious, particularly as our visualized edge is usually based on structurally obvious vegetation. It is more realistic to view the edge

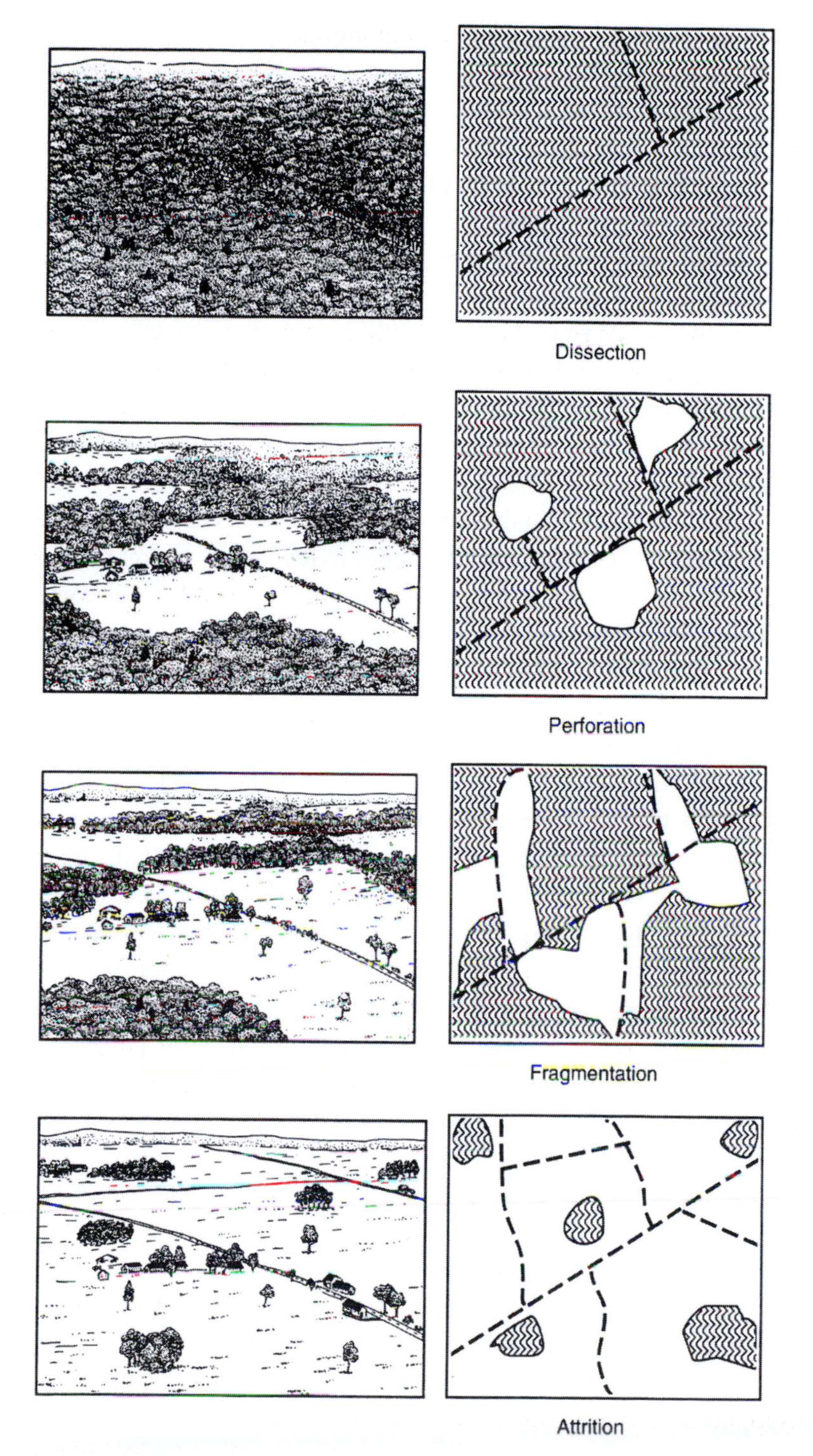

5.1 Fragmentation of the landscape begins when a road is built across the natural landscape, thereby *dissecting* it. Next, as agriculture and settlement begins along the road, the landscape then becomes *perforated*. As more lands are converted to agriculture, the disturbance patches coalesce, thus separating natural ecosystems, and causing *fragmentation* of the landscape. Finally, as more of the natural patches are converted, they become smaller and farther apart, this being the process of *attrition*. (From Hunter, 1996, following the terminology of R. T. T. Forman, with kind permission of Blackwell Publishing.)

more as a membrane that is semi-permeable to different species, as in the case of certain African butterflies (Pryke and Samways, 2001). This then invokes a variegation model rather than one of fragmentation. The range of behavioural and ecological responses will increase as more species are added to the model, as each species has a characteristic response to a particular boundary type (Figure 5.2).

Additionally, sex (Matter, 1996; Angelibert and Giani, 2003), age (Lawrence and Samways, 2001) and morph (Peterson and Denno, 1997) will all affect individual responses. The accumulated effects of these individual movements can be illustrated by reaction-diffusion models. However, as Fry (1995) points out, these diffusion models need also to consider the physical and biological properties of the landscape. As landscapes are heterogeneous, there will be different movement rates in different habitat patches. Where there is great movement of many individuals of many species, a landscape linkage or corridor comes about. But in reality a movement corridor for one species is not necessarily a corridor for the next (Wood and Samways, 1991) (Figure 5.3).

5.2.2 *Patch selection*

The differential selection of patches by different species, and even by different individuals, has important implications for insect diversity conservation. Edges as we see them may not be edges for particular individuals or species, or even interactions. In a woodland patch in England, natural enemies and other mortality factors of the Holly leaf-miner *Phytomyza ilicus* contributed differently to the edge and the interior, which arose from the interaction between microclimate, adult movement and host plant quality (McGeoch and Gaston, 2000). Furthermore, edges have various degrees of permeability for particular species (Wiens and Milne, 1989; Duelli *et al.*, 1990; Ims, 1995) (Figure 5.4). This in turn depends on the contrast between the patch and matrix, which also depends on the mobility of the patch organisms.

Flight-limited species, such as the brachypterous bush cricket *Metrioptera bicolor*, can be highly vulnerable to local extinction as remnant patches become smaller and more isolated (Kindvall and Ahlén, 1992) (Figure 5.5) resulting, as with the closely related species, *Platycleis fedtshenkoi azami*, in southern France, in regional extinction (Samways, 1989b).

In the case of the Apollo butterfly *Parnassius apollo*, the population acts as a patch population where adults mix over the whole area, but successful reproduction can only take place in discrete host plant patches. Occurrence on a host plant patch is restricted by the area size of the host plant patch and the configuration of nectar patches (Brommer and Fred, 1999). Similarly, densities of the tropical butterfly *Hamadryas februa*, which is a good colonizer, are constrained by lack of appropriate host plants as well as by emigration from isolated, smaller patches (Shahabuddin *et al.*, 2000). Having said that: How small is too small for

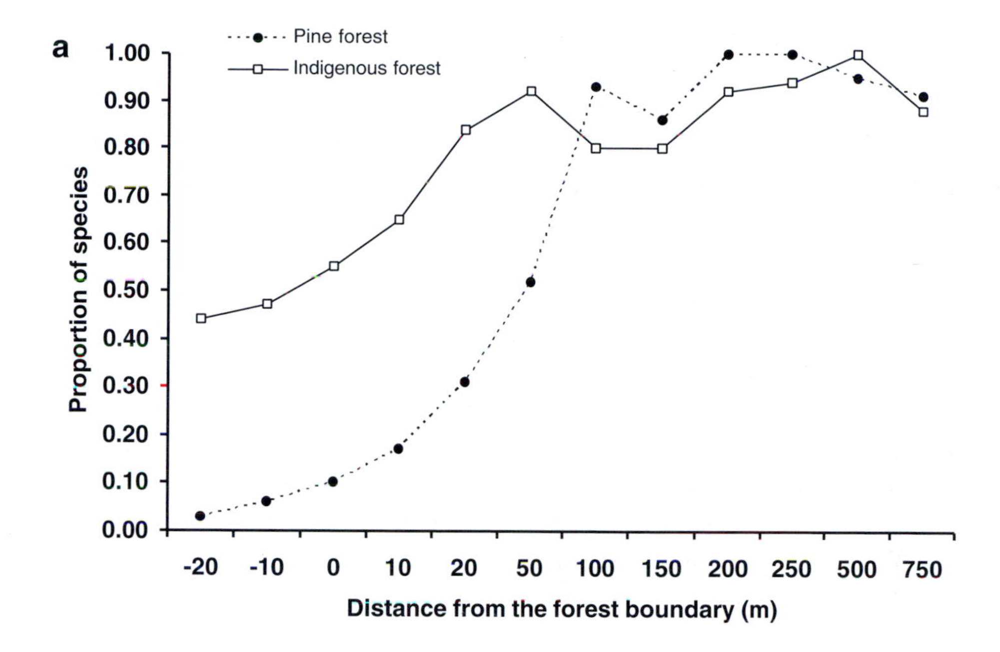

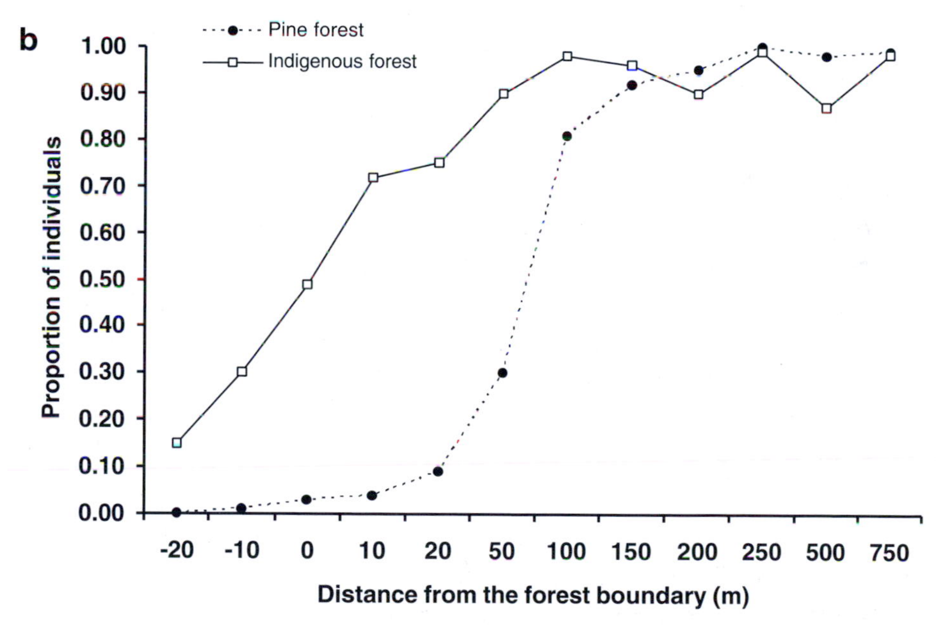

5.2 Response of African butterflies to tree boundaries along the edge of natural grassland linkages. The boundary is designated as '0'. Both proportion of species (a) and proportion of individuals (b) penetrate into natural forest boundaries to a much greater extent than into alien pine-tree boundaries (minus values in graph). In other words, pines present a much harder boundary than natural forest. Furthermore, the edge effect of pines into the grassland has a much greater depressing effect on both species and individuals than does natural forest (positive values in graph). (Redrawn from Pryke and Samways, 2001. Copyright 2001, with permission from Elsevier.)

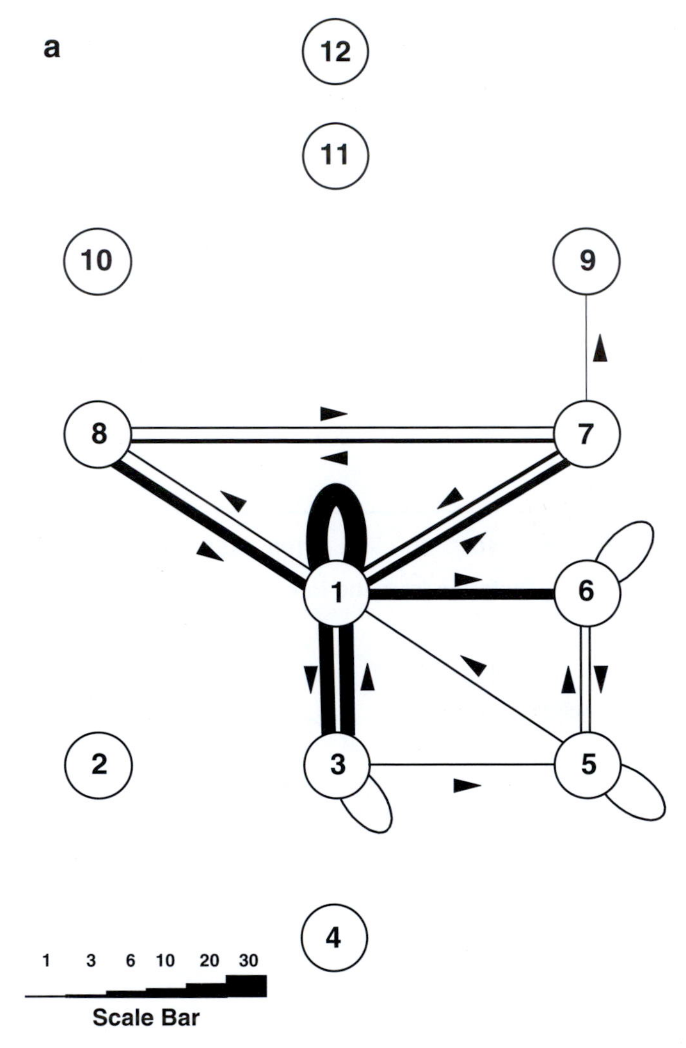

5.3 Flight response of two African butterflies (a) *Papilio dardanus* and (b) (see next page) *Acraea encedon*, illustrating their different responses to different features of the landscapes (represented by the numbers 1–12) such as clumps of bushes, lawn, pond. Each species moves across the featured landscape in its own characteristic way. *P. dardanus* principally follows a course from edge of trees (2) to pond edge (1), to reservoir inlet (8) etc. *A. encedon* frequently ventures out over the pond water but returns to the edge (1), and frequently flies back and forth over the lawn (1–3). (From Wood and Samways, 1991. Copyright 1991, with permission from Elsevier.)

small animals? Abensperg-Traun and Smith (1999) addressed this question using four arthropod species and suggested that even very small Australian remnant woodlands on farms may play an important role in sustaining certain species, either as stepping stones for dispersing individuals or for providing adequate habitat in the long term. But caution is required because other species in other

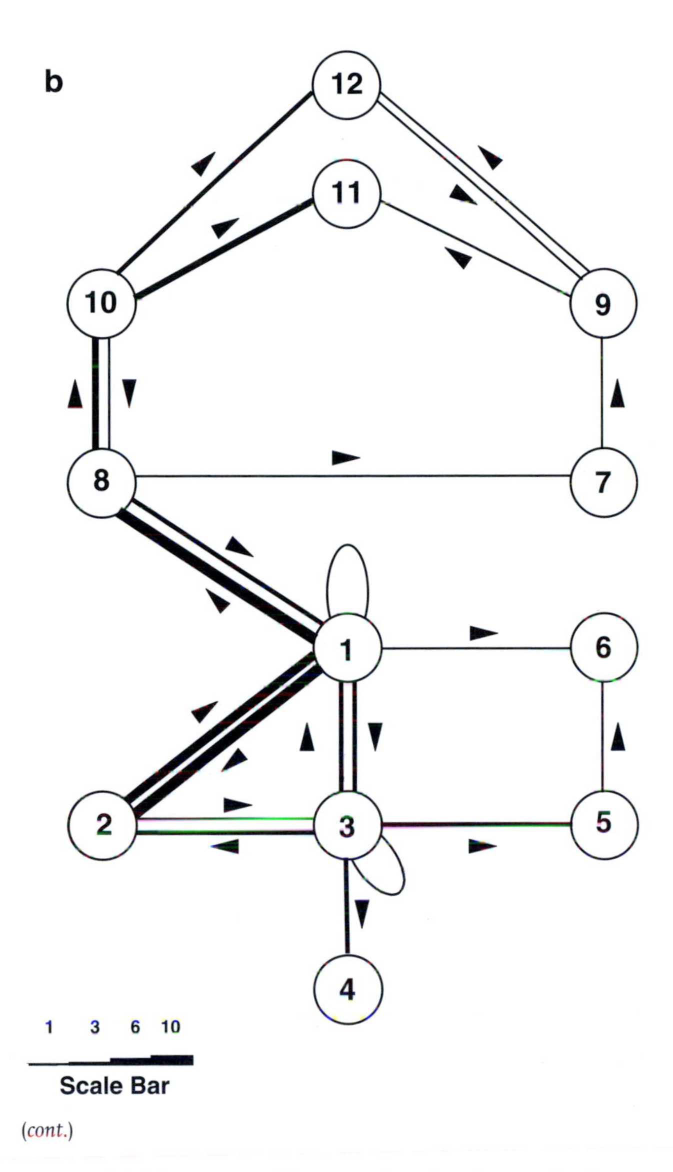

5.3 (*cont.*)

ecosystems are highly susceptible to reduction in patch size, particularly when associated with reduction in patch quality.

5.2.3 *Movement between patches*

Some sedentary species can be remarkably capable of moving across unsuitable habitat gaps. In the North American prairies, the flightless leafhopper *Aflexia rubranura* and the sedentary moth *Papaipema eryngii* readily crossed habitat gaps as wide as 36 m and 25 m respectively (Panzer, 2003).

Some individuals in a population may move considerably longer distances than most (Thompson and Purse, 1999). In the case of the Glanville fritillary butterfly *Melitaea cinxia*, females emigrating from habitat patches were on

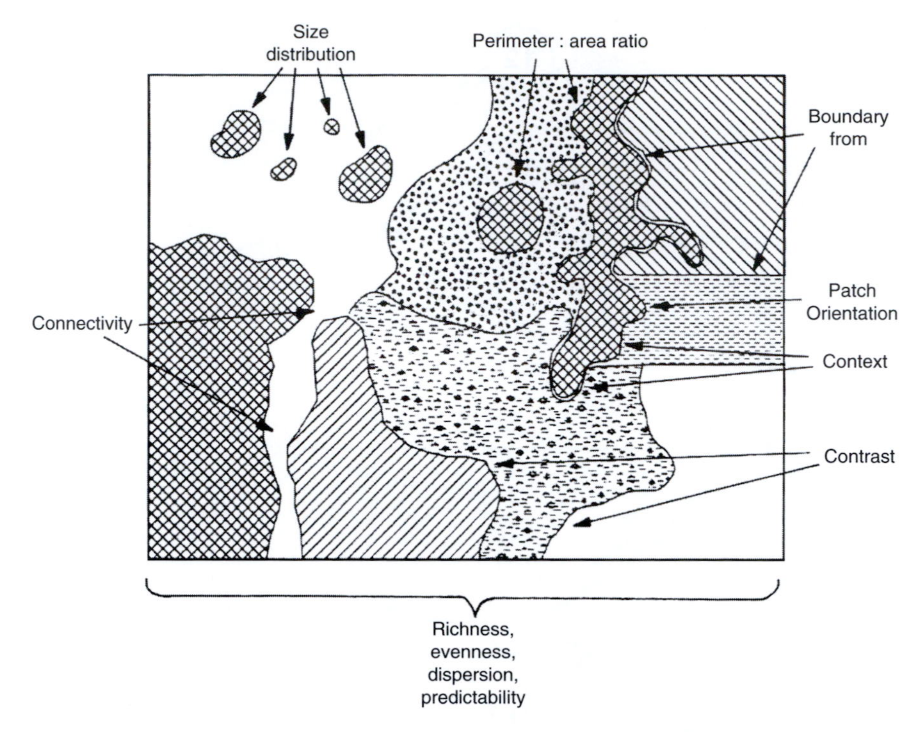

5.4 Spatial structures of a hypothetical landscape, with descriptors used in landscape ecology. (From Wiens *et al.*, 1993.)

average larger than females that stayed. Interestingly, this led Kuussaari *et al.* (1996) to conclude that conserving an isolated butterfly population is more successful in an area with physical barriers to migration than in an open landscape, although it is essential to have sufficient nectar sources and a large enough patch size. In the case of the Comma butterfly *Hesperia comma*, there was only 18% emigration from 5.7 ha patches compared to 100% emigration from 0.01 ha patches (Hill *et al.*, 1996). Generalizations, however, are difficult, because in the case of the French butterfly *Proclossiana eunomia*, there is a hierarchical nested metapopulation structure with much inter-habitat movement, which ensures population cohesion at the regional scale (Néve *et al.*, 1996).

Such inter-habitat movement appears also to be the case for Dutch carabids (De Vries *et al.*, 1996). This metapopulation structure also underpins survival of moths on Finnish islands, but as well as island characteristics, various traits among species, and even the sexes, influence the levels of migration (Nieminen, 1996). Interestingly, migration was lowest when sea surface temperatures were low, which emphasizes yet again that the matrix can be as important as the patch itself in maintaining diversity (Franklin, 1993). Furthermore, patch quality can be as important as patch size. Among such qualities are improved microhabitat

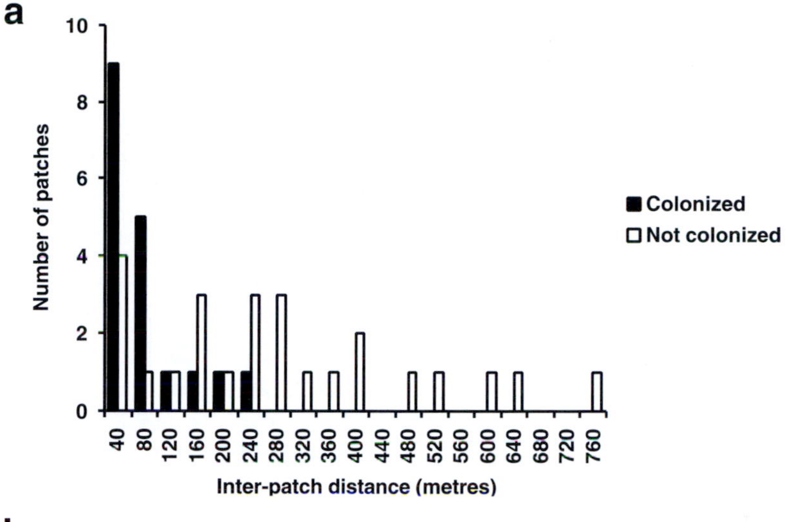

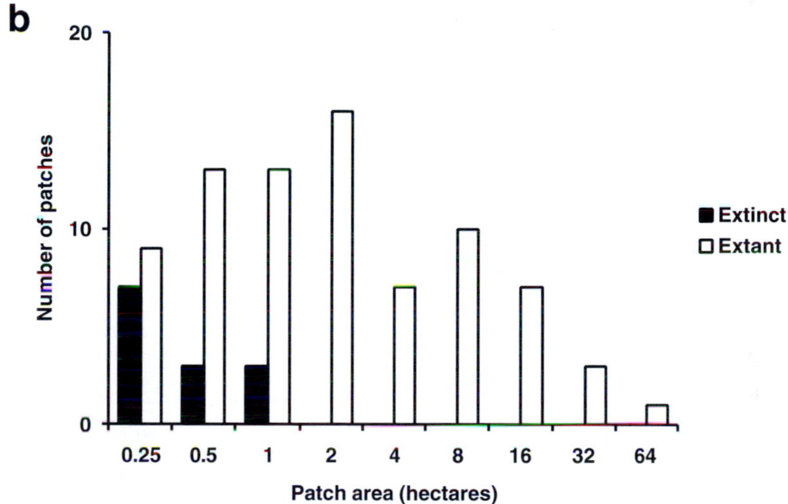

5.5 (a) Frequency distribution of inter-patch distance of habitat patches that were colonized or not colonized by the bush cricket, *Metrioptera bicolor*. Nearest patches were colonized most frequently. (b) Frequency distribution of different-sized habitat patches with extant or extinct populations of *M. bicolor*. Smallest patches saw the most extinction. (Redrawn from Kindvall and Ahlén, 1992, with kind permission of Blackwell Publishing.)

for parasitoids with increasing patch complexity (Marino and Landis, 1996) (Figure 5.6).

On balance, what we have learnt from insect behavioural responses to landscape change suggests that, for maximal insect diversity conservation the aim is to reduce the contrast between patch and matrix. Put another way, this means creating soft edges around patches and their linkages. From information on

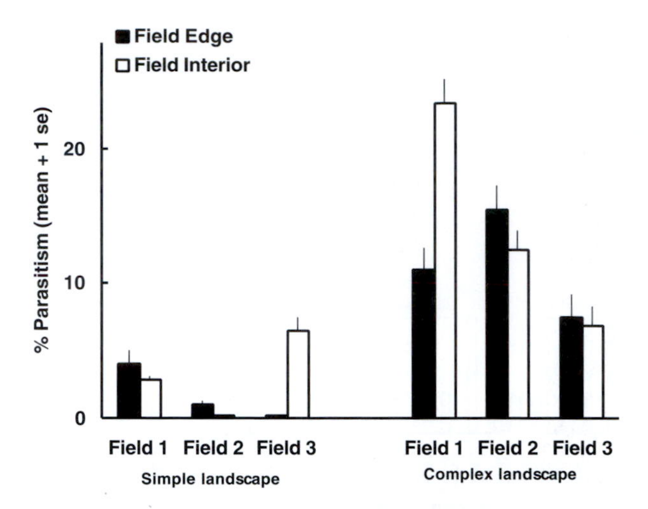

5.6 Mean percentage parasitism of the armyworm (*Pseudaletia unipunctata*) 5 m from a hedgerow ('field edge') and 90 m from a hedgerow ('field margin') in simple and complex landscapes in southern Michigan, USA. Complex landscapes support much higher levels of parasitism than the simpler ones, both at the edges and interiors. (From Marino and Landis, 1996.)

tropical forest butterflies (Mallet and Gilbert, 1995; De Vries *et al.*, 1997; Willott *et al.*, 2000) this also means maintaining unlogged refuge areas within a logged forest to act as source habitats.

5.3 Population response and local extinction

Occasionally, natural populations of insects go extinct. This is the background noise against which we must measure the pressure of threatening processes. Extinction of a species is the brief moment when the last individual of a last population passes from life to death. This is rarely observed but there do appear to be two butterfly species in South Africa that may have gone extinct naturally in historical times (Henning and Henning, 1989).

The road to extinction is loss of populations and break up of metapopulations through reduction, isolation and transformation of habitat. Patch size can be a critical factor. When coupled with connectivity indices, it is the best predictor of patch occupancy by the Wart-biter bush cricket, *Decticus verrucivorus* (Hjermann and Ims, 1996). Small patches (> 250 m^2) had an extinction frequency of 0.75, whereas larger patches (> 600 m^2) had a value of > 0.25 (Figure 5.7). Similarly, another bush cricket, *Metrioptera bicolor*, tended to become extinct in small (< 0.5 ha) and relatively isolated (> 100 m) patches (Kindvall and Ahlén, 1992). Patch quality is additive upon patch size and isolation. For *D. verrucivorus*, the probability of patch occupation was also positively influenced by other, 'quality' factors such as increasing slope to the south, increasing amounts of the plant

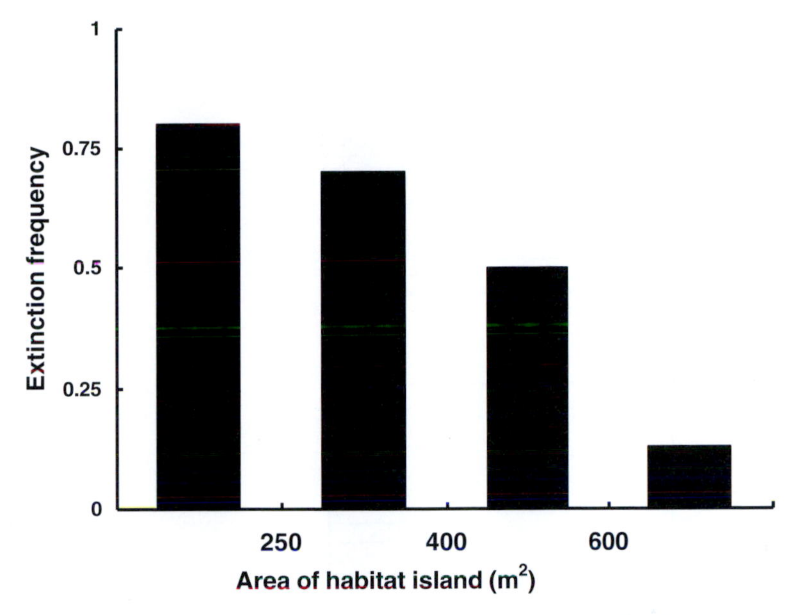

5.7 Although the extinction frequency of the Wart-biter bush cricket *Decticus verrucivorus* becomes less with increasing size of habitat patch, other 'quality' factors, such as slope, plant composition and vegetation architecture also play a role. (Redrawn from Hjermann and Ims, 1996, with kind permission of Blackwell Publishing.)

Achillea millefolium and decreasing fraction of vegetation lower than 10 cm, in addition to habitat area and connectivity (Hjermann and Ims, 1996).

Studies on the Grayling butterfly *Hipparchia semele* in Europe indicated that for oceanic islands, area followed by isolation were the most important predictors of long-term occupancy (R. L. H. Dennis *et al.*, 1998). Populations on small islands suffering habitat loss were highly vulnerable because the aggregated archipelagos could not function as a metapopulation.

Temporal factors can be additive upon spatial ones. Tilman *et al.* (1994) point out that even moderate levels of habitat transformation and fragmentation can cause time-delayed, but deterministic extinction, especially of once-common species. Some large-sized habitat-specialist insects have been particularly affected. The American burying beetle *Nicrophorus americanus* is large, forages widely, needs big carcasses and needs to bury them in deep, loose soils with a substantial litter layer. Inevitably, it has suffered through a cascade effect in the food web, with its large mammal resource having declined (Lomolino and Creighton, 1996).

Tilman *et al.*'s (1994) models suggest that our current habitat destruction will not be fully felt until many years in the future, even 400 years forwards, which is what they term an 'extinction debt'. Superior competitors will be affected most which suggests that there will be long-term, as yet unseen, effects on ecosystem

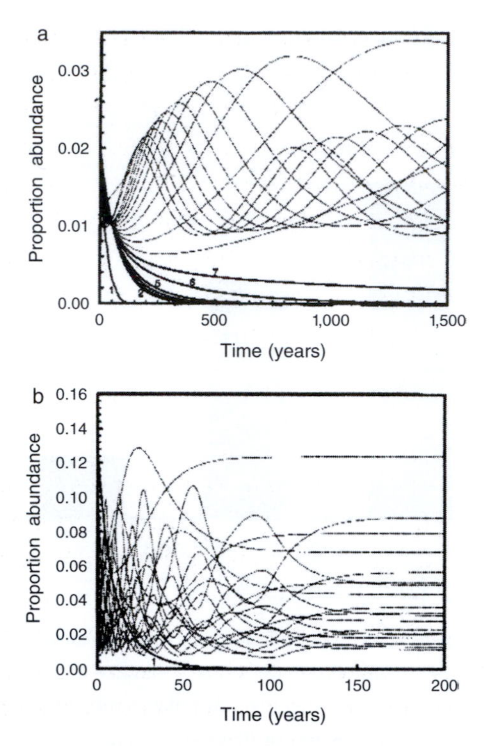

5.8 Species dynamics after habitat destruction. (a) Destruction of one-third of a habitat led to the extinction of the seven best competitors (solid curves). The 13 inferior competitors (dotted curves) persisted. Parameters were chosen to give geometric abundance series in a virgin habitat, that mimics a tropical forest, and with the best competitor occupying 3% of sites. Lower mortality rates would lead to slower extinctions. (b) Same as (a), except that the best competitor occupied 20% of the virgin habitat sites, which mimics a temperate forest. Now only the best competitor (solid curve) was driven extinct by habitat destruction, and the remaining 19 species (dotted curves) persisted. (From Tilman *et al.*, 1994.)

function, which will negatively affect the future biotic richness of the planet (Figure 5.8).

5.4 Community response and long-term prognosis

5.4.1 *Impact of landscape fragmentation on specialist and on common species*

What is the long-term prognosis for insect diversity given the fragmentation, attrition of remnant patches and the formation of landscape mosaics? What glimpses does current research give us? Besides the type of disturbance, the effect on assemblages also depends on severity, extent and regularity of the disturbance, with moderate disturbance often generating diversity (Connell, 1978; Petraitis *et al.*, 1989). But fragmentation, and its consequences, is relatively permanent. It may be viewed as a continuous disturbance, because it is essentially negentropic, with energy input into agriculture and urbanization maintaining the patch-matrix, or more often, the landscape mosaic. Inevitably this leads to loss, for example, of butterflies whether in North America (Kochér and Williams, 2000), Britain (Cowley *et al.*, 1999) or the Mediterranean islands (Dennis *et al.*, 2000) from long-term agriculturally induced patch attrition. While specialist species are often the first to go locally extinct (Kitahara and Fujii, 1994), we cannot assume that 'common' species are safe. The Small copper butterfly *Lycaena phlaeas* has decreased its area of occupancy by 90.6% in 96 years in North Wales, with many other species showing major declines, and seven species going regionally extinct (Cowley *et al.*, 1999).

5.4.2 *Temporal and spatial scales*

Appreciation of both temporal and spatial scales is essential. Certain British butterflies that prefer sunlit grassland have benefited from the clearance of the postglacial wildwood over the last 6000 years (Dennis, 1977), relegating certain species such as the Black hairstreak butterfly *Satyrium pruni* to remnant areas of mature woodland (Thomas, 1991). This pattern appears to be repeated for other taxa in Europe, e.g. carabids (Andersen, 2000), and in the tropics, where intact, large remnants of original forest are nodes where complete preservation is essential (Fermon *et al.*, 2001).

Understanding species diversity in local assemblages requires knowledge of processes acting at larger spatial scales, including determinants of regional species richness and spatial turnover of species (Caley and Schluter, 1997; Sax and Gaines, 2003). However, a metapopulation-like perspective might well overlay a purely interspecific–interaction perspective or a purely regional perspective, suggesting that recruitment limitation may be more important, even on a local scale (Tilman, 1997). This botanical perspective is supported by Basset's (1996) findings on tropical insects, where most of the variation in local species richness could be predicted from local processes (i.e. food resources, and abundance of natural enemies), and not from regional processes.

In temperate areas, regional processes and vagility may be more important. Indeed, movement of carabids between Baltic Sea islands can be surprisingly great, and colonization success appears to depend on many factors, including availability of suitable habitat, competitive superiority, survival ability during dispersal and island arrival sequence (Kotze *et al.*, 2000). Scattered islands accumulate species at a faster rate than closely grouped islands, with more mobile, macropterous species readily colonizing distant islands (Kotze *et al.*, 2000), as do powerful-flying moths in the same Finnish archipelago (Nieminen and Hanski, 1998). Many of the carabid specialists are large and wingless, and in the absence of fragmentation would probably have little need to found new populations in a continuous forest, whereas dispersal and founding of new populations is necessary for heathland species (De Vries *et al.*, 1996).

Habitat destruction also affects interactions between insect hosts and parasitoids. Parasitism by four species of parasitoids that attack the Forest tent caterpillar *Malacosoma disstria* is significantly reduced or enhanced depending on the proportion of forested to unforested land. Each of the parasitoid species responds to the mosaic at four different spatial scales that correspond to their relative body sizes (Roland and Taylor, 1997).

5.4.3 *Response to patch quality*

Moderate habitat fragmentation and ecological stress may be more pronounced in tropical systems than in temperate ones, and may result in a greater proportional loss of local biodiversity in the tropics (Basset, 1996). This includes changes in size of the faunal components, with the larger dung and carrion beetles being lost with increasing forest fragmentation (Klein, 1989). Stress also depends on susceptibility of the insect fauna. In the case of *Melipona* bees, species succumb to deforestation in stages: (1) forest clearing, when colonies are killed from the impact of the nest cavity hitting the ground and breaking apart, or by human harvesting of the accessible colony; (2) fire survival stage, when having survived felling, the colony's survival now depends on how well it can close itself off from the outside and resist the heat of the fire; and (3) recuperation and long-term survival stage: having survived the fire the colony must rebuild its nest architecture (Brown and Albrecht, 2001).

The impacts of fragmentation also depend on quality of resources remaining in the patch (Collinge *et al.*, 2003; Pryke and Samways, 2003). Ant assemblages are more depauperate in patches with a long history of severe disturbance, especially if small in size, than historically less-disturbed patches (Mitchell *et al.*, 2002). No matter how well a species can disperse, it must have, in the case of butterflies, the appropriate host plants in the patches (Shahabuddin *et al.*, 2000). These may exist, even in small remnant patches, which can act as stepping-stones (Thomas, 1995), at least for the more vagile species (Usher and Keiller, 1998). But for most species, small fragments do not provide suitable long-term conditions, so that over time, such fragments become depauperate (Bolger *et al.*, 2000; Lövei and Cartellieri, 2000).

Although there are records of increased beetle species richness (Barbalat, 1996) in logged temperate forest clearings, this is a perforation situation (i.e. the disturbance is a patch in a forest matrix) and not the converse of fragmentation/attrition (i.e. remnant patch in felled matrix). A forest gap can mimic natural tree fall. Indeed some species need tree gaps, like the neotropical grasshopper, *Microtylopteryx hebardi*, which interestingly is flightless, suggesting that tree fall is sufficiently frequent to forgo fast, airborne search and locate (Braker, 1991).

There is considerable variation between different taxonomic groups (Abensperg-Traun *et al.*, 1996) across forest patch boundaries. These differences are linked with differences in functional activity of the taxa (Kotze and Samways,

2001), which in turn means that ecosystem processes will be differentially affected, with the more specialized interactions suffering the most from fragmentation (Didham *et al.*, 1996).

In an experimental study in Europe, abundance of only the most sessile and specialized groups (leafhoppers and wingless aphids) was affected by plant diversity, and the effect was mostly indirect and mediated by changes in plant biomass and cover (Koricheva *et al.*, 2000). Although the plant species richness and plant functional groups were important determinants of insect diversity, it was plant species composition (i.e. which plant species were present) that was the most important factor (Koricheva *et al.*, 2000), indicating that insect diversity conservation depends heavily on plant diversity conservation. This has also been confirmed in studies of arthropods in Azores (Borges, 1999) (Figure 5.9) and Scotland (P. Dennis *et al.*, 1998) (but see also Chapter 8).

5.4.4 *Predictions from models*

Can we predict what level of species loss there will be with fragmentation? Pimm (1998) and others have used the species–area relationship to predict this. Like many ecological models, the predictions have built in assumptions that are not necessarily obeyed by organisms. Among these assumptions is that extinction estimates assume that species have uniformly distributed range requirements and a minimum abundance level required for persistence. This means that as the total number of species is reduced through habitat destruction, an increasing proportion of species are lost relative to increasing attrition of habitat. The species that are lost are the least abundant ones, and so, as the total available habitat decreases, the species–area relationship can be used to predict species lost (Figure 5.10).

Ney-Nifle and Mangel (2000) have refined this methodology and have shown that habitat loss can change the species–area relationship and consequently the number of species predicted to go locally extinct. This is because spatial effects such as geographic range distribution, minimum variable ranges, edges and type of habitat transformation have an effect. These model assumptions appear to be supported in real life. Ney-Nifle and Mangel's (2000) model also indicates that the geometry of fragmentation also plays a role, and its effect is dependent on the proportions of aggregated species, aspects of geographic ranges and form of habitat destruction, as well as the sampling protocol (Figure 5.11).

This information is essential to the model because how many species are lost depends on where the habitat is removed, and because species' ranges rather than abundance determine which species are lost. Additionally, real world situations occur between the two limiting cases of abundance and range requirements. Species with small ranges are also usually rare within them. In practice, this means that, as well as species lost from habitat destruction, some species will survive in the transformed area but will, in time, go extinct because they are

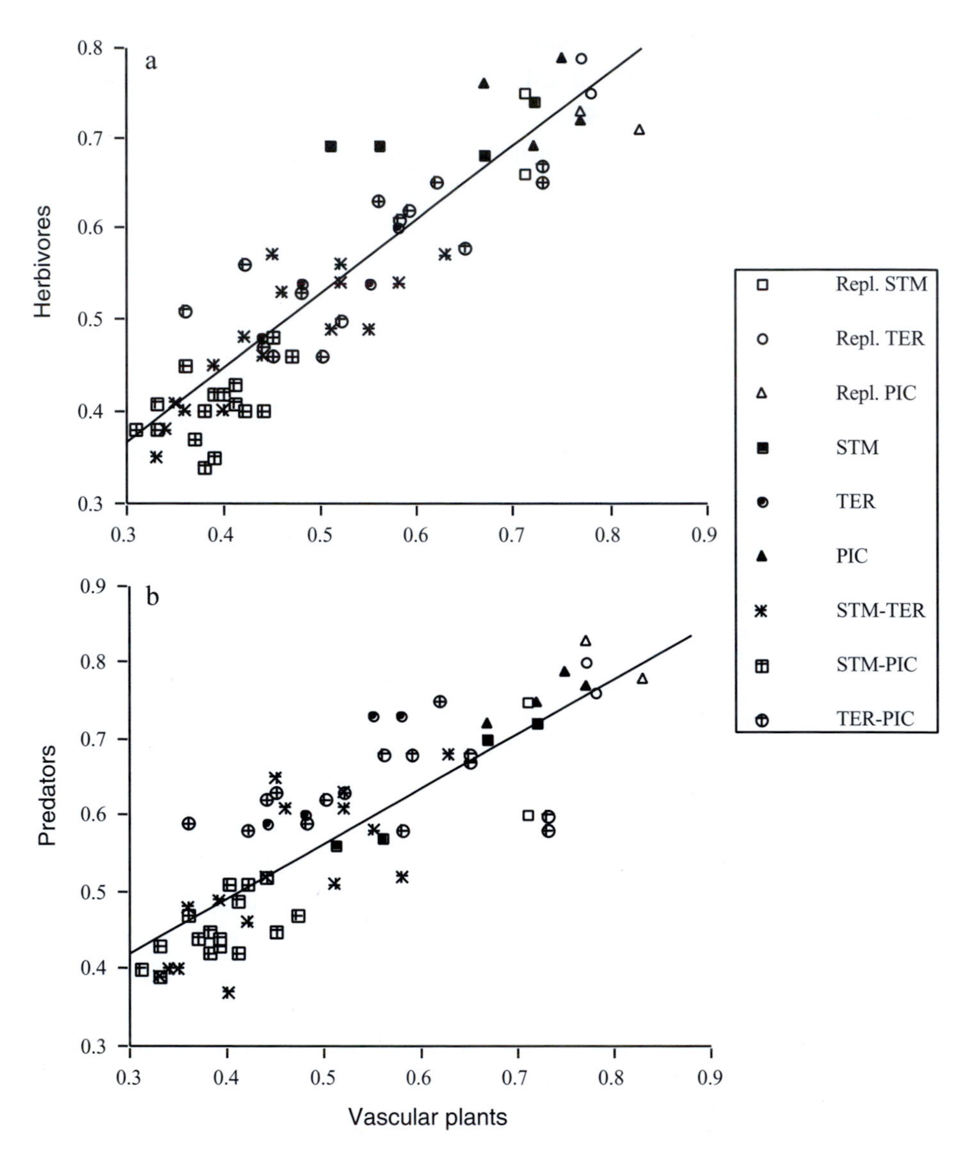

5.9 In the Azores, not only does arthropod herbivore diversity (Sörensen's Index of similarity) increase with vascular plant diversity (a), but so does arthropod predator diversity (b). The symbols represent different sites. (From Borges, 1999).

not sufficiently numerous. What the model does not predict however, is that certain other species will benefit from the transformation and become more abundant. This was shown experimentally in grassland plots in Switzerland, where fragmentation was beneficial to some plants via decreased competition intensity along the fragments as retreats, and because some animals may use fragments as retreats between foraging bouts into the disturbed area. The fragmentation had the most adverse affect on the rare specialists, especially butterflies (Zschokke *et al.*, 2000).

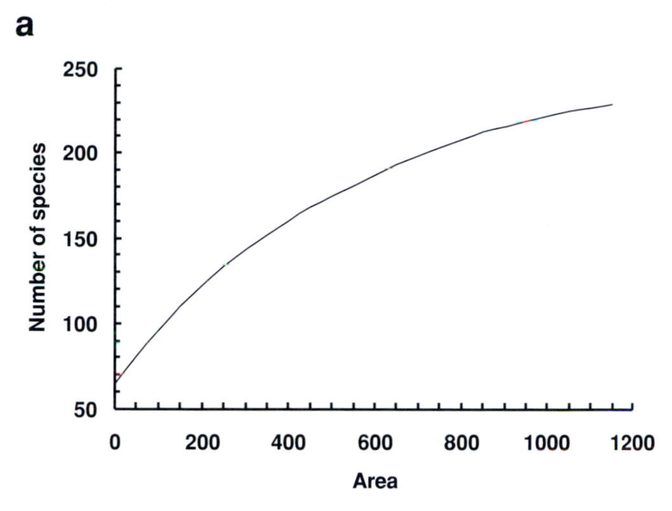

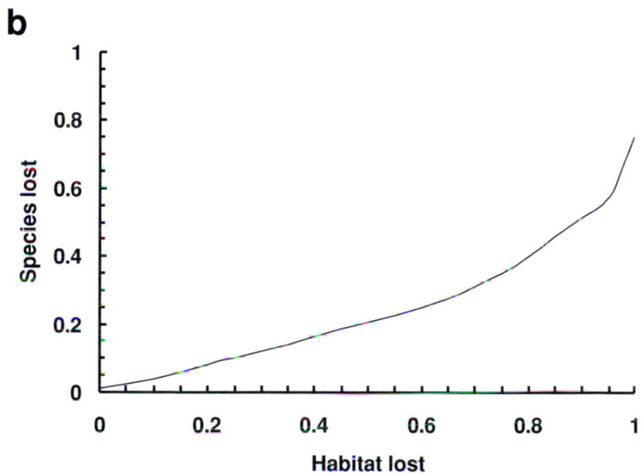

5.10 (a) The species–area relationship (SAR), where the number of species increases with area ($S = CA^z$; S, number of species; C, intercept; z, exponent, which in this case is 0.23). (b) The SAR has been used to predict extinction based on habitat reduction. As the total number of all species is reduced during habitat destruction, an increasing fraction of species falls below the minimum number needed to persist. With this model, how many species are lost depends solely on how much habitat is removed, not on where the habitat is removed. Species abundances determine which species are lost and thus as area decreases, the SAR can be used to predict species loss. (Redrawn from Ney-Nifle and Mangel, 2000, with kind permission of Blackwell Publishing.)

5.4.5 *Maintaining nodes of natural habitat*

Distilling these various findings, it is essential to maintain nodes of natural forest and other habitats, which may, in part, involve topographic considerations (Samways, 1990; Basset, 1996), particularly at the lower latitudes. The nodes should be as large as possible to avoid edge effects (Laurance and Yensen, 1991) yet small nodes also have partial value (Usher and Keiller, 1998;

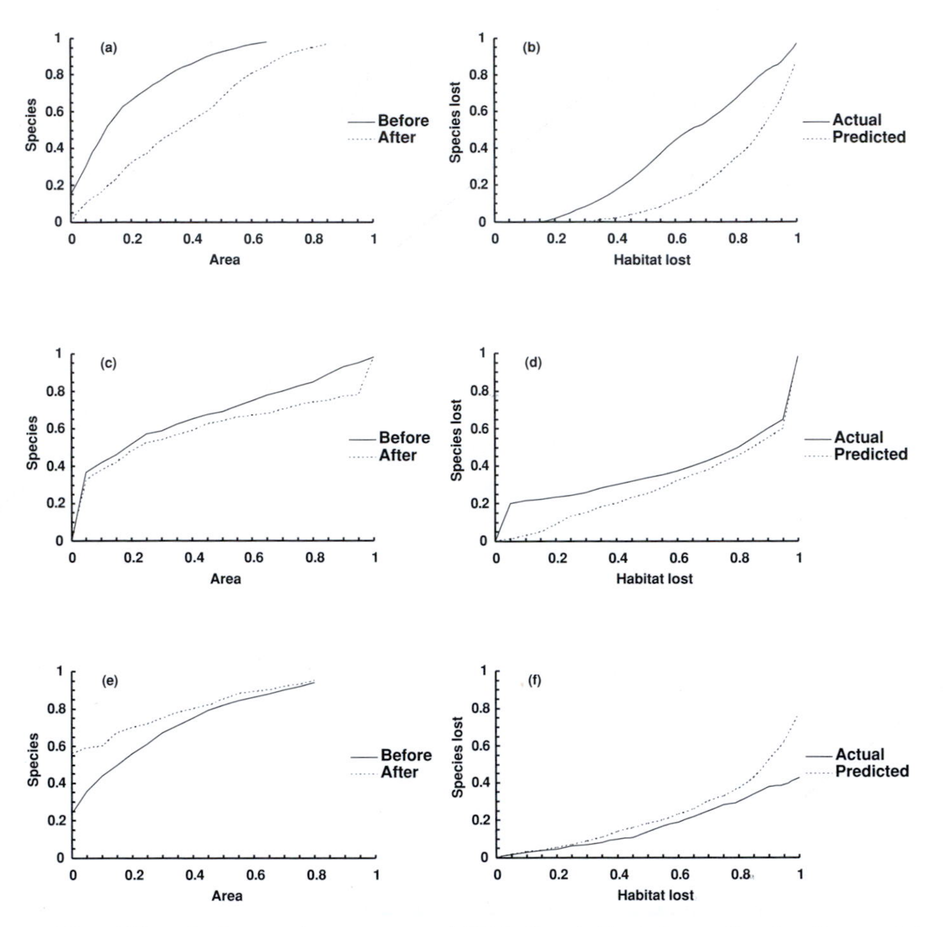

5.11 Although the species–area curve (SAR) before habitat destruction is often used to predict species loss after habitat destruction, assumptions must be clearly stated. Spatial effects such as geographic range distribution and minimum viable ranges, edges and fragmentation can change the SAR. In particular, the larger the fraction of aggregated species (those having a restricted range), the more likely a change in the SAR with habitat destruction. Shown here is the prediction of species lost with habitat loss: (a, c, e) SAR before and after habitat destruction, and (b, d, f) the fraction of species predicted to be lost (based on the original SAR) as a function of habitat loss and the actual fraction (based on a new SAR) depending on the geometry of habitat loss, such as rectangular cuts of tropical forest (a, b), rectangular cuts with habitat fragmentation (c, d) and square cuts (e, f), the latter of which had a severe impact. (Redrawn from Ney-Nifle and Mangel, 2000, with kind permission of Blackwell Publishing.)

Magura *et al.*, 2001). These nodes must cater for specialist species and specialist interactions in particular. These nodes also need to be linked with corridors (Pryke and Samways, 2001, 2003). Other nodes may have selective and careful logging. These local networks then need to be linked into a larger, regional network that maintains evolutionary as well as ecological processes (Erwin, 1991).

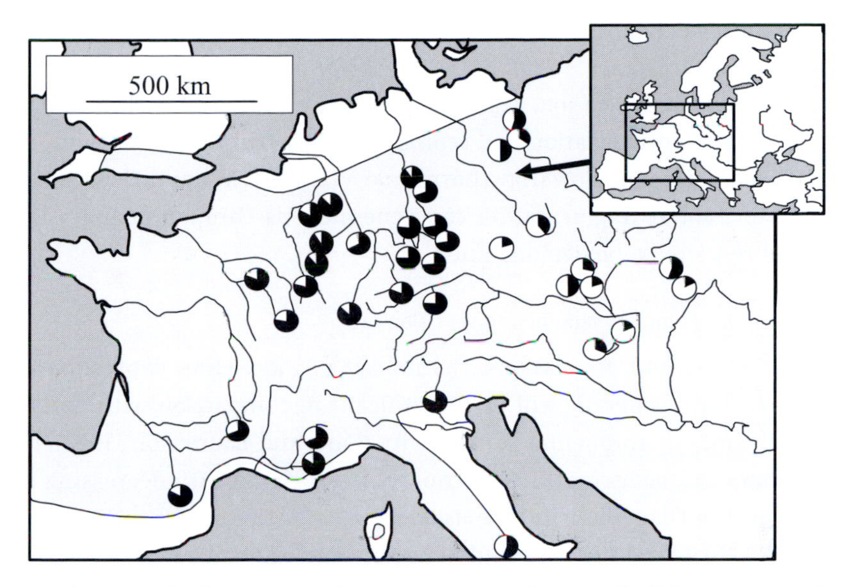

5.12 Distribution of allele frequencies of the Chalk-hill butterfly *Polyommatus coridon* (allele 3: black; allele 5: white). Allele 3 dominated in the south-western lineage and allele 5 in the north-eastern lineage. There is a loss of genetic diversity from south to north with both lineages, reflecting the decline of diversity during the postglacial expansion. The lines represent the actual geographic range (extent of occurrence) as given by Tolman and Lewington (1997). (From Schmitt and Seitz, 2001, with kind permission of Blackwell Publishing.)

5.5 Genetic changes

5.5.1 Marginality

Marginal populations i.e. those on the edge of geographical ranges, have been observed to go locally extinct and to re-establish, especially during climatic adversity and salubriousness respectively (Samways, 2003c). During the expansion phase, although seemingly advantageous for a species, there tends to be reduced genetic heterogeneity (Schmitt and Seitz, 2001) (Figure 5.12). It is important to understand these genetic processes because, in the face of human transformation of the landscape, for a species to exist in the long-term, it must be able to respond adaptively. The anthropogenic landscape and biogeographical squeezing process has led to many species now having reduced geographical ranges, and in many cases, isolated populations.

Genetic divergence in populations isolated by human activity may be termed anthropovicariance, which Williams (2002) considers as a special case of speciation. This can come about because stress from anthropogenic impacts increases mutation/recombination, decreases gene flow and leads to more phenotypic variation, all of which induce evolutionary change (Hoffmann and Hercus, 2000). Anthropovicariance thus produces novel evolutionarily significant units. In turn, this requires debate as to whether it is of genuine conservation concern. Besides these rapid vicariance events, insect diversity conservation must also consider rapid sympatric speciation. Traits that are genetically tightly linked and which contribute to specialization and reproductive isolation can lead to rapid speciation, even with gene flow (Hawthorne and Via, 2001; Via and Hawthorne, 2002). Whether habitat fragmentation can generate this through changes in habitat quality remains to be demonstrated.

5.5.2 *Importance of maintaining gene flow*

In natural populations, the impact of inbreeding depression on population survival generally will be insignificant in comparison with that of demographic and environmental stochasticity (Caro and Laurenson, 1994). Nevertheless, there can be increased extinction risk associated with decreasing heterozygosity in the Glanville fritillary *Melitaea cinxia* in the field (Saccheri *et al.*, 1998) (Figure 5.13). Larval survival, adult longevity and egg-hatch rate were all adversely affected by inbreeding, and appear to be the fitness components underlying the relationship between inbreeding and extinction. It seems that the Glanville metapopulation maintains a high genetic load, making it susceptible to inbreeding depression. Selection against deleterious recessives exposed by localized inbreeding may be relatively inefficient owing to drift within, and gene flow among, neighbouring local populations that carry different deleterious alleles (Saccheri *et al.*, 1998).

Evidence is accumulating that the effects of inbreeding depression as a result of anthropogenic disturbance might be more widespread than formerly thought, with plant dynamics (on which many insects depend), as well as insect dynamics being affected (Keller and Waller, 2002). Thus there is now the need for practical demonstration that or, in the field it is necessary to retain and even enhance gene flow among populations with fragmented habitat patches. This is borne out by Couvet's (2002) models which show that for mainland populations, the rate of decrease of viability with reduction of gene flow is not uniform and becomes increasingly higher when the number of migrants per generation is below one (Figure 5.14). In turn, this corresponds to a migration rate equal to the inverse of population size.

Couvet (2002) presents an interesting perspective on island populations. The viability of individuals in the island population depends also on the gene flow *within* the source, mainland population. This is because low gene flow within the mainland population leads to a higher frequency of deleterious mutations, including among the immigrants to the island population. The frequency of deleterious mutations among the migrants then has a strong influence on the frequency of such mutations in the island population. Consequently, the positive effects that could result from enhanced gene flow depend on a combination of (1) number of immigrants per generation to the island population, along with;

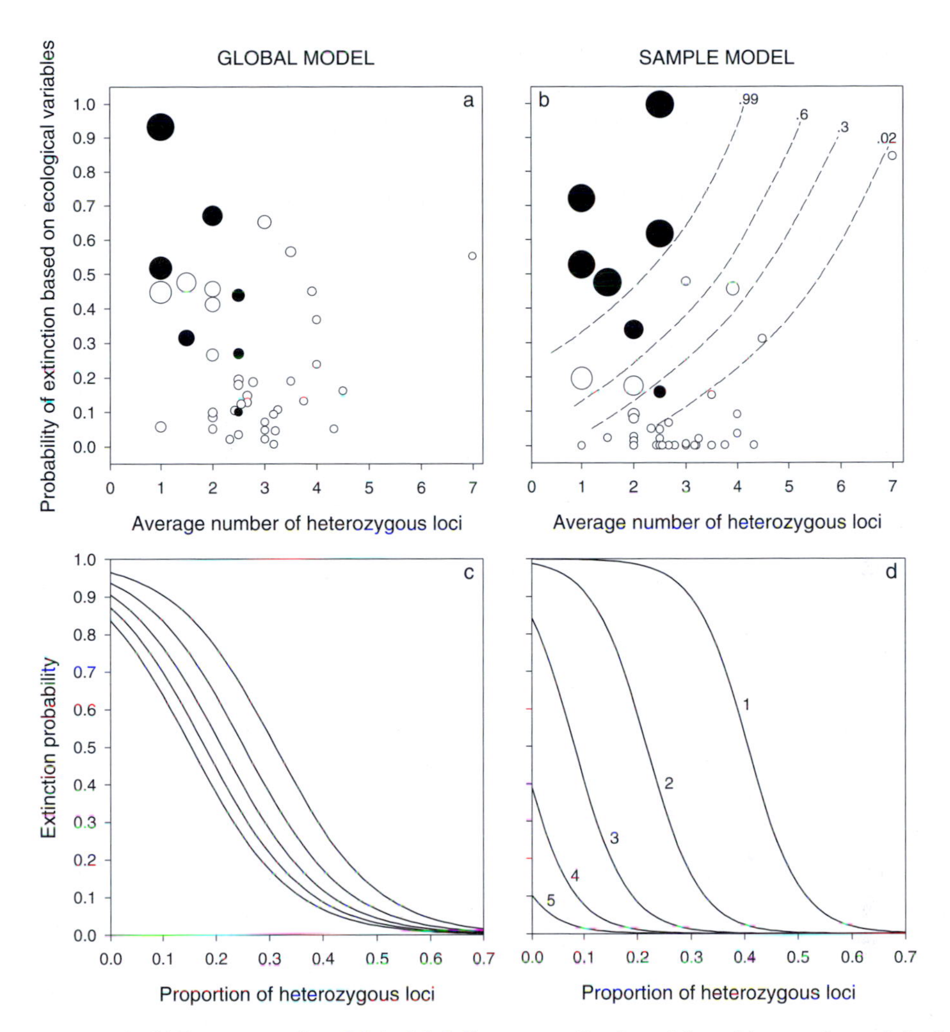

5.13 (a, b) Two types of model (a: 'global', or generalized model; and b: sample model) of populations of the Glanville fritillary butterfly *Melitaea cinxia*, showing the observed average number of heterozygous loci in extinct (black) and surviving (white) populations. (c) The probability of extinction predicted by the models without heterozygosity compared with the observed heterozygosity. (d) The probability of extinction predicted by the full model, including heterozygosity (proportional to circle size). For the sample model, isoclines for the extinction risk predicted by the model are drawn, including ecological factors and heterozygosity. These figures illustrate that both the ecological factors and heterozygosity influence the extinction risk. (b, c) Risk of local extinction and heterozygosity predicted by the global and sample models. Model predictions are shown for local population sizes of 1–5 larvae groups, fixed at the lower quartile value of change in regional density and the lower quartile value of meadow habitat area in the global model, and fixed at a further regional density and median flower abundance in the sample model. (From Saccheri *et al.*, 1998.)

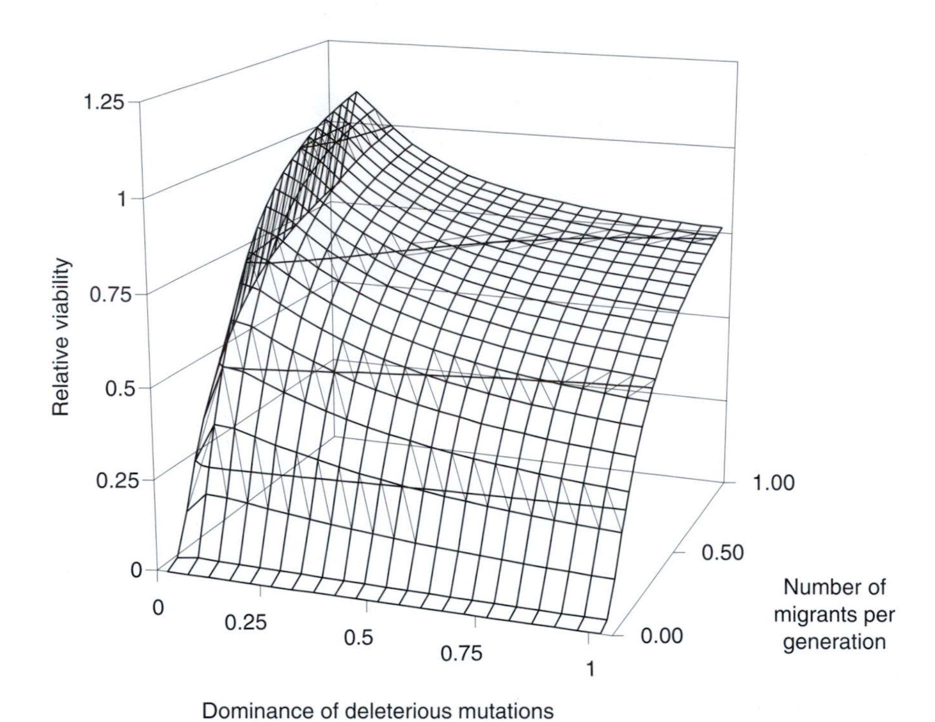

5.14 Expected viability in a mainland population relative to that in an effectively infinite population as a function of the number of migrants per generation, and the dominance of deleterious mutations. (From Couvet, 2002, with kind permission of Blackwell Publishing.)

(2) relative number of immigrants per generation to the mainland population; and (3) relative viability of the island population (Figure 5.15). In other words, low gene flow among mainland populations will significantly impair viability in all island populations connected to them.

5.6 Summary

Loss of habitat through landscape fragmentation and attrition is having a devastating effect on insect diversity worldwide. The problem lies also with knock-on effects. For example, landscape transformation goes hand in hand with increased invasion by alien species. There is also increased filtration of insect movement by the subdivided landscape, resulting in the isolation of populations. Not all insects respond to these changes in the same way. Different species, and even different individuals, are affected in different ways. One species may lose its food plant while another may have its dispersal route blocked. As the range of variation of responses by insects to this huge worldwide landscape change is so immense, generalizations are difficult. What is emerging, however, is that we need to maintain as much wild land in as big as patches as possible. Nevertheless,

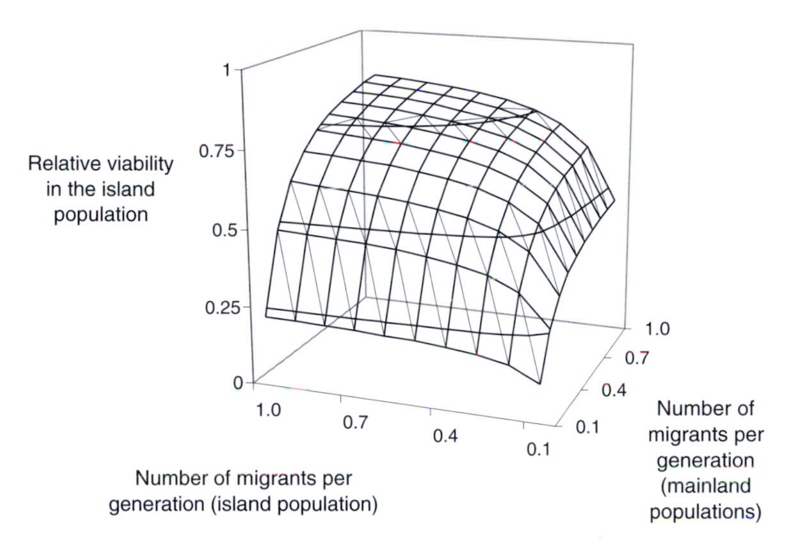

5.15 Expected viability in an island population relative to that in an effectively infinite population as a function of the number of migrants in the island and in the mainland populations. (From Couvet, 2002, with kind permission of Blackwell Publishing.)

even small remnant patches can have some insect diversity value, especially if there is reduced contrast between patch and matrix. In other words, it is necessary to make the transformed landscape, as close as possible, structurally, compositionally and functionally, to the untransformed one. This encourages residence and inter-patch movement, and the maintenance of metapopulation dynamics.

Caution says that we must be wary of today's results, because only with time, perhaps over centuries, will the full effects of this landscape transformation be seen. Although rare species will suffer most, it appears that some currently common ones will also be affected. Tropical insects, which are often highly specialized and may have low vagility, are likely to be pressurized the most, as their home habitat patches become smaller and more isolated. Associated with this is the likelihood that more specialized interactions (parasitism, pollination etc.) are likely to succumb first.

It is becoming clearer that insect diversity conservation is heavily dependent on maintenance of habitat patches both large in size and high in quality. It is not just the extent of fragmentation and attrition that is of concern, but also where it occurs and its spatial geometry. With these transformations, most specialists will decline, while some generalists will benefit.

Evidence is accumulating that untransformed patches, which we may call nodes, should be connected as much as possible using near-natural linkages. For insects, this involves consideration of a whole range of habitat characteristics to which the many insect species are highly sensitive. These linkages have long-term genetic benefits, as well as short-term, demographic ones. They, are, in effect, increasing the habitat options, and hence genetic viability, of populations leading to survival security into the future.

The aim of this chapter has been to review insect responses (behavioural, ecological, evolutionary) to both short- and long-term changes in the land mosaic. It is important to identify the types, aspects and intensity of changes in land mosaic that pose genuine threats to insect diversity. Once we know the status of these threats, only then can we proceed with appropriate management (Part III).

6 Threats from invasive aliens, biological control and genetic engineering

In the month of August, 1862, a nest of the common WASP (*Vespa germanica*), was taken near Brighthampton, and handed over to Mr Stone, who has long been in the habit of experimenting upon these insects . . . The nest was very much damaged by carriage, and Mr Stone took it entirely to pieces, placing one or two small combs inside a square wooden box . . . He then fixed the box in a window, so as to allow the insects free ingress and egress through a hole in the back.

J. G. Wood (1876)

Insect imports, for better purpose, not merely as cabinet specimens but as living objects of admiring interest, we should indeed most gladly welcome from China . . .

Acheta domestica (1849)

In many obscure shipyards Long-horned Beetles have been found crawling, transported with timber from out of the virgin woods of North and South America; a dozen such were known and described by Stephens.

A. H. Swinton (*circa* 1880)

6.1 Introduction

We are entering the Homogenocene, the new geological era where biota the world over are becoming increasingly the same. This is because of establishment of organisms in foreign lands through human agency, accidental or deliberate. Such invasive alien organisms are now considered as a major threat to biological diversity, ecological integrity and ecosystem health. Among these invasive organisms are alien plants that establish, spread and outcompete indigenous plants. This changes the composition, structure and function of natural ecosystems and, in turn, has a major impact on indigenous insects. Invasive alien insects and vertebrates also pose a threat to indigenous insects, mostly through direct competition or predation, especially on oceanic islands.

Biological control is a multifaceted activity. Use of indigenous predators or parasitoids to control pests is generally benign to local insect diversity. Classical biological control, on the other hand, where alien natural enemies are introduced to control an alien pest, carries risks for some indigenous, non-target organisms. The biggest concern is that once introduced and established, it is virtually impossible to recall a natural enemy.

There is also concern for introduction of alien pathogens, deliberate or accidental. The problem is that the pathogen is small and not readily seen or traced. As with the possible impact of insect natural enemies on non-target insects, the impact of pathogens on non-targets is difficult to detect. The impact can easily happen quickly and out of human sight, as we cannot constantly monitor all ecosystems simultaneously.

Genetic engineering is the introduction of alien genes as opposed to whole organisms. To date, this has had little influence in the field of insect diversity conservation, although it may in the future. As with any deliberate introduction, great caution and circumspection is required before introduction, whether of a whole organism or of a gene.

Where will this unabated homogenization of the world's biota all end? How stoppable is it? There is a huge ethical and practical dilemma that is not easily soluble. Agriculturalists, and ultimately consumers, seek biocontrol as an alternative to pesticides for crop protection. Yet ecological integrity is being compromised.

In the case of invasive alien organisms, we are simply not doing enough to stop them.

6.2 Invasive alien plants

The magnitude of the invasive alien plant problem is highlighted by the fact that 23% of the plant species in the USA (Pimentel *et al.*, 2000) and 47% in New Zealand (McNeely, 1999) are aliens. The problem is more than just

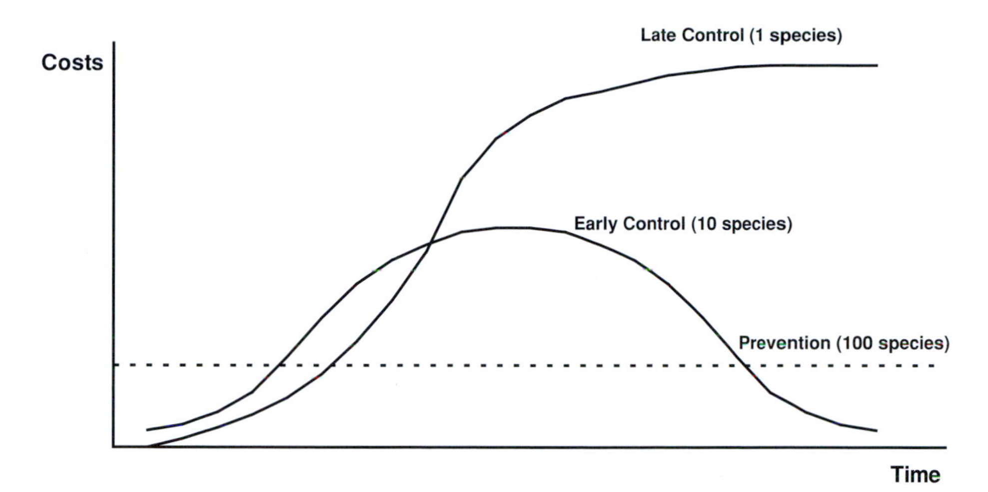

6.1 Illustrated trade-off between target specificity and cost of control of invasive alien organisms for three control strategies. Late control of a single species that actually becomes invasive will be less costly in the short run, but most costly in the long run, than prevention measures for a much larger number of species, many of which may never invade. (Redrawn from Naylor, 2000.)

numbers of species. Besides some aliens displacing indigenous plants, about 700 000 ha of US wildlife habitat are being invaded every year, with some species over-running entire ecosystems, with adverse effects on hydrology (van Wilgen *et al.*, 1996), as well as on biodiversity. The financial losses incurred and costs of control are enormous, with the USA spending $100 million per year to control aquatic weeds alone (Pimentel *et al.*, 2000). Both Samways (1999a) and Naylor (2000) have emphasized the importance and long-term benefits of predicting and preventing invasions rather than dealing with them afterwards (Figure 6.1).

Our knowledge of how invasive alien plants affect insect diversity and conservation is very limited. On South Georgia island in the South Atlantic, body size of the beetle *Hydromedion sparsutum* was smaller in areas where alien grasses predominated (Chown and Block, 1997). This mirrors the situation in the arid Northern Cape Province of South Africa where thickets of the alien tree *Prosopis glandulosa* reduced dung beetle species richness and disfavoured larger, heavier and rare species (Steenkamp and Chown, 1996). Invasive alien ragwort *Senecio jacobaea* in New Zealand reduced saprophytic microarthropod abundance immediately adjacent to the flowering plants (Wardle *et al.*, 1995). However, the effect of the alien was varied and complex and affected various interactions and various indigenous species differently. Similar findings come from South Africa, where indigenous insect diversity was altered but not necessarily impoverished under alien vegetation. Both abundances and species composition changed as a result of alien vegetation (Samways *et al.*, 1996).

Sun-loving dragonflies are excluded when invasive alien trees shade out the habitat (Samways and Taylor, 2004). The odonate assemblage under the alien canopy converges on that of indigenous forest, but where alien plants, such as bramble, do not radically change light conditions, even localized endemic species appear to survive perfectly well (Kinvig and Samways, 2000). This seems to be a general phenomenon, so long as water conditions remain suitable, as similar results were found on the islands of Mayotte (Comoros) (Samways, 2003a) and Silhouette (Seychelles) (Samways, 2003b). Similarly, reduced insolation from introduced dense-canopy trees such as pines can reduce acridid diversity in adjacent grassland, both in Australia (Bieringer and Zulka, 2003) and South Africa (Samways and Moore, 1991).

The effect of invasive alien plants is not always detrimental to insect diversity. Moderate levels of aquatic invasives in South Africa can encourage certain odonate species, but these are habitat generalists with wide geographical distributions (Stewart and Samways, 1998). Similarly, in Argentina, grasshopper assemblages were not adversely affected by alien plant species (forbs, dicots and sown grasses), with densities generally being higher in such disturbed areas, especially of potential pest species (Torrusio *et al.*, 2002).

On Mediterranean Islands, alien *Opuntia maxima* cactus provides refuge for tenebrionids in the absence of suitable indigenous vegetation (Cartagena and Galante, 2002). Clearly though, the effects of invasive as well as planted vegetation involves a complex interaction of factors. In a South African study, there was marginally lower arthropod species richness in alien compared with indigenous vegetation. But there was also a different assemblage of species in the two vegetation types, with certain species being negatively or positively sensitive to the vegetation type. Certain species however, were restricted to the indigenous vegetation (Samways *et al.*, 1996).

An emerging research field is the effect that invasive alien plants can have on ecosystem processes (Rejmánek, 1999) and as a stress factor for the indigenous insect species. As invasive alien plants are largely an irreversible feature of the landscape, it is likely that the stress will generate evolutionary change, perhaps in a way outlined by Hoffmann and Hercus (2000) (Figures 6.2 and 6.3; Table 6.1).

A further consideration is that many invasive species have lag periods where the impact comes long after the invasion. This is especially so in plants which may remain scarce for decades before suddenly impacting. There are various population, environmental and genetic reasons why this happens (Crooks and Soulé, 1999). The point, as regards insect diversity conservation, being that the pressures will continue into the future not only from new plant appearances but also from those that emerge from their lag period.

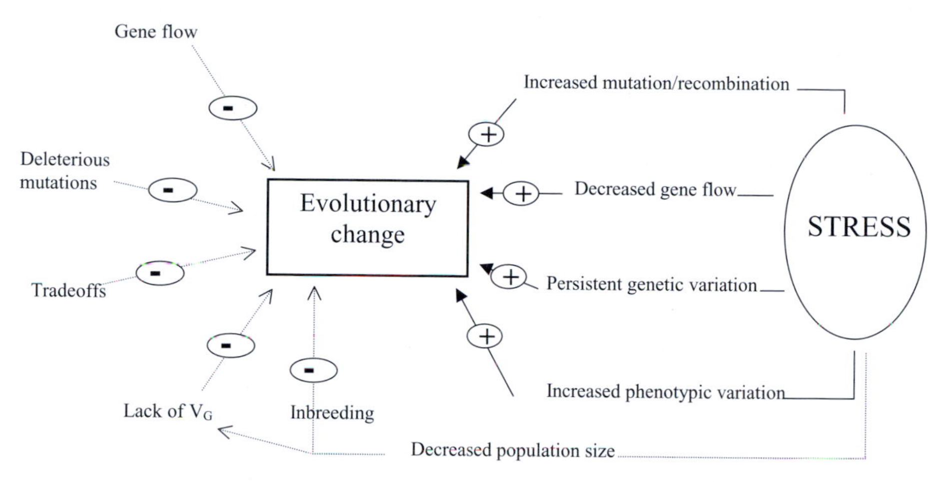

6.2 Outline of the processes that can have positive (+) or negative (–) impacts on rates of adaptation in a population, and the impact of stress on these processes. V_G is genetic variance. (From Hoffman and Hercus, 2000. Copyright, American Institute of Biological Sciences.)

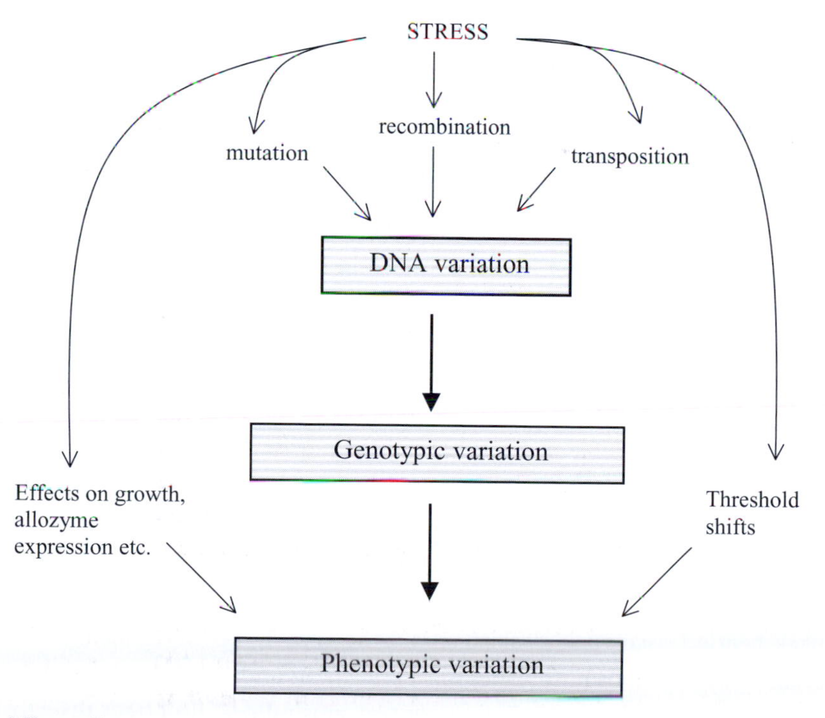

6.3 Effects of stress on variation. Genetic variation can be produced following increased rates of mutation, recombination, and transposition. Stress can also increase the expression of variation at the phenotypic level by lowering thresholds for the expression of traits, by influencing growth or metabolic flux, or by other processes. (From Hoffmann and Hercus, 2000. Copyright, American Institute of Biological Sciences.)

Table 6.1. *Possible ways in which stressful periods may have had a creative role in evolution by causing extinction events (modified from Hoffmann and Parsons, 1997)*

Decrease in predation pressure allows the establishment of novel genotypes; subsequent radiations following renewed predation pressures.
Decrease in competitive interactions enables previously non-competitive species to survive; evolutionary radiations following renewed competition.
Clearing of ecological space provides habitat for new adapted forms.
Stress induces expression of increased genetic and phenotypic variability, increasing rates of evolutionary change.
Reduction in gene flow allows new combinations of genotypes to persist.

6.3 Invasive alien vertebrates

The impact of vertebrates on insect diversity can be direct, such as eating insects, or indirect, by changing the insects' habitat. Such impacts are often synergistic with others, such as invasive alien plants and deterioration of the soil layer. These impacts have been magnified on islands where small area, lack of population restraints on the alien vertebrates, coupled with intrinsic vulnerability of some island insects (because of small population size, large physical size and naïvity) have mediated a reduction in island insects.

High tourism pressure coupled with impact of introduced turkeys, hens, rabbits and rats have had a major impact on the number of tenebrionid beetles surviving on eastern Iberian Mediterranean islands (Cartagena and Galante, 2002). Indeed, rodents can have a phenomenal appetite for indigenous fauna, with mice (*Mus musculus*) on Marion Island consuming up to 194 g ha^{-1} of invertebrate larvae and adults, weevil larvae and adults, earthworms, spiders and flies. Weevil adults alone were consumed at such a rate that nearly six times the annual average weevil (*Ectemnorhinus* spp.) biomass was consumed over a year (Smith *et al.*, 2002). No wonder the Black rat *Rattus rattus* eliminated the Lord Howe Island stick insect *Dryococelus australis* on that island (Wells *et al.*, 1983; Priddel *et al.*, 2003).

Much of the predatory activity of introduced agents goes undetected, or at least unresearched. What is clear is that species like the cane toad *Bufus marinus*, which was introduced into Australia in 1935 to control larvae of the Grey back beetle *Dermolepida albohirtum* and the Frenchi beetle *Lepidota frenchi* in sugarcane, and the mosquito fishes *Gambusia affinis* and *G. holbrooki* which have been introduced in many countries to control mosquitoes, do affect insect diversity. What we do not know, however, is which insect species are affected and by how much, and whether it matters for their long-term survival.

In a study of alien trout on an indigenous Australian dragonfly *Hemicordulia tau*, Faragher (1980) found that despite high levels of predation on the last three instars, the long-term survival of the dragonfly was not in jeopardy, owing to its opportunistic dispersal ability over a large geographical area. This contrasts with the situation in Hawaii where various poeciliid fish, including *G. affinis*, had a negative impact on the indigenous *Megalagrion* spp. damselflies (Englund, 1999).

6.4 Invasive alien insects

6.4.1 *The scale of the problem*

There has been, over the years, a major spread of insects across the world through human agency. This is partly because insects are excellent stowaways, particularly Diptera, but also Hemiptera, Hymenoptera, Lepidoptera and Coleoptera on aircraft, and Lepidoptera and Odonata on ships (New, 1994). As mentioned in Chapter 4, Hawaii accumulates 20–30 new insect species per year (Beardsley, 1991) and Guam 12–15 new species, facilitated by the huge amount of human traffic (Schreiner and Nafus, 1986). Even on the remote South Atlantic Gough Island, out of the 99 species of winged insect species recorded on the island, 71 are established introductions (Gaston *et al.*, 2003).

Probably only a small percentage of insects arriving in foreign lands establish (Williamson, 1996), with many remaining local and relatively rare, such as the moth *Phyllonorycter messaniella* around cities in southeast Australia (New, 1994). Nevertheless, on Gough Island most human landings may lead to the arrival of at least one alien. What is concerning is that these rates of introduction of new insect species are estimated to be two to three orders of magnitude greater than background levels for Gough Island, an increase comparable with that estimated for global species extinctions (many of which occur on islands) as a consequence of human activities (Gaston *et al.*, 2003).

6.4.2 *The special case of invasive alien ants*

Invasive alien ants in particular have been devastating to many ecosystems across the globe (Holway *et al.*, 2002). This is especially so on islands. Hawaii, which was originally devoid of any social insects, has been heavily invaded by ants, with the Big-headed ant *Pheidole megacephala* and other invasive ants having had a major adverse effect on indigenous insects (Wilson, 1996) and spiders (Gillespie, 1999). At the present time, the crazy ant *Anoplolepis gracilipes* is radically changing ecosystem processes on Christmas Island, 350 km south of Java, while the invasive Red imported fire ant *Solenopsis invicta* has locally reduced indigenous USA ant species by 70% and arthropods in general by 30% (Schmidt, 1995). On islands in the Seychelles, *A. gracilipes* has radically reduced components of the ground invertebrate fauna (Figure 6.4) and that of the tree canopy (Hill *et al.*, 2003). In Texas, *S. invicta* locally excluded four endemic ant species,

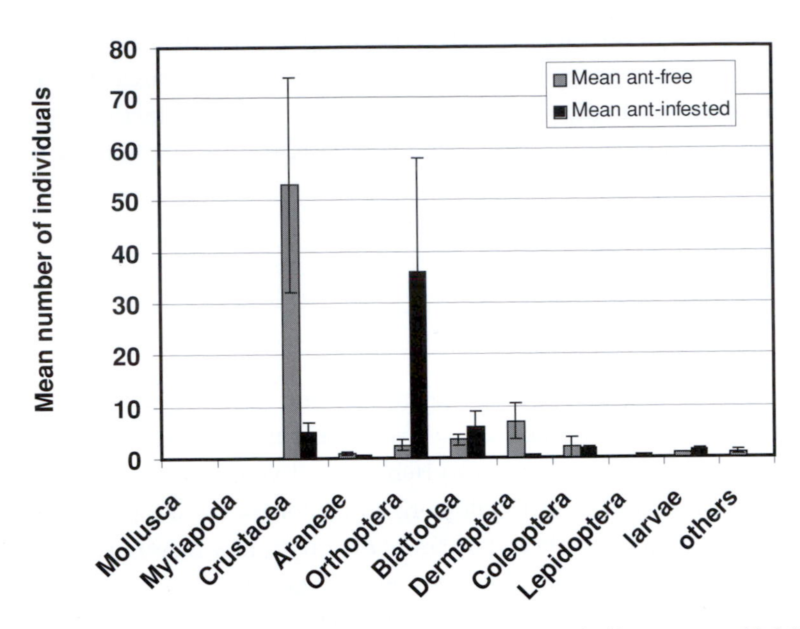

6.4 Taxonomic composition of pitfall trap assemblages, excluding ants, on Bird Island, Seychelles, where the invasive alien ant *Anoplolepis gracilipes* has and has not invaded. The large figure for Orthoptera in the ant-infested area is made up of one species, *Myrmecophilus* sp., which is ant tolerant, as are the cockroaches. (Redrawn from Hill *et al.*, 2003, with kind permission of Kluwer Academic Publishers.)

although when this invasive ant is controlled with bait treatments, there is some recovery of the indigenous ant assemblage (Cook, 2003).

Even some continental ecosystems are being modified by alien ants. The fynbos of South Africa, renowned for its botanical diversity, is being modified by the alien invasive Argentine ant *Linepithema humile*, which pre-empts seed burial by indigenous ants (Bond and Slingsby, 1984). It is such a highly invasive ant that it can exclude indigenous ant species (Donnelly and Giliomee, 1985) and reduce invertebrate diversity as a whole (Human and Gordon, 1997). Indigenous ant species that are tolerant of *L. humile* tend to have small body sizes, small colonies, are rare and/or forage at times that do not compete with foraging Argentine ants (Carpintero *et al.*, 2003). Although *L. humile* tends to follow watercourses and lines of disturbance, particularly in Mediterranean-type ecosystems (de Kock and Giliomee, 1989; Holway, 1995; Human *et al.*, 1998), it has now been discovered penetrating indigenous evergreen forest (Ratsirarson *et al.*, 2002). Part of the reason for the pioneering invasive success of *L. humile* is that queens need only as few as ten workers for the population to establish successfully and grow quickly (Hee *et al.*, 2000). During the pioneering, invasive stage, colonies can go through a genetic bottleneck that reduces genetic diversity. This, in turn, leads to reduced intraspecific aggression among spatially separate nests, and leads

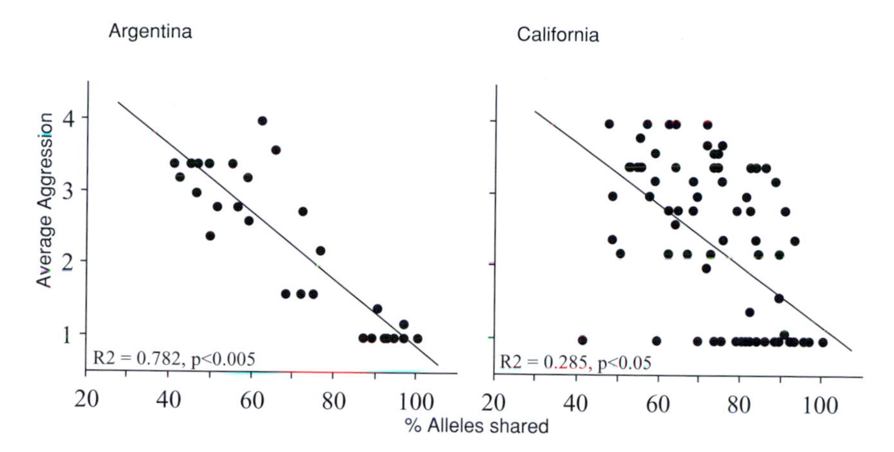

6.5 Relationship of aggression between nests of the Argentine ant, *Linepithema humile* and their genetic similarity. Aggression between pests was measured by using a behavioural assay ranging from 1 (no aggression) to 4 (high aggression). (From Tsutsui *et al.*, 2000.)

to the formation of supercolonies which dominate other, sympatric ant species (Tsutsui *et al.*, 2000) (Figure 6.5).

We must be careful about generalizing on the effects of invasive alien ants. Although Human and Gordon (1997) report that the Argentine ant negatively affects spiders, possibly through competition for prey, Bolger *et al.* (2000) found spiders most abundant and diverse in the smaller, older habitat fragments, and where the Argentine ant was most abundant. This and other variances in results suggest that the situation is very complex and may well depend on aspects of microhabitat, including microclimate, microarchitecture and particular biological features of, and interactions between, various taxa.

In Texas, where carrion resources were scarce, *S. invicta* drastically altered decomposer community composition and the process of succession (Stoker *et al.*, 1995). Even when resources were abundant, the fire ant still significantly altered population levels of various fly and beetle taxa. This mirrors the situation in decomposing fruit, where the decomposing community was similarly altered (Vinson, 1991). Although indigenous fire ants (*Solenopsis geminata*) also affect the decomposing community in carrion, the point is that *S. invicta* is competitively aggressive to the point of totally excluding other ant species (Stoker *et al.*, 1995).

In the artificial ecosystem Biosphere 2, the Crazy ant *Paratrechina longicornis* eventually became totally dominant, feeding almost exclusively on homopteran honeydew. Such a mutualistic spiral between an increasing ant population and increasing homopteran population is not unusual, being a frequent feature in agriculture. What is interesting however, is that the only invertebrates thriving in Biosphere 2, besides the ant and the homopterans, were either species with

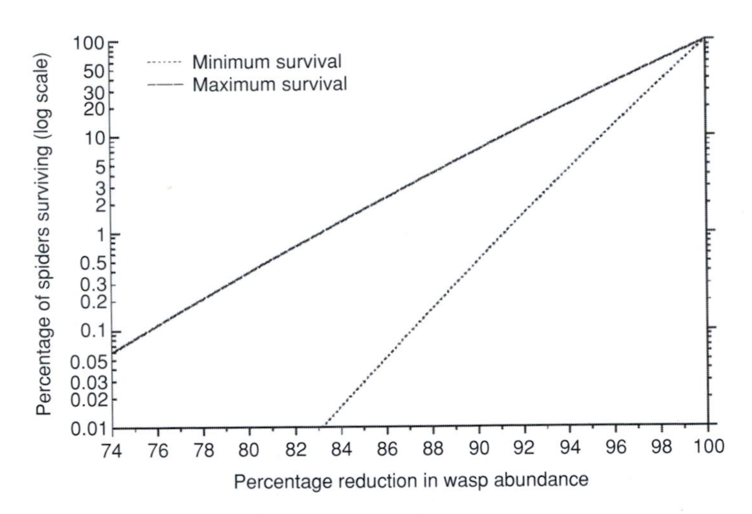

6.6 The Common wasp *Vespula vulgaris* has invaded New Zealand and is reducing Garden orb-web spiders (*Eriophora pustulosa*). The graph shows the percentage of orb-web spiders predicted to survive at the end of the wasp season given varying percentage reductions of wasp numbers (assuming constant reduction over the entire season). (From Toft and Rees, 1998, with kind permission of Blackwell Publishing.)

effective defences against ants (well-armoured isopods and millipedes) or tiny subterranean species that can escape ant predation (mites, thief ants and springtails) (Wetterer *et al.*, 1999).

Moller (1996) points out that only a few interrelated characters may be important in determining invasiveness. Some intercorrelated characteristics may increase while allowing invasiveness to come from the flexibility that sociality itself allows. The morphological size range of the different castes tunes each to different tasks. For example, bigger workers forage farther or are confined to food defense roles. The most polymorphic ant species are more successful, partly due to feeding flexibilities, with general ant individuals being able to team up to corral and kill prey, carry it back to the nest, and guard it from competitors during the journey. This means that they can kill and retrieve a greater range of food types, either as a group or as individuals. The result is close tracking of temporal and spatial changes in resources, and of dangers from competitors or natural enemies (Moller, 1996).

Ants are not the only hymenopterans that can reduce indigenous insects. Honeybees (*Apis mellifera*), for example, appear to exclude some indigenous bees in Tasmania (Goulson *et al.*, 2002).

6.4.3 *Past and future effects of invasive insects*

One concerning feature as regards the impacts of invasive alien insects is that current research may already be on a depauperate fauna. In an experimental

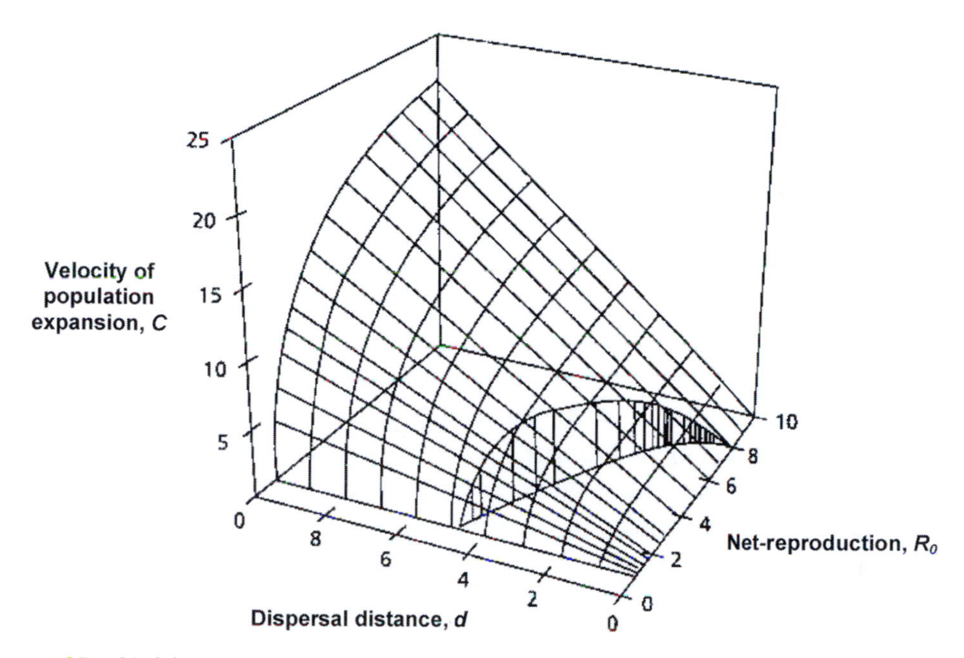

6.7 Models enable us to improve the understanding of the invasion process. Model behaviour demonstrates that invasion of uniform areas is a linear function, whereas invasion of non-uniform areas is non-linear. In the non-uniform areas, as with most landscapes, expected invasion rates initially increase with fraction of suitable area, reach a maximum, and subsequently decline with decreasing suitable habitat. The position of the maximum is determined by dispersal risk as a function of the suitable area. The fraction of suitable area most favourable for the invader's progress depends on its dispersal and mortality rates, which determine the dispersal risk. Here the invasion rate (C) is plotted as a function of both the net rate of reproduction (R_0) and dispersal distance (d). (From Hengeveld, 1995, with kind permission of Kluwer Academic Publishers.)

study of the effect of the alien Common wasp *Vespula vulgaris* on orb-web spiders in New Zealand, Toft and Rees (1998) showed that the effect is so severe that the probability of a spider surviving to the end of the wasp season was virtually nil, and that invertebrate taxa most vulnerable to wasp predation may already have been removed from the indigenous *Nothofagus* forests during the 40 years of wasp occupation (Figure 6.6).

Considerable effort is currently going into predicting how invader species might spread (e.g. Tribe and Richardson, 1994). But the difficulty of prediction using analytical modelling is that we cannot make fully confident estimates before a potential invader has been introduced and actually spreads. Nevertheless, models can show how variables interrelate, which practical difficulties of the measurements should be solved, and where an invasion is likely to stop (Hengeveld, 1999) (Figure 6.7).

6.5 Risks of introducing insect natural enemies

6.5.1 *Risks of classical biological control of insects*

The risks of classical biological control (CBC) (the deliberate introduction of foreign control agents against foreign pests) for indigenous insect faunas have been known for many years (Samways, 1988; Howarth, 1991). The risks are no less today because new pests are constantly appearing in new lands. In response, new biocontrol agents are being translocated between nations, resulting in global homogenization of faunas. Although establishment of an alien pest is unintentional biotic contamination, dealing with the problem using foreign biocontrol agents can be viewed as intentional contamination (Samways, 1997d). This raises profound ethical issues. In particular, the introduction of alien organisms offends our moral concept of *place*, because it violates a deep cultural, historical, and aesthetic context (Lockwood, 2001).

In practical terms, the risks of biocontrol have to be weighed against the advantages. But how do we measure these risks? And who has the advantage, and others, presumably, a disadvantage? The source of the conflict lies in the very nature of biological control agents – they persist, spread and evolve (Lockwood *et al.*, 2001). One of the risks is that biocontrol agents are rarely specific (monophagous), with only 7 of the 45 species of chalcidoid parasitoids (considered as among the most host-specific biocontrol agents) introduced into South Africa being monospecific (Prinsloo and Samways, 2001). This is potentially hazardous, particularly on tropical islands such as Guam (Nafus, 1993). Barratt *et al.* (1996) have recorded the parasitoid *Microctonus aethiopoides*, introduced for control of the weevil forage pests *Sitona discoideus* and *Listronotus bonariensis*, attacking 13 non-target indigenous weevil species in New Zealand. Furthermore, one of the unintended hosts of *M. aethiopoides* is *Rhinocyllus conicus*, a weed biocontrol agent introduced into New Zealand to control the weed, Nodding thistle, *Carduus nutans* (Barratt *et al.*, 2001).

On La Réunion island, Indian Ocean, after the parasitoid *Tamarixia dryi* eliminated the alien citrus psylla *Trioza erytraea*, it switched to the indigenous host *T. litseae* (*eastopi*) but did not eradicate it (Aubert and Quilici, 1983) (Figure 6.8). On Fiji, the tachinid fly *Bessa remota* was introduced to control the endemic coconut moth, *Levuana iridescens*. The moth went from a serious agricultural pest to a threatened species in only 2 years, and shortly after, was thought to be extinct. Another non-target moth *Heteropan dolens* became extinct at the same time. The parasitoid continues to exist on Fiji parasitizing only non-target species (Howarth, 2001). This may not, however, be the whole story, as Sands (1997) makes a strong counter-argument including the fact that *L. iridescens* may not be endemic to Fiji, and not even extinct there or elsewhere.

The risks to non-targets are not necessarily just an island phenomenon. The tachinid parasitoid *Compsilura concinnata*, which was introduced into continental

6.8 The African citrus psylla *Trioza erytraea* has the dubious distinction of being one of the few alien insects locally and completely eradicated by an introduced biological control agent, in this case by an introduced parasitoid (*Tamarixia dryi*) on the Indian Ocean island of La Réunion.

USA several times in an attempt to control 13 pest species was found to attack non-target saturniid moths and may have been responsible for their decline (Boettner *et al.*, 2000).

6.5.2 *Risks of weed biocontrol*

In Canada, the weevil *Larinus planus* which was introduced to control the alien thistle *Cirsium arvense* has been found to attack an indigenous *Cirsium* thistle yet is not having much impact on seed production in the target thistle. This suggests that this weevil entails a high risk-to-benefit ratio, with other evidence suggesting that this is not an isolated case (Louda and O'Brien, 2001).

One of the greatest concerns in weed biocontrol is the unintentional arrival of the South American moth *Cactoblastis cactorum* in Florida. It feeds on prickly pear cacti, *Opuntia* spp., and has been hailed as a great success in controlling prickly pear in Australia. It has now spread to South Carolina where it is attacking several *Opuntia* spp. (Hight *et al.*, 2002). Concerns are that it will spread to the *Opuntia*-rich areas of the USA and Mexico. Biological control using introduced natural enemies might cause an upset in the balance between indigenous closely related moths and their natural enemies and cause secondary pest problems. An approach is to release large numbers of genetically altered (sterilized or partially sterilized) individuals of the moth to ensure that when matings occur in the field, a significant proportion of these will involve sterile insects, thus resulting in greatly reduced viability of the F_1 generation from these matings, and stimulate the build up of effective natural enemies (Carpenter *et al.*, 2001).

6.5.3 *Positive effects of biological control*

Not all aspects of CBC are harmful to biodiversity. The alien pest scale insect *Aonidiella aurantii* in South Africa can infest indigenous African trees such as *Trichilia* spp. and *Rhus* spp. Not only is it suppressed by the introduced parasitoid *Aphytis melinus* and the invasive ladybird *Chilocorus nigritus* but also by the highly effective and indigenous *Aphytis africanus*. In the case of the *Juniperus bermudiana* forests of Bermuda, which were being devastated by the scale insects *Carulaspis minima* and *Lepidosaphes newsteadi*, it appears that the introduced ladybird *Rhyzobius lophanthae* and an unknown parasitoid may have prevented the demise of the juniper. A similar situation occurred on St Helena where the indigenous gumwood *Commidendrum robustum* was being threatened by the alien scale insect *Orthezia insignis*. The introduced ladybird *Hyperaspis pantherina* has had a major impact on the scale, and the last stands of the tree have recovered (IIBC, 1996).

6.5.4 *Risk–benefit analysis of biological control*

The problem with detecting any adverse action of biocontrol agents is that besides being small and cryptic, their role in local extinctions and even total extinctions is rarely likely ever to be definitively recorded. This is perhaps the reason behind why some biocontrol proponents have suggested that there is little evidence for agents having done any harm. Additive upon this is the fact that any harmful effects of biocontrol agents is likely to be synergistic with other impacts, such as landscape fragmentation. This was shown to be the case for four species of parasitoids attacking the Forest tent caterpillar, *Malacosoma disstria*. Parasitism was significantly reduced or enhanced depending on the proportion of forested to unforested land (Roland and Taylor, 1997).

Clearly there is a need for much more risk–benefit analysis, especially as classical biological control is an economically effective and arguably a necessary method of pest insect and mite control. The problem is that there are risks from over-regulation, and in the case of Hawaii, 'The current atmosphere of bureaucracy and over-regulation is stifling the science and the practice of biocontrol to the detriment of both agriculture and native Hawaiian ecosystems' (Messing and Purcell, 2001).

Although there has been a strong call for much more screening of biocontrol agents (Samways, 1997d; Follett and Duan, 2000; Prinsloo and Samways, 2001) the reality of doing this is extremely difficult and uncertain. Not only can recently founded insect populations show genetic change (Thomas *et al.*,

2001) but screening can be overly conservative. The polyphagous egg parasitoid *Ooencyrtus erionotae*, despite earlier concerns, turned out to be host-specific to the Banana skipper *Erionata thrax* in Papua New Guinea (Sands, 1997). In the laboratory, the ladybird *Coccinella septempunctata* will feed on the eggs and young larvae of the threatened North American lycaenid butterfly *Erynnis comyntas* when its aphid prey is unavailable. But this has not been observed in the field (Horn, 1991).

We also need to know much more on when, where and how hosts invade and expand as a result of being released from natural enemies (the 'Enemy Release Hypothesis') (Keane and Crawley, 2002) which underpins biocontrol philosophy, in that classical biological control aims to put back the key mortality factor. This is an alternative (but not necessarily a mutually exclusive one) to the 'Niche Opportunity Hypothesis' where invasions are exploiting new niches open to them (Shea and Chesson, 2002). One of the key ways forward is to fully document the invasion biology of a natural enemy that has already been introduced into one area before it is introduced into another.

In the case of a newly discovered natural enemy, it is essential to test for possible attack on a range of closely related indigenous species to the target host. Results of pre-screening trials also need to be widely disseminated, so that different countries can benefit from the findings. This would enable us as a whole to get a better picture of which natural enemies carry a relatively low and which carry a high risk of switching to non-targets. However, assigning risk status to natural enemies requires great care, although it is clear that insect predators in particular carry a very high risk (Van Lenteren *et al.*, 2003) (Figure 6.9). Such risk assessments need to take into account population growth, interaction strengths and ecological interactions as well as the more familiar behavioural and developmental attributes, as evidence is accumulating that the interaction between biocontrol agents and indigenous species is more complex than originally thought (Louda *et al.*, 2003).

6.6 Risks of introducing insect pathogens

6.6.1 Risks from pathogens

Certain fungi (e.g. *Metarhizium anisopliae*), protoctistans (e.g. *Nosema locusteae*), bacteria (e.g. *Bacillus thuringiensis*), viruses (e.g. specific to some lepidopterans) and nematodes (e.g. *Steinernema feltiae*) have been employed in particular localities against specific target pests. These are pathogens that originally occurred naturally but have since been cultured, particular strains selected, and prepared as products that are sprayed onto the crop. An extension of this is to incorporate genes of entomopathogens into crop plants. The genes for producing the crystal proteins (Cry toxins), which are an insecticidal component of

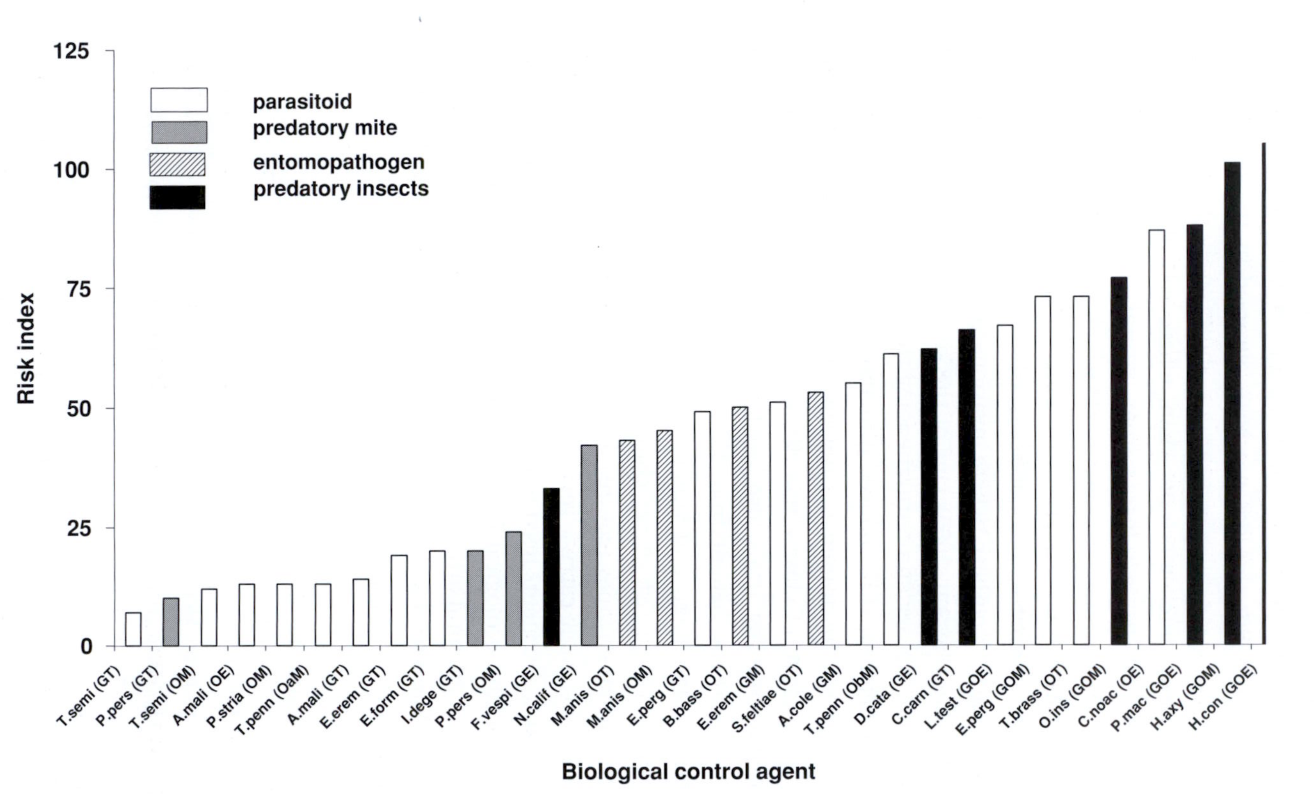

6.9 Risk indices of commercially available inundative and classical control agents released in European and Mediterranean countries. Names of organisms: first letter of genus name, three or more letters from species name. Letters in parentheses: a = risk index after release, b = risk index before release, G = greenhouse, O = open field, M = Mediterranean climate, T = Temperate climate, E = across Europe. (From Van Lenteren *et al.*, 2003, in which the full species names are given. With kind permission of Kluwer Academic Publishers.)

B. thuringiensis, have since been incorporated into various crop plants as a form of protectant against lepidopteran, coleopteran and dipteran pests.

Howarth (2001) has overviewed the impacts of pathogens, and points out that little has been done to ascertain their effect on non-targets. Indigenous species that have never been exposed to commercial pathogens may be particularly vulnerable, and in the new environment with new susceptible hosts, the pathogens might evolve new virulence. Furthermore, the availability of alternate hosts may support high populations of disease agents, potentially resulting in greater environmental impacts. It is essential now to monitor their non-target effects more closely. *B. thuringiensis israelensis* for example, which is widely used in mosquito control, has been recorded to cause mortality of mayfly and dragonfly larvae, as well as of other aquatic insects (Zgomba *et al.*, 1986).

Miller (1990) found that although populations of most non-target Lepidoptera recovered after the third year of spraying *B. thuringiensis*, species richness did not fully recover during the 3 years of the study. This result has been supported by other studies. While the mycoinsecticide *Metarhizium anisopliae* has been hailed as a breakthrough for control of the Brown locust *Locustana pardalina* in southern Africa, there are risks for non-target grasshoppers and the dependent birdlife as well as for various hymenopteran species (Samways, 2000a).

6.6.2 *Past impacts*

The challenge is to ascertain what damage may already have been done by application of pathogens many years ago when effects on non-targets were little considered. A further, vexing problem is knowing how many and to what extent pathogens have been introduced inadvertently with insect and other arthropod agents for weed biocontrol. Kluge and Caldwell (1992) have emphasized that some moths such as *Pareuchaetes* spp., which are potential biocontrol agents of the weed *Chromolaena odorata*, can carry enormously high densities of the microsporidian pathogen *Nosema* sp. (a protoctistan) in their faeces. This is cause for considerable concern in the target area, with no record of how non-target lepidopteran species are being affected. Although only 4.5% of fully grown larvae of the Winter moth (*Operophtera brumata*) in Britain die from microsporidian infection, this is likely to be a natural 'balance', with much higher figures likely in the case of new, virulent strains and physiologically naïve hosts. Outbreak of an apparently new strain of entomopathogenic fungus *Cladosporium oxysporum*, whose origin is unknown, eliminated local populations of Homoptera species (Samways and Grech, 1986).

Clearly, much more caution is needed with introduction of pathogens, especially on small islands, or close to special habitats. It seems, on balance, that again 'we know not what we do'. To compound the issue, certain pathogens are being genetically modified, especially insect baculoviruses, so that they kill the insect pest before the economic threshold is reached. The assessment of any risks attached to releasing genetically modified baculoviruses must focus on the host range of the virus. Some baculoviruses are host specific but others kill a range of host species. Tests on partially susceptible lepidopteran species showed that a ground-dwelling cutworm species was far more at risk from infection from genetically modified baculovirus than a foliar-feeding species of equivalent susceptibility (Cory, 2002). As the deployment of genetic technology develops, the risks to the maintenance of insect diversity will increase. This is such a major development that it now deserves a section of its own.

6.7 Risks of genetic engineering

6.7.1 *Scale of the challenge*

The role of genetically modified organisms (GMOs), particularly transgenic plants, has become a contentious issue in the public arena (Hails, 2000). The point is whether a plant containing a transgene (an implanted gene) can result in a negative impact on organisms or systems relative to the status quo. The impact of a transgenic crop could be direct, through increased invasiveness, or indirect, via alteration of agronomic practices. Ecological risks might be classified into three groups: (1) those concerning genome organization within the plant; (2) the escape of transgenes into wild relatives (the likelihood and the consequences); and (3) the impact on non-target species in the wider ecosystem. The edges of these classifications are fuzzy, with the genomic location of a transgene, for example, influencing the rate of gene flow through hybridization (Hails, 2000).

A further aspect of this debate is whether the use of transgenic insect-resistant crops results in less of the harmful insecticidal products and procedures being applied, so reducing overall environmental impact (Hope and Johnson, 2002). Currently available transgenic pest-protected plants appear to require fewer pesticide inputs, potentially reducing direct impacts on non-target species. For example, the Environmental Protection Agency in the USA estimates that the annual benefit to maize growers is $38–$219 million, coming from reduced application of broad-spectrum insecticides in addition to other factors (Ortman *et al.*, 2001). A sobering note, however, is the likely development of resistance to transgenic toxins, emergence of secondary pests following changes in pesticide regimes, and possible changes in community dynamics caused by removal of crop-feeding herbivores from fields. All these could undermine the potential for transgenic insect-resistant crops to provide a medium- to long-term solution to pest control problems. As regards resistance, to date it appears that only the Diamondback moth *Plutella xylostella* has developed resistance to transgenic plants with *Bacillus thuringiensis* insecticidal toxins, but this is unlikely to be a one-off case in the future (Wright, 2002).

6.7.2 *Importance of experiments*

Hope and Johnson (2002) emphasize that maintenance of insect diversity in agricultural ecosystems may be an important factor, both for long-term success of crop production and for the conservation of biodiversity in and around farmland. Insects carry out many important and indeed crucial functions, including pollination and maintenance of soil fertility, and are a food source for vertebrate predators. The point is that novel pest control measures should not compromise the long-term viability of agricultural ecosystems (Hails, 2002). In response to these concerns, there has been an emphasis on conducting high-quality detailed experiments which are discussed in a rational and constructive fashion (Poppy, 2002). Combining knowledge of insect ecology, behaviour and pest management can help drive research in this area, rather than just acting as a quality control of technology developed by others. In response to this call for scientific light rather than emotional heat, some large-scale experiments have been set up where plants and invertebrates are being sampled using sound protocols.

The impact of transgenes on non-targets was first highlighted when it was found that a high concentration of transgenic *B. thuringiensis* maize pollen (i.e. genes for insecticidal toxins, originally from a strain of *B. thuringiensis* and incorporated into maize plants but also expressed in the pollen) experimentally applied to *Asclepias curassavica*, a foodplant of the Monarch butterfly *Danaus plexippus*, caused significant mortality of its larvae (Losey *et al.*, 1999). This has been verified in field trials (Jesse and Obrycki, 2000). Predictions are that the effects of transgenic pollen on *D. plexippus* is likely to occur over 10 m from transgenic field borders, with the highest larval mortality within 3 m of the transgenic fields. However, it has been suggested that this will not pose a risk to these butterflies on a national scale (Ortman *et al.*, 2001).

These viewpoints emphasize the importance of large-scale experiments as have taken place under the Farm Scale Evaluations (FSE) in Britain (Firbank, 2003). The FSE found 27% fewer butterflies in genetically modified herbicide-tolerant (GMHT) beet fields, and 22% fewer in GMHT spring rape fields with 24% fewer around their margins. Woiwod (2004) has pointed out that these findings must be put in context. Firstly, the FSE was about weed management regimes and not GMO technology. If future herbicide-tolerant crops were to be created by conventional plant breeding, similar results would apply. Secondly, arable fields, by virtue of their intense cultivation, are not suitable for butterflies anyway. It is almost as if such fields are devoid of suitable habitat whatever way they are modified. For butterfly conservation, the real issue takes place outside the fields, and in Britain an effective way forward is to create wide, flower-rich, permanent margins around fields. But genetically modified (GM) crop technology has a more pernicious side to it. As Woiwod (2004) emphasizes, an area the size of Wales is being cleared annually in Amazonian Brazil to grow GM-free soya for the European market, with an inevitable devastating effect on insect diversity.

6.7.3 *Effects of genetically modified organisms on natural enemies*

There are some other concerns, and among them is the effect of GMOs on insect and mite natural enemies. Current transgenic crops expressing *B. thuringiensis* toxins are only effective against some lepidopteran, coleopteran and dipteran pests. This means that the effects of *B. thuringiensis* transgenic crops are likely to have differential impacts on different natural enemies, especially on their feeding behaviour. *B. thuringiensis*-toxin uptake by the aphid *Rhopalosiphum padi* and the spider mite *Tetranychus urticae* had no negative effect on their neuropteran predator *Chrysoperla carnea*, in contrast to the deleterious situation where the same predator was fed lepidopteran larvae (*Spodoptera littoralis*) reared on *B. thuringiensis*-maize (Dutton *et al.*, 2002). Interestingly, Jeanbourquin and Turlings (2002) found that the lepidopteran parasitoids *Cotesia marginiventris* and *Microplitis rufiventris* did not distinguish between odours emanating from transgenic or isogenic (without *B. thuringiensis* genes) plants.

Although to date the overall effect on natural enemies has been largely neutral, there have been reports of some negative effects on some parasitoids in the field, including a 30–60% reduction in *Macrocentris cingulum* parasitoids (Obrycki *et al.*, 2001). This contrasts with the situation where potato plants engineered to express snowdrop lectin, and being fed on by the Tomato moth *Lacanobia oleracea*, resulted in a positive synergism between the moth's parasitoid (*Eulophus pennicornis*) and the transgenic crop (Bell *et al.*, 2002). This is not the case with other similar transgenic potato pests. Adult *Aphelinus* spp. parasitoids of the aphid *Myzus persicae* were affected physiologically by the insecticidal construct 'GNA', although in different ways (Couty *et al.*, 2002).

Hawes *et al.* (2003) showed that in general the biomass of weeds was reduced under genetically modified herbicide-tolerant management of some British field crops. This change in resource levels then had a knock-on effect on higher trophic levels. Generally, herbivores, pollinators and natural enemies changed in abundance in the same direction as their resources, while detritivores increased in abundance in the genetically modified crops. Furthermore, the effect of the genetically modified crops was not only dependent on type of crop but also on phenology and ecology of the insect taxa concerned (Haughton *et al.*, 2003), which inevitably will have repercussions on food chains.

6.7.4 *Effects of genetically modified organisms on pollinators and soil organisms*

In the case of pollinators, Marsault *et al.* (2002) found little difference among a variety of bees and flies to transgenic herbicide-tolerant, oilseed-rape compared with an isogenic variety. The only significant difference between control and transformed plants was for honeybees which were more abundant on the control. Ramsay (2002) has pointed out that insects, especially honeybees and bumblebees, are responsible for most oilseed rape gene flow, with each insect species contributing its own pattern to this genetic flux. Gene flow in this plant consists of a small element from wind-dispersed pollen, intense gene flow over short distances reflecting short hops by pollinating insects, less intense gene flow spreading as a fine network over bee colony foraging ranges, and very low levels of gene flow from free-living beetle pollinators dispersing over extremely long distances. However, the absolute magnitude of such gene flow depends on a range of factors, from weather, landscape, and pollinator populations to the relative sizes of sources and sinks, and fertility characteristics of receiving flowers.

A further consideration is the effect of transgenic *B. thuringiensis* toxins on soil organisms. The information to date is contentious (Obrycki *et al.*, 2001), and highlights again how much more research is needed. There is also a need for bioassays to test individual components of interactions. Lövei *et al.* (2002) found the generalist wolf spider *Pardosa amentata* to be a suitable model predator for screening transgenic wheat, while Haughton *et al.* (2003) suggest Collembola, bees and butterflies as indicators of the use of genetically modified herbicide-tolerance in British field crops. Brooks *et al.* (2003) found that the effects of herbicide management of genetically modified herbicide-tolerant British spring crops, had a very variable effect on soil surface-active invertebrates, depending on higher taxon and even between species. Nevertheless, there were consistent increases in abundance of detritivore Collembola and some of their predators, possibly through enhancement of the habitat quality (such as fungal food resources) of the Collembola.

6.7.5 *Do genetically modified organisms affect management activities?*

The real challenge for insect diversity conservation relative to GMOs is determining what are really the direct impacts, and to what extent they are synergistic with other impacts such as fragmentation. In a converse sense, we also need to know whether any adverse effects inhibit conservation management activities. Do, for example, GMOs reduce the effectiveness of corridors of indigenous vegetation between agricultural fields? In particular, do they reduce the effectiveness of pesticide-untreated margins of fields (conservation headlands) or does the transgenic crop benefit these headlands because less broad-spectrum insecticides, and herbicides, are required? These questions are beginning to be addressed, with initial results suggesting the effects vary substantially from one genetically modified herbicide-tolerant crop to another (Roy *et al.*, 2003). Butterflies were particularly sensitive to differences, possibly due to lower nectar supplies in genetically modified crop margins.

6.7.6 *Value of mathematical models*

Hillier and Birch (2002) are developing mathematical models to assist in this risk assessment. One model explores the impact of changes to parameters associated with food quality in a theoretical food chain. Conditions are presented under which changes to parameters describing the food quality of the crop and pest are related to the trophic levels above (pest and predator respectively) so as to predict long-term environmentally adverse consequences. A second model concerns pest adaptation to a transgenic insecticidal crop. Analytical results from the model can be used to (1) suggest measures to control the rate of pest adaptation and (2) explore the potential impact of effects on the rate of adaptation. Hillier and Birch (2002) point out that such models might be used as a first step in risk assessment, and warn of direct and indirect effects on managed ecological communities.

6.7.7 *Risks and synergies*

An interesting aspect of GMOs is development of possible control strategies for the Argentine ant, *Linepithema humile* (Tsutsui *et al.*, 2000). With the discovery of the association between genetic variability and intraspecific aggression, the introduction of new alleles into introduced populations was thought to possibly increase genetic differentiation to a level sufficient to trigger intraspecific aggression. This has now been shown not to happen in reality, with individuals from less genetically diverse colonies attacking individuals from more diverse colonies (Tsutsui *et al.*, 2003). A way to overcome this problem would be to introduce genetically different males, which could disperse into established colonies, mate with virgin queens, and thereby infuse new alleles. There are still risks involved, and increasing genetic diversity could undermine future control strategies designed to exploit genetic homogeneity of the pest populations, such as introduction of alleles that confer resistance to potential biocontrol agents (Tsutsui *et al.*, 2003).

The field of GMOs and their effects, good and bad, are in their infancy. One challenge will be deciphering what is purely an effect of a GMO and what is partial or synergistic. This is the same challenge that faces proponents of classical biological control and of use of entomopathogens.

6.8 Summary

Invasive alien plants are now a major component of many floras around the world. The effect on ecosystem processes and health of this invasion is locally enormous. The impact on insect diversity is variable and complex. There is generally an impoverishment of the indigenous insects where alien plants establish. Some predaceous insect species are highly tolerant of the change in plant assemblage, although not necessarily of the new plant structure, which may shade the habitat.

Little solid evidence is available on how alien vertebrates, whether fowl, fish or mouse, have impacted on insect diversity. Yet some of these vertebrates clearly have an enormous appetite for some insects, and there is strong circumstantial evidence that island insect faunas in particular have been radically affected.

Some insects themselves are invasive, with tramp ants being among the worst offenders. Although some insect introductions have been accidental, others were deliberate. Classical biological control, where insects and other organisms are introduced to control a target pest, is a highly contentious issue, largely because we do not know exactly the magnitude of the harm that the practice does. Furthermore, it depends on our viewpoint on the trading off of economic benefit derived from biological control versus the environmental risks. Evidence is accumulating that there are clearly risks to non-target species being attacked, making it essential now to screen more astutely before new introductions are made.

Introduction of insect pathogens also carries risks. In the past, these risks were not apparent and not recorded. Much more circumspection is now required. The same may be said of genetically modified organisms (GMOs). In this case however, at least bioassays, large-scale trials and mathematical modelling are under way from the beginning. Like biological control, there are economic benefits but some environmental risks. It is important now to sharpen up on exactly what those risks are their magnitude and their repercussions. Evidence so far suggests that the risks are relatively small and localized, and involve localized changes of certain insect species' population levels. However, trade issues associated with GM crops are a pernicious threat to insect diversity. There is a dire need for much more research, particularly as the effects of GMOs are highly likely to be synergistic with major impacts such as habitat fragmentation, invasive aliens and global climate change.

7 Global climate change and synergistic impacts

My occasional summers spent among those dark romantic lochs that indent the deep depressions in the picturesque clay-slate mountains of the Western Highlands of Scotland, first brought me face to face with the great problem of the influence exerted by climate over our fauna, when the gloomy glen and heathery hill disclosed the existence of species unknown in the genial south. The dark Scotch Argus Butterfly fluttering in the shady bushes, the globular papery nest of the Tree Wasp hung at the rushing burnside – no less than the sooty aspect assumed alike by the Garden Moth and Braeside Butterfly . . .

A. H. Swinton (*circa* 1880)

It is only change that is at work here

R. Wilhelm (1964)

7.1 Introduction

Anthropogenic global climate change is upon us. It is likely to continue for many decades yet. Global climate change is a major consideration for insect diversity, both directly and indirectly. Insects, being ectotherms and sensitive to temperature changes, are theoretically likely to respond directly to elevated temperatures by shifting their geographical ranges closer to the poles, or to higher elevations. There is evidence that some species are indeed changing their geographical ranges in this way. In turn, those in the tropics may come under increasingly severe stress, being adapted to a narrow range of temperatures and often with nowhere else to go.

It can be difficult in global climate change research to determine cause and effect, as so many ramifying variables are involved. Also, having one earth, we have a sample size of only one. For this reason, it is difficult to be sure of the accuracy of geographical range-change predictions until they actually happen.

The effects of global climate change on individual insects, populations and communities are many, varied and variable. Also, global climate change is more than simply about temperature. Weather patterns, rhythms and intensities will also change. These large-scale effects are also strongly synergistic with other, local effects, such as pollution, landscape fragmentation and attrition. The large-scale synergistic and adverse changes can make a mockery of reserve selection and maintenance, and even of small-scale conservation efforts in general. The field of global climate change is an immensely complex one, that will become increasingly important in every aspect of our lives. Not only will wild populations and communities change, but so will insect diversity associated with crops. In short, all aspects of insects' and other biota's well-being, whether viewed from a utilitarian or deep ecology standpoint, will change. The evolutionary resourcefulness shown by insects to changing climates in prehistorical times will now be tested to the full.

7.2 Ecosystem response to global climate change

7.2.1 Changes taking place

The concentration of atmospheric CO_2 has been increasing since the mid-nineteenth century. The continued increase in greenhouse gas concentrations are predicted to have a major effect on the world's climate on a timescale of decades to centuries. Global mean surface temperatures have increased 0.6 °C since the late nineteenth century, and by 0.2–0.3 °C over the past 40 years, with the most recent warming being greatest over continents between 40°N and 70°N. The 0 °C isotherm in tropical latitudes (15°N–15°S) rose in elevation by about 110 m in the 1970s and 1980s, and there has been a decrease in the diurnal temperature range. Predictions are for continued climate warming by 1.4–5.8 °C during the twenty-first century (IPCC 2001).

Climate models predict an increase in global mean precipitation, but some regions will be drier. There has been an increase in precipitation over land by about 1% over the last century. This increase has been mostly in the northern hemisphere, with the USA experiencing a 10% increase, as well as more extreme precipitation events than in the past. Meanwhile, parts of the subtropics and tropics have been drier. The length of the snow season and amount of snow in the Swiss Alps has decreased substantially since the mid-1980s.

Atmospheric CO_2 concentration rises in winter and drops in summer, mostly in response to seasonal changes in terrestrial vegetation growth. Since the early 1960s, the amplitude of this oscillation has increased 20% in Hawaii and 40% in the Arctic, with the lengthening of the growing season in the northern hemisphere. There has also been an increase in conifer plantation yields as well as an acceleration of turnover rates and biomass of tropical trees. Hughes (2000) points out that future climate change might be partially mitigated if CO_2 gas emission is already promoting forest growth, which, in turn, is sequestering carbon.

As a result of stratographic O_3 (ozone) depletion, solar UV-B reaching sea level is increasing, having an effect on plants. This enhanced UV-B can damage plants' DNA, change membrane lipid composition, affect photosynthesis in various ways and inhibit cell expansion (Rozema *et al.*, 1997). Plant morphogenetic responses include decreases in plant height, leaf length and leaf area, as well as an increase in leaf thickness and axillary branching. Related to these are altered leaf angle, plant architecture, canopy structure, emergence, phenology, senescence and seed production. Inevitably, there are direct and indirect effects on ecosystems and on insect diversity. This is still a little-researched field, partly because the experimental approach is difficult, as the effect is over a large geographical area.

Evidence is accumulating however that community changes are taking place as a result of the enhanced greenhouse effect (Hughes, 2000; Post *et al.*, 2002) (Figure 7.1). In Antarctica, there has been greater seed germination and seedling survival, while a shift in plant species to higher elevations is being widely reported.

7.2.2 *Direct effects of global climate change on insects*

Increased levels of CO_2 are likely to result in greater carbon:nitrogen ratios in plant tissues, which may stimulate greater feeding activity in some insects. An additional effect of increased CO_2 is its influence on water use by plants, which results in closure of the stomata which reduces transpiration. This, in turn, results in higher leaf surface temperatures and consequently a reduction in relative humidity which could make conditions less favourable for the development of some herbivorous insect species (Porter, 1995).

This change in communities is having pronounced direct and indirect effects on the invertebrate fauna of Gough Island in the sub-Antarctic (Chown *et al.*, 2002). With warmer, drier conditions, house mice have had a major impact on their prey, including increasing their food-resource base by widening the number of indigenous insect species eaten. The warming and drying has favoured introduced invertebrates. An increase in temperature on the islands is likely to

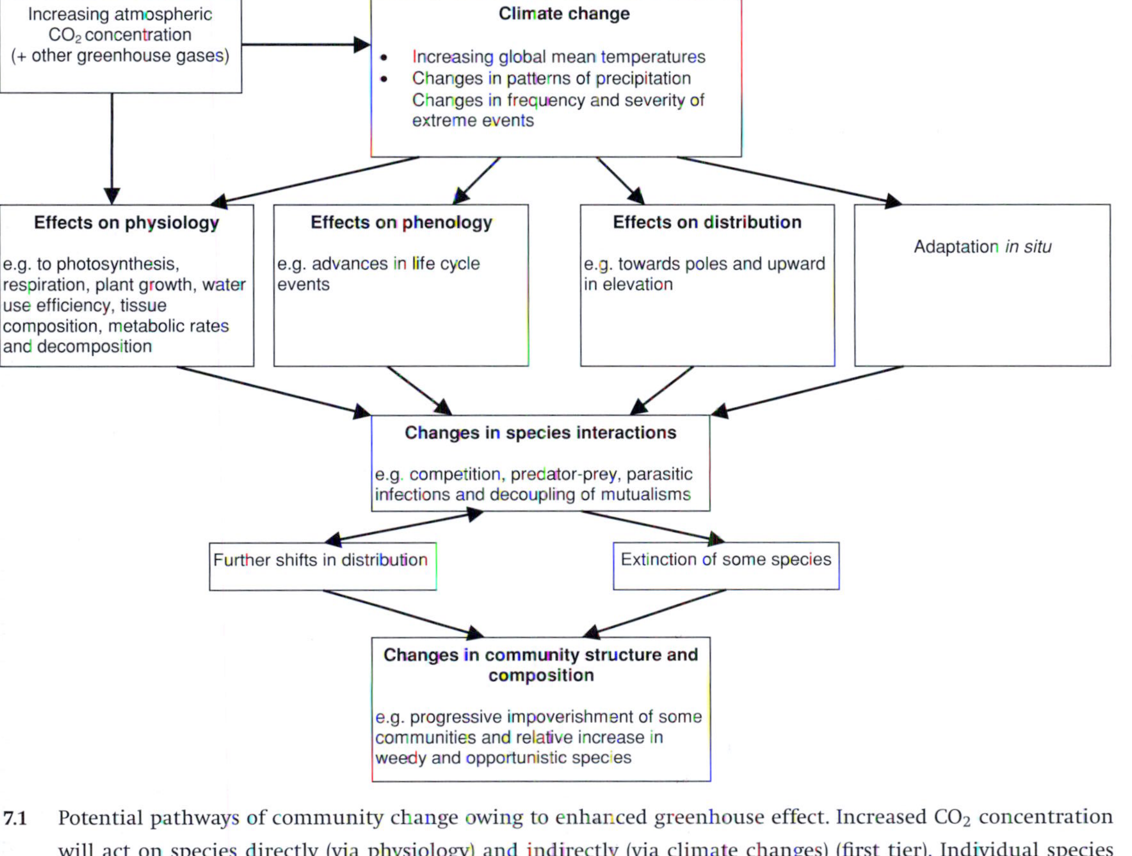

7.1 Potential pathways of community change owing to enhanced greenhouse effect. Increased CO_2 concentration will act on species directly (via physiology) and indirectly (via climate changes) (first tier). Individual species might potentially respond in four ways (second tier), resulting in changes in species interactions (third tier). These changes might then lead either to extinctions or to further shifts in ranges (fourth tier), ultimately leading to changes in structure and composition of communities. (From Hughes, 2000. Copyright 2000, with kind permission from Elsevier.)

affect the life cycles of both indigenous and alien species. But given the rapid generation times and overlapping generations of most alien species in comparison with indigenous species, it is possible that the aliens have a reproductive advantage.

As well as changes in plant chemistry and architecture, climate change is also likely to affect phenologies of various components of the community. For British butterflies, climate change may be positive for some species (particularly lowland, southerly species) through an increase in flight-dependent activities such as mate location, egg laying, nectaring, predator evasion and dispersal (Dennis and Shreeve, 1991). However, droughts associated with climate change may have a negative effect. Evidence is beginning to accumulate that phenologies are beginning to change. The first appearance of British butterflies in the summer has advanced significantly for 13 species, with the Orange tip butterfly *Anthocharis cardamines* and the Red admiral *Vanessa atalanta* advancing by 17.5 and 36.3 days respectively, over the period 1976–98 (Roy and Sparks, 2000) (Figure 7.2). Similarly, the moth *Orthosia gothica* was flying a month earlier in Britain in 1995 than it was in 1976 (Woiwod, 1997).

7.2.3 *Effects on higher trophic interactions*

Besides climate change affecting plant–insect herbivore interactions, it is also likely to alter higher trophic interactions, especially those between insect herbivores and their predators, parasitoids and pathogens. Although at first sight these interactions are difficult to research, they become tractable to some extent with long-term data sets (Harrington *et al.*, 1999). Nevertheless, there is still a need to understand far more fully how the interactive components of global climate change (e.g. elevated temperatures along with more extreme droughting and higher UV-B) will affect or are affecting trophic cascades. The approach will need to be multidisciplinary because the first studies on climatically induced forest disturbance, for example, affect herbivore and pathogen survival, reproduction, dispersal and distribution. Indirect consequences of disturbance from herbivores and pathogens include elimination of nesting trees for birds and negative effects on mycorrhizal fungi (Gehring *et al.*, 1997; Ayres and Lombardero, 2000).

Not all effects will necessarily be knock-on through the food web. Masters *et al.* (1998) found that leafhoppers had larger populations irrespective of the response of the vegetation in a drought/rainfall climate change experiment, although in the wild, natural enemies might well change the outcome. Finding a simple generalized outcome, however, is not going to be easy, because experiments on insect herbivores in elevated CO_2 grow more slowly, consume more plant material, take longer to develop and suffer heavier mortality (Figure 7.3). This stimulated Watt *et al.* (1995) to emphasize the need for studies on the combined

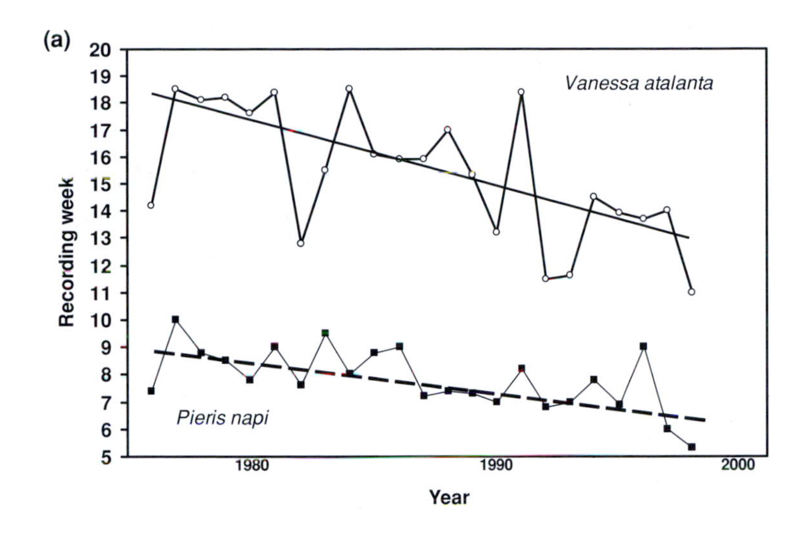

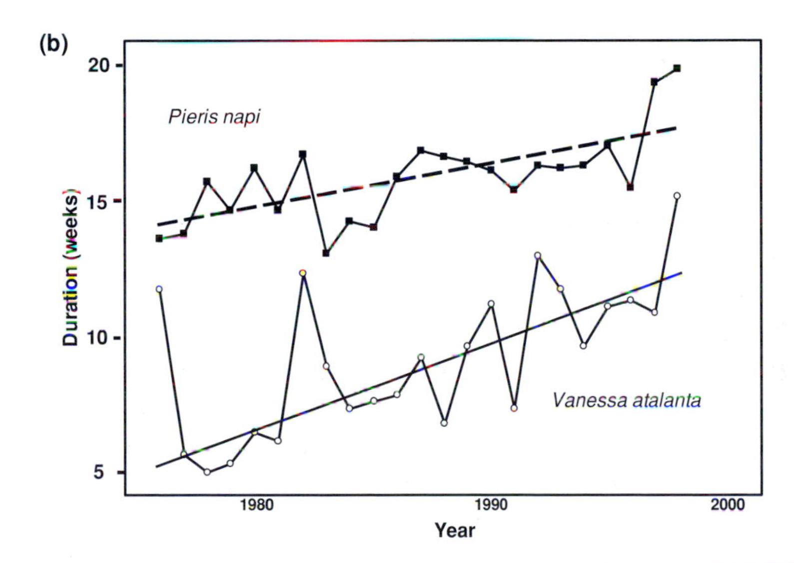

7.2 The first appearance of most British butterfly species has been earlier in recent years and the duration of the flight period longer. This is highly significant for the Red admiral *Vanessa atalanta* and the Green-veined white *Pieris napi*. (a) First appearance and (b) duration of flight period. (From Roy and Sparks, 2000, with kind permission of Blackwell Publishing.)

effects of factors such as CO_2, drought, temperature, plant nutrient status and pollutants such as ozone.

Some interactions, at least between plant and insect herbivore, may remain in step with climatic warming. Winter moth *Operophtera brumata* larvae have, so far, remained in step with oak budburst (Buse and Good, 1996) and the Orange tip butterfly with one of its foodplants, garlic mustard *Alliaria petiolata* (Sparks and Yates, 1997).

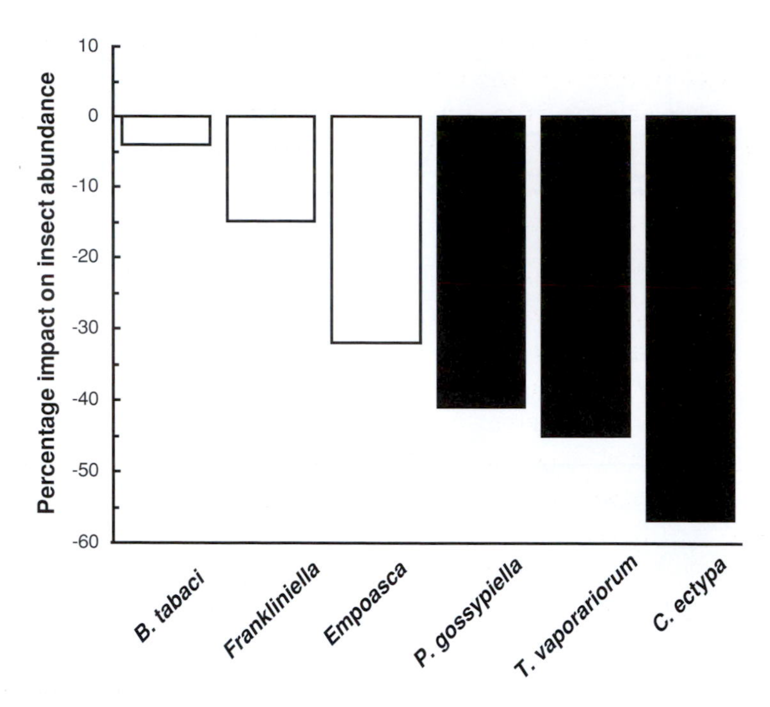

7.3 Elevated CO_2 levels decrease the abundance of various insect herbivores, and statistically significantly in the case of three species (solid columns). *Bemisia tabaci* is a whitefly, *Frankliniella* a thrips, *Empoasca* a leafhopper, *Pectinophora gossypiella* a moth, *Trialeurodes vaporariorum* a whitefly and *Chaetocnema ectypa* a beetle. (Adapted from Watt *et al.*, 1995. Copyright 1995, with permission from Elsevier.)

However, among aphids, interactions may be too numerous and complex for a mechanistic approach to provide valuable insights within an acceptable time frame (Harrington *et al.*, 1995). This is why the studies on sub-Antarctic islands such as Gough are so valuable, where the communities are small, physically limited and composed of relatively few interactions. A similar argument can be made for controlled experiments, such as those on *Drosophila* where different temperature regimes favoured one species over another (Davis *et al.*, 1995). But we must be aware that islands will not give all the answers. On a continental scale, an increase in the year to year variation in minimum winter temperatures is expected to favour more northerly outbreaks of the Southern pine beetle *Dendroctonus frontalis* but could reduce more southerly outbreaks (Ungerer *et al.*, 1999).

Increased warming would most likely increase the diversity of insects at higher latitudes. Because insects typically migrate much faster than trees, many temperate tree species are likely to encounter non-native insect herbivores that previously were restricted to subtropical forests (Dale *et al.*, 2001). These geographical range shifts are likely to shuffle communities as each species responds differentially, even though the changes in ranges may be relatively small

(Peterson *et al.*, 2002). Exactly how communities will re-assemble is difficult to predict. Currently, there are far too many unknowns associated with translating global-scale climate changes to population and community-level linkages and dynamics, with the likelihood also of chaotic reactions.

7.2.4 *Effects on medical insects*

Insect vectors of human disease will also be affected by changes in climate (Lines, 1995). Increases in temperature may alter the distribution of an arthropod-borne pathogen either by affecting the distribution of the vector, or by accelerating the development of the pathogen within the vector. Where the vector is long-lived in comparison with the development period of the parasite, the main question is how the distribution of the vector will be changed. Temperature changes may also affect the parasite directly, in particular by reducing the maturation time in the vector. This will be particularly important with short-lived vectors such as mosquitoes (Rogers and Randolph, 2000) in the highlands of east Africa, Madagascar, Nepal and Papua New Guinea, where temperature is a limiting factor. While malaria may well return to parts of the former USSR, it is unlikely that it will return to Western Europe, because the European vectors are not susceptible to tropical strains of the parasite, and it appears that the European strain of *Plasmodium falciparum* is extinct.

7.3 Changes in species' geographical ranges

7.3.1 *Imperfections of generalized climate change models*

Considerable attention has been given to how species might expand or change geographical ranges with global climate change. Yet, predictions of geographical range change with global climatic change have two variables. The first is an untested, implicit assumption that we can accurately predict any species range prior to climate change. The second is the one normally considered, and that we can predict changes according to changes in physical climatic variables. Using ladybird biocontrol agents (*Chilocorus* spp.), Samways *et al.* (1999) showed that the first point cannot be assumed, simply because climate is not always the only overriding feature determining whether a species will establish or not. Other determinants, such as localized response to microclimate, phenology, host type and availability, presence of natural enemies and hibernation sites play a varying role over and above climate in determining whether a species will establish at a new locality. Nevertheless, climate alone did give correct predictions of geographical range for about a quarter of the species, but for one species its accuracy was zero, with another four less than 50%.

A priori modelling of species' geographical ranges is an art rather than a science, and shown by the ladybird study which used falsifiable evidence. The weakness of a priori climatic modelling is being borne out with studies on actual range changes now that anthropogenic climatic change modelling has begun. Hill *et al.* (2002) pointed out that 30 out of 35 British butterflies have not tracked recent climate changes because of lack of suitable habitat, despite the fact that in the past 200 years and in a theoretical future without landscape change most of these butterflies are predicted to increase in abundance (Roy *et al.*, 2001). Warren *et al.* (2001) have also indicated that for most British butterfly species, the widespread alteration and destruction of natural habitats means that newly available, climatically suitable areas are too isolated to be colonized or do not contain suitable habitat.

In view of Coope's (1995) findings, that prior to human impact, insects have tracked climatic changes, we might be tempted to predict that insects will likewise track the recent climate changes rather than adapt on site, habitat limitation aside. This may happen in the case of highly mobile species rather than sedentary species (Warren *et al.*, 2001), the differences of which are magnified by the fragmented landscape acting as a differential filter (Ingham and Samways, 1996). We may also be in for some surprises because insects can be remarkably genetically variable and adapt quickly. This is 'contemporary evolution' (Stockwell *et al.*, 2003), with insect resistance to chemical pesticides being an example. Where populations are large enough, and thus avoiding risks pertaining to small populations (Frankham *et al.*, 2002), there may be rapid adaptive evolution. This assumes however, that there are not too many synergistic impacts, which would require a bigger adaptive hurdle of contemporary, multiple adaptation.

It is difficult to separate the effects of global climate change from that of landscape change. Nevertheless, evidence is accumulating that for some mobile generalist butterflies (Dennis, 1993; Parmesan, 1996, Parmesan *et al.*, 1999; Warren *et al.*, 2001) and for dragonflies (Aoki, 1997; Ubukata, 1997) in the northern hemisphere there is a poleward shift in geographical ranges. Geertsema (2000) cautions however, that some herbivorous insect range expansions, at least in the southern hemisphere, that have been cited as a result of climate change are more parsimoniously explained by host plant switching or newly available food sources.

7.3.2 Variation in response between species

Clearly there is much variation in response to climate from one species to the next. In Britain, the Chequered skipper butterfly *Carterocephalus palaemon* has disappeared from England and is now restricted to Scotland, while the Peacock butterfly *Inachis io* has expanded its distribution within the core of its range but is less abundant at its range margin in northern Scotland. The range margin of the Comma butterfly *Polygonia c-album* has moved northwards more than 170 km in the last three centuries (Hill *et al.*, 2002). Mobile and generalist species are moving along a broad front, while others are restricted mainly because of being habitat specialists. Butterfly species that are at the southern limit of their geographical ranges in Britain, or have montane distributions, have become extinct at low elevation sites in the southern parts of their ranges and colonized sites at higher elevations (Hill *et al.*, 2002). Although with some species, their northern range margins have shifted more than their southern margins (Parmesan *et al.*, 1999). It also appears that others are compensating by moving up in elevation rather than in latitude.

This differential shift in range changes among species results in a change in species composition of local sites, as seen among microlepidoptera in The Netherlands (Kuchlein and Ellis, 1997). This has serious implications for Red Listing of species. Kuchlein and Ellis (1997) suggest that because the species composition of communities changes in time, there is little point in monitoring individual species to assess the conservation status of specific ecosystems. This applies particularly to threatened species which, as well as usually being rare habitat specialists in a fragmented landscape, are generally also range restricted, making them ill-equipped to adjust to climate change through movement.

7.4 Synergisms and future perspectives

Much of the recent research on global climate change has aimed at deciphering the effects of the industrial age from background abiotic fluctuations on the one hand, and the effect of other anthropogenic disturbances, particularly landscape fragmentation, on the other. The reality, however, is that the future of insect diversity rests on the synergisms between all the various human-generated changes. To date, we are only getting glimpses of the ramifications of climate change. Although information is coming forward at the level of plant primary producers, little is known of the interactions between stresses on individual plants, and especially of multiple environmental stress interactions at the level of ecosystem, which includes insect communities. Nevertheless, the lattice models of Travis (2003) show sharp thresholds for both habitat availability and rate of climate change below which a species rapidly becomes extinct. During climate change, the habitat threshold occurs sooner, meaning that the synergism between climate change and habitat loss might be disastrous. Habitat specialists, especially those with poor dispersal abilities, are likely to be affected most (Figure 7.4).

Vagaries of species' behaviour may make it difficult to accurately model how some species' range will change, particularly at the sub-regional down to landscape spatial scale, given the response of ladybirds on a global experiment basis (Samways *et al.*, 1999). Surprises are likely too, with the Brown argus butterfly *Aricia agestis* in Britain using an alternative hostplant *Geranium molle*, enabling it to inhabit new localities (C. D. Thomas *et al.*, 2001). This contrasts with the fact that many other species are not adaptable. In Britain, 89% of the habitat

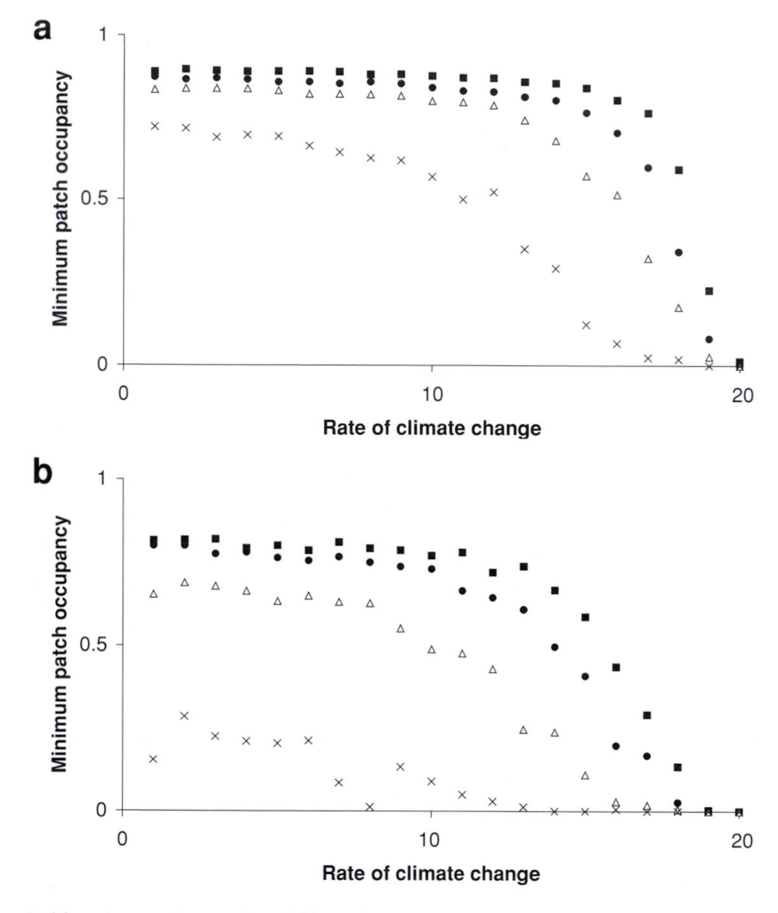

7.4 Habitat loss reduces the ability of a species to survive climate change for a more generalist (a) and a specialist (b) species. The squares show the results for no habitat loss, circles with 20% loss, triangles 40% and crosses 60%. (Redrawn from Travis, 2003, with kind permission of the Royal Society.)

specialist butterflies have declined in their distribution size since the 1970s, compared with only 50% of the more vagile habitat-generalist species (Warren *et al.*, 2001) (Figure 7.5).

British moths seem to be suffering a similar fate to the butterflies (Woiwod, 2003), with an overall decline during the 1950s of about 70% in the total number of larger moths, and many once-common species have either become very rare or have disappeared. This decline has probably been the result of rapid agricultural intensification involving both landscape change and widespread adoption of herbicides and insecticides. Not all British moth species have fared so badly, with the Ruby tiger *Phragmatobia fuliginosa* having increased in abundance (Figure 7.6).

At this point, it is not clear how climate change and landscape change are interacting to cause the declines but a pattern is emerging that indicates that insect diversity is being dramatically affected, at least in Britain, as a result of

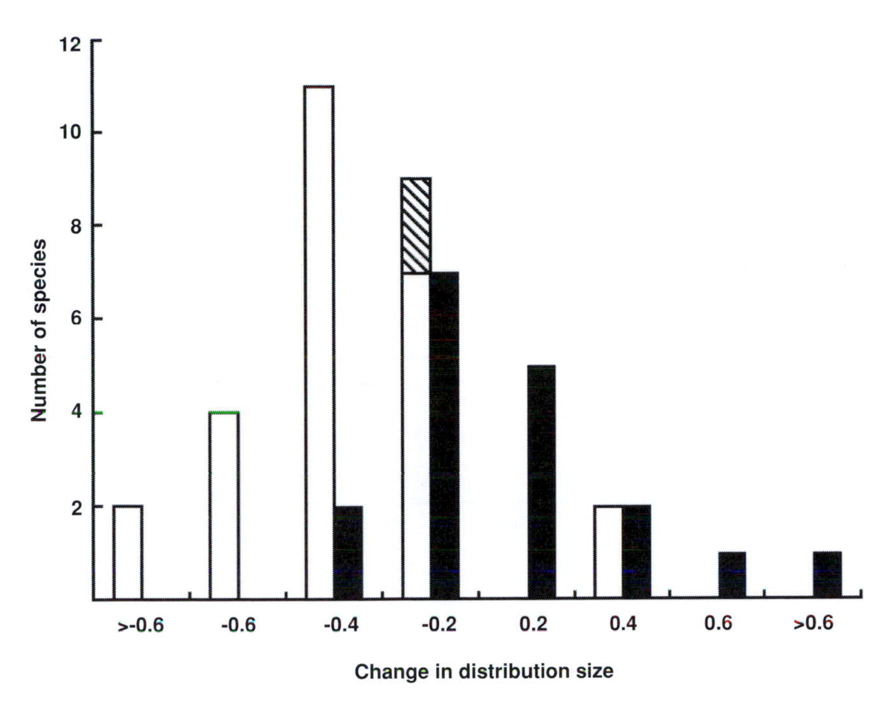

7.5 Proportional changes in geographical distribution sizes of British butterflies between 1970–82 and 1995–99. Sedentary habitat specialists (white) have declined the most, while mobile specialists (hatched) have also declined. Wider-countryside species (black) have not fared so badly, with several increasing their ranges. (Redrawn from Warren *et al.*, 2001.)

synergistic human-induced impacts. For some insect taxa, this might arise from increased competition, among many other factors. Predicted rises in the North and Baltic Seas, which will be in step with a possible 20–30 cm increase in mean sea level by 2050, are likely to have a profound influence on carabid beetles and spiders (Irmler *et al.*, 2002), with increased competition for food resources, reduced abundances and greater risk of local extinction from frequent heavy flooding.

Clearly, global climate change is a major consideration when planning for future insect diversity conservation. Kuchlein and Ellis (1997) consider it even more important than changes in landscape use, at least in The Netherlands. The reality, however, is that across the globe the various manifestations of human impact will compound to various and varying degrees, and it is these synergisms that we need urgently to assess.

An increase in global temperature of 4–5 °C will make the world warmer than at any time in the last one million years. Warming will thus lead to novel ecological scenarios as current trophic interactions are modified or decoupled and new ones formed. While we can generalize by saying that an increase in 1 °C will raise the rate of respiration of plants and the decay of organic matter in soils

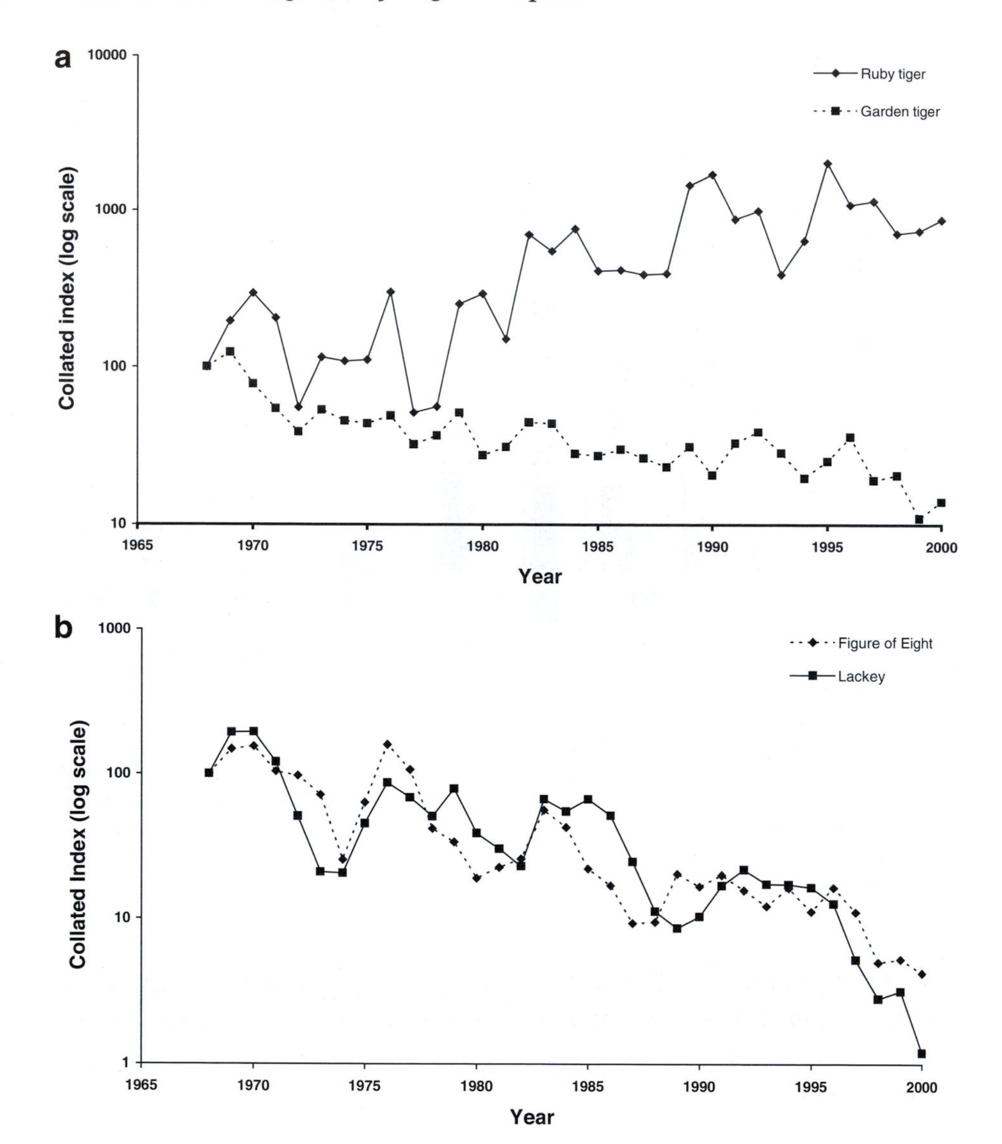

7.6 British national trends in moth abundances from the Rothamsted Insect Survey. (a) While the Ruby tiger moth (*Phragmatobia fuliginosa*) has increased in abundance over the last few decades, the Garden tiger (*Arctia caja*) has declined steadily (above). (b) The Lackey (*Malacosoma neustria*) and Figure of Eight (*Diloba caeruleocephala*) moths have shown a particularly severe decline in recent years. (From Woiwod, 2003.)

by 10–30%, and that organisms will need to move up mountains by 170 m to stay at the same temperature, or to move up in latitude by 150 km, these figures do not take into account all the abiotic and biotic interacting factors, which make up changing habitats and landscapes. This has been shown for butterflies (Warren *et al.*, 2001; Hill *et al.*, 2002) and ladybirds (Samways *et al.*, 1999). Nevertheless, based on responses so far, it is possible to estimate how British butterflies, for example, are likely to change their geographical distribution sizes

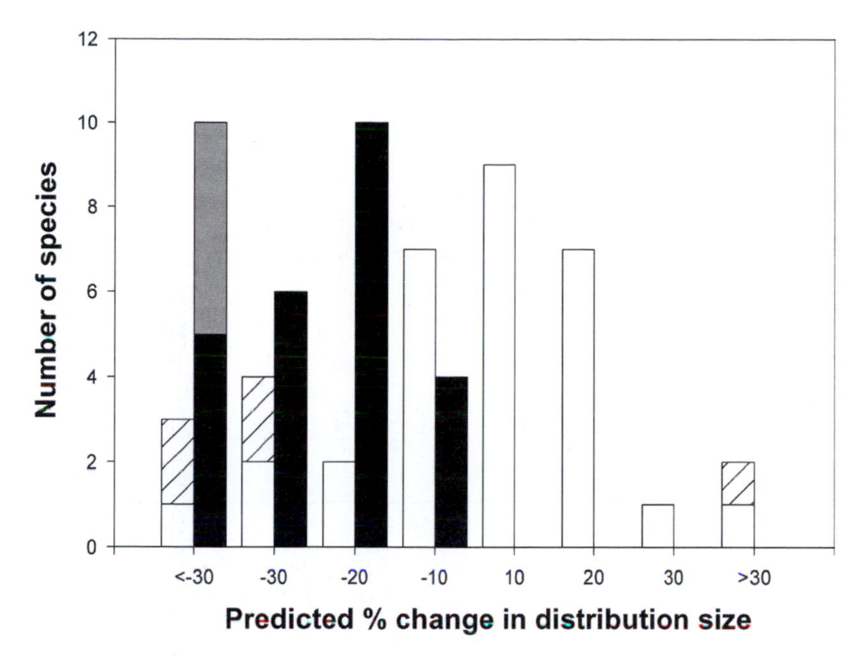

7.7 Predicted changes in size of European butterfly geographic ranges for a subset of 35 species for the period 2070–2099. The histogram illustrates two sets of predictions assuming that (i) all the species have perfectly tracked climate changes (open bars (hatched bars are northern species)), and (ii) the 30 species that have failed to respond to recent climate change also fail to colonize newly available northern areas (black bars (grey bar, northern species)). (From Hill *et al.*, 2002, with kind permission of the Royal Society.)

in the future (Figure 7.7). This is clearly going to be an important exercise to undertake on many other taxa and many interactions relative to future predictions (Figure 7.8). It is a major field of research that permeates all other aspects of insect diversity conservation.

7.5 Summary

Global climatic change is upon us, with a 0.6 °C increase during the twentieth century. Precipitation and droughting events are changing, with more extreme events becoming increasingly regular. The amplitude of seasonal cycles of CO_2 concentration has increased, with lengthening of the growing season in the northern hemisphere, and an acceleration of turnover rates and biomass of tropical trees. UV-B has also increased, and is changing plant structure and function. Some plant species are already moving to higher elevations in response to warming.

These global climate changes are favouring generalist, mobile insect species, and, on sub-Antarctic Gough island, it is the invasive alien species that are

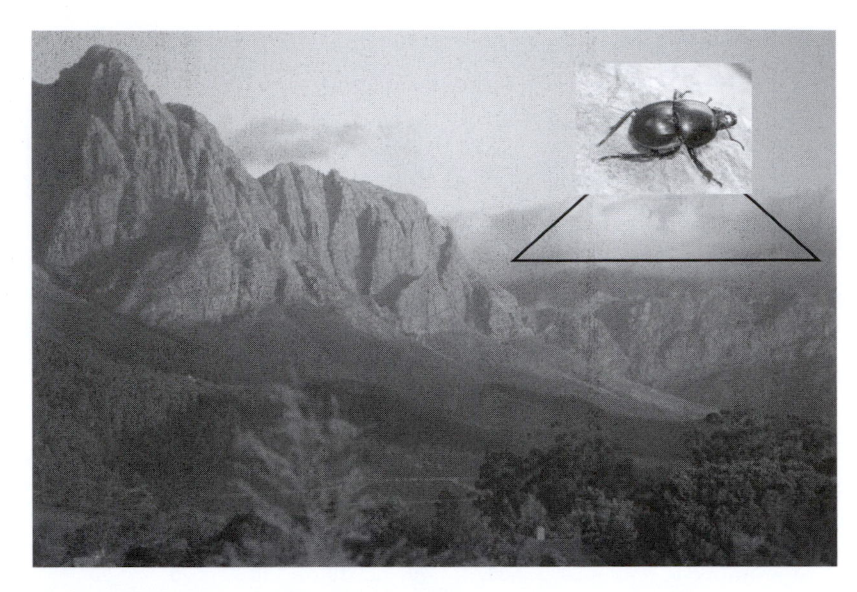

7.8 Global climate change is of great concern for montane specialists, such as the rare, endemic, South African stag beetles (*Colophon* spp.) which live on mountain peaks. These beetles depend on condensation and cloud formation from strong south-easterly winds. Without adaptation to elevated global temperature, there is nowhere else for them to go. And if cloud-formation patterns change, they would be under desiccation stress as well. Pictured here is *Colophon stokoei* which has its home in the Hottentots-Holland mountains, near Stellenbosch. (Photo of beetle by courtesy of Henk Geertsema and Alicia Timm.)

benefiting most. As plants change in their architecture and chemistry, so do insect abundances and assemblage composition. Additionally, warming is resulting in butterflies appearing earlier in the summer.

Evidence is accumulating that the effects of global change are ramifying all aspects of food webs, with some insects keeping in step with their host plants and others not, particularly in lowlands. As insects are more mobile than trees, high latitude trees are likely to encounter novel insects from lower latitudes. This, in turn, leads to more competitive interactions, with natural enemies also modifying outcomes. Insect-borne diseases such as malaria are expanding to higher elevations and to higher latitudes.

Geographical range expansions are now being experienced, with butterflies in the northern hemisphere moving farther northwards, and up in elevation, while dropping out in their southern margins. Modelling such changes is compounded by the fact that there are various aspects of biology, such as hibernation sites for ladybirds, that mean species range changes will not be a simple, straightforward response. These changes will be further compounded by landscape changes which have a filtering effect on insect movement, more so for sedentary habitat-specialists than for mobile habitat-generalists. The outcome is that species composition, such as moths in The Netherlands, at certain sites is changing. This makes single reserve selection only a temporary solution.

The real problem with global climate change is that it involves multiple stressors and synergisms between impacts. This makes accurate modelling of future scenarios extremely difficult. Evidence from British Lepidoptera indicates that climate change has been a major interacting force with landscape fragmentation and disturbance in reducing abundances and geographical ranges of many (but not all) species. The prognosis for the future is that the insect diversity that we are encountering today at any one site will not be the same for our grandchildren.

III Conserving and managing insect diversity

A blending of anthropocentric and biocentric values continues to be vital. These duties toward nature involve analysis of ecosystem integrity and evolutionary dynamism at both scientific and philosophical levels; any responsible environmental policy must be based on plausible accounts of ecosystems and a sustainable biosphere. Humans and this planet have entwined destinies. We now envision an Earth ethic beyond the land ethic.

Holmes Rolston III (2000)

In our hands Themis placed
the Scales of Law and Order.
The power to protect,
and a prophecy for destiny.
Would the scales be empty
or filled with life?
Silent faces stare-
waiting, pleading, asking
only for the right to be.
And when you look up my child,
what will you see?
In our hands was placed earth's eternal life.
On silent pages, the sun ticks away time.
The Sands of the ages
running through the
hourglass of our existence . . .
When you ask me my child
'What, Daddy, will you leave behind for me?'
For you 'I will open the gate . . .'

8 Methods, approaches and prioritization criteria

There is scope aplenty for innovative enquiry at a time when life's abundance and diversity appear poised for both an unprecedented reduction and an unprecedented revolution.

Norman Myers (1996b)

8.1 Introduction

In the final analysis, conservation biology is a management exercise, the success of which is measured in terms of the amount, quality and health of the biological diversity maintained. We discussed the 'why' we are doing this in Chapter 1. We now consider the act of doing insect diversity conservation.

Broadly, the doing of conservation has two components. The first is research, or the finding out. The second is the practical implementation, management and assessment of success of the conservation initiative. In some respects, these two components have different quality controls. Research is measured in terms of scientific rigour, while implementation and management is assessed on feasibility, ergonomics and what has actually been conserved. Research of course underpins implementation and management, although at times it seems oblivious as to actually what can be achieved in practice. Conversely, management needs to better know what research findings are out there, and how these findings might improve management. Clearly, communication between researchers, managers, policy makers, regulatory authorities and educationalists is essential. In this regard, insect diversity conservation differs little from other areas of biodiversity conservation, only that insects are vast and varied yet little known, and therefore their diversity conservation requires a particular type of articulation.

This chapter is essentially about research, especially the broad approaches used to identify and prioritize geographical areas and networks that give the greatest chance for biodiversity to survive into the future. It is not intended to be a 'how to do it' chapter, for which there are some excellent texts that are referenced here. This chapter underpins the landscape management perspectives in later chapters.

8.2 Towards an 'Earth ethic'

Conservation biology, being a young and vigorous science, is undergoing considerable methodological change. It seems at times that we need to reflect more deeply on what actually we are trying to achieve. Such reflection needs, in turn, to be clear on the ethical angle from which we are coming (Chapter 1). Expressions like 'conserve the world for future generations' is distinctly utilitarian and does not consider any aspects of deep ecology. Also 'maintaining biodiversity at current levels', is biologically and practically unrealistic, as inevitably there are going to be more extinctions before a levelling out occurs.

Biodiversity has been in constant flux with changing geophysical and atmospheric conditions in the past, and is likely to suffer from repercussions from anthropogenic climate change in the future. Nevertheless, it is essential that we muster optimism. Probably what conservation biologists are trying most to do is to reduce extinction rates. These are occurring both through direct impacts on certain species, and, more pervasively, through changes in ecosystem functioning, often because of multiple stressors, or synergistic effects.

For meaningful conservation action, we need to be clear on our goals. While these depend on the spatial and temporal scales under consideration, there are two factors that underpin all that we do. The first is that we ourselves are changing. Our consciousness and unstoppable inventiveness for changing the world around us now involves tinkering even with our own genome. Culture has now become an evolutionary path in its own right (Blackmore, 1999), and this self-manipulating genome the driving force (Samways, 1996d). This new evolutionary direction poses interesting questions. Selfish genes now can be even more selfish. Human genes no longer will be selected by nature, but by culture, by their own genomes. This is likely to be a profound new mutualism, and will be our parting from wild nature. The sheer speed of the changes about to take place, as the super-selfish genome increases in magnitude, will leave no future for many organisms. This prompted Rolston (1994) to ask: Do we have the moral tools to match our technological ones?

Increasingly, there may be a sector of human culture that no longer needs wild nature, and the humans that do may increasingly rely on 'wild nature' from virtual reality. The variety of life will then become increasingly impoverished, with wild forms being mostly micro-organisms, plants, invertebrates and fungi, with the big animals largely gone (Dixon, 1990). In view of the recent analysis indicating that 9% of the earth's permanent ice has disappeared each decade since the late 1970s, the fate of the polar bear, for example, is of great concern.

We urgently need to maintain every 'cog and wheel'; the so-called art of intelligent tinkering (Leopold, 1953). This involves maintaining as many interactions and interaction strengths, as well as individuals and populations, as possible, to retain the status quo of current ecosystems. But this returns us again to the inevitability of change. Besides changes to our own genome, which generally seems to have a minimal environmental ethic, and natural changes to the earth's climate, there is the looming, difficult-to-reverse juggernaut of human-induced climate change. Whether Gaia has the correctional fortitude to attend these issues we do not know, and is currently at the hands of the human consciousness.

So we do what we can, after firstly prioritizing our goals as best we can. At the large, global scale, we can highlight those areas richest in biodiversity (Myers *et al.*, 2000). As these areas largely hug the tropics, it does risk neglecting other geographical areas, although it is tropical forest ecosystems and the oceans that are much of the powerhouse for planetary maintenance. Defining our conservation goals at the global level involves considerable international co-operation (Sandalow and Bowles, 2001). At the sub-regional and landscape scales, consultation from the 'bottom up', involving all human communities from the start of the planning process, appears to be a vital ingredient for conservation success (e.g. Rozzi *et al.*, 2000).

Refining our conservation goals therefore is about discussion and collaboration. Besides, we are all in the same biotic boat which must be seaworthy and rowed with rhythm if we are to get anywhere. This is the concept of 'shared values' of Norton (2000) and the 'Earth ethic' of Rolston (2000).

8.3 Identifying geographical areas for conservation action

There are several initiatives to identify and to prioritize conservation areas across the world at a gross scale. Myers *et al.* (2000), using species endemism and degree of threat to plants, mammals, birds, reptiles and amphibians, identified 25 such areas or 'hotspots' (Figure 8.1). Their argument is that to stem global species extinctions, conservation efforts should best be focused on these areas, which together contain as many as 44% of all species of vascular plants and 35% of all species in the four vertebrate groups in a total of only 1.4% (formerly 11.8%) of the land surface of the Earth.

Had insects been included in the analyses, it is likely that there would have been similarities and differences, depending on which taxa were included. There are no dragonflies, for example, endemic to the Succulent Karoo, yet the neighbouring Cape Floristic Region is rich in endemic species, and has many highly threatened species (Samways, 2004). Other taxa, as far as we know, are likely to show a reverse trend to this. We can, however, never prejudge a situation when it comes to insects. In this general geographical area, a new insect Order, the Mantophasmatodea (related to the Grylloblattodea) has recently been discovered (Klass *et al.*, 2002; Picker *et al.*, 2002) (Figure 8.2).

Myers *et al.*'s (2000) foundation requires complementary approaches. The first of these is that with the rapid rate of global climate change, these 25 hotspots are likely to be severely impacted. This is clearly the case, for example, with the Succulent Karoo and the Cape Floristic Region (Rutherford *et al.*, 1999, 2000) (Figure 8.3), which require a dynamic approach. Such an approach has been outlined by Hannah *et al.* (2002) as a 'Climate change-integrated conservation strategy' (CCS), which although tailored to individual regions, has five key elements: (1) regional modelling of biodiversity response to climate change; (2) systematic selection of protected areas with climate change as an integral selection factor; (3) management of biodiversity across regional landscapes, including core protected areas and their surrounding matrix, with climate change as an explicit management parameter; (4) mechanisms to support regional co-ordination of management, both across international borders and across the interface between park and non-park conservation areas; and (5) provision of resources, from countries with the greatest resources and greatest role in generating climate change to countries in which climate-change effects and biodiversity are highest. To adequately respond to the uncertainties posed by climate change, the provision of resources will be required on a much larger scale than has occurred in the past.

An insightful complementary approach to Myers *et al.* (2000), and against a background of global climate change, is division of the world into ecoregions

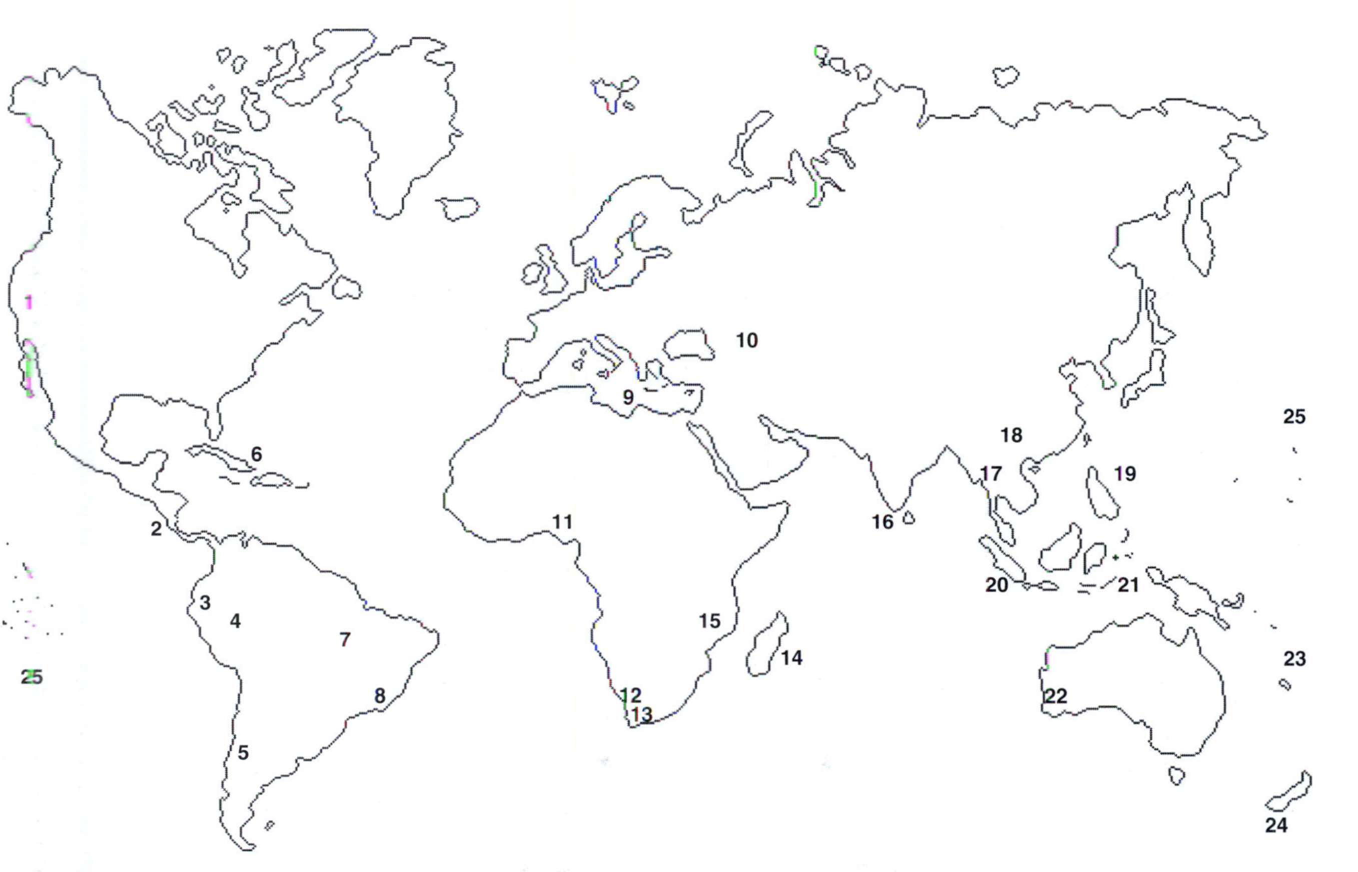

8.1 The 25 hotspots in the world where there are exceptional concentrations of endemic species that are undergoing exceptional loss of habitat. As many as 44% of all species of vascular plants and 35% of all species in four vertebrate groups are confined to these 25 hotspots which comprise only 1.4% of the Earth's land surface. (Based on Myers *et al.*, 2000.)

8.2 The discovery of a completely new insect Order, the Mantophasmatodea, emphasizes just how little we still know of insect diversity. Pictured here is a female nymph of *Austrophasma rawsonvillensis* from the southern Cape of South Africa. (Photo by courtesy of Mike Picker.)

(i.e. areas with characteristic abiotic and biotic features) for conservation planning. This has been done by the World Wildlife Fund through extensive collaboration with biogeographers, taxonomists, conservation biologists and ecologists (Olsen and Dinerstein, 1998). A total of 233 ecoregions have been identified and each allocated a conservation status: CE, critical or endangered; V, vulnerable; or RS, relatively stable or intact. While it is easy to criticize this approach as being subjective, it does nevertheless, like the hotspots, enable us to identify areas

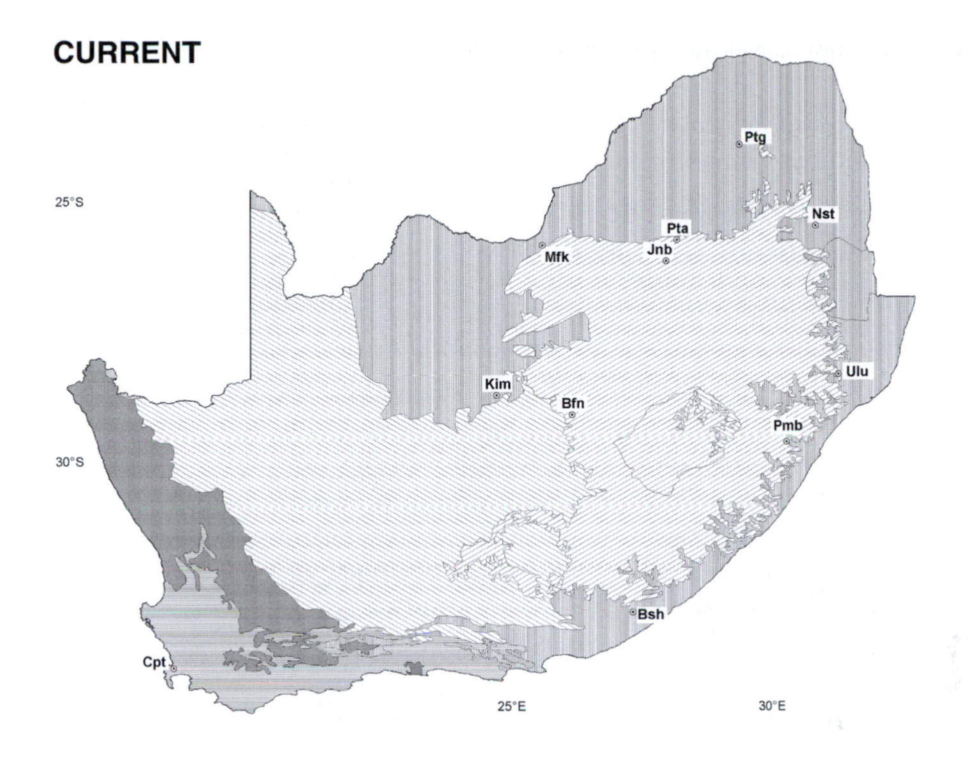

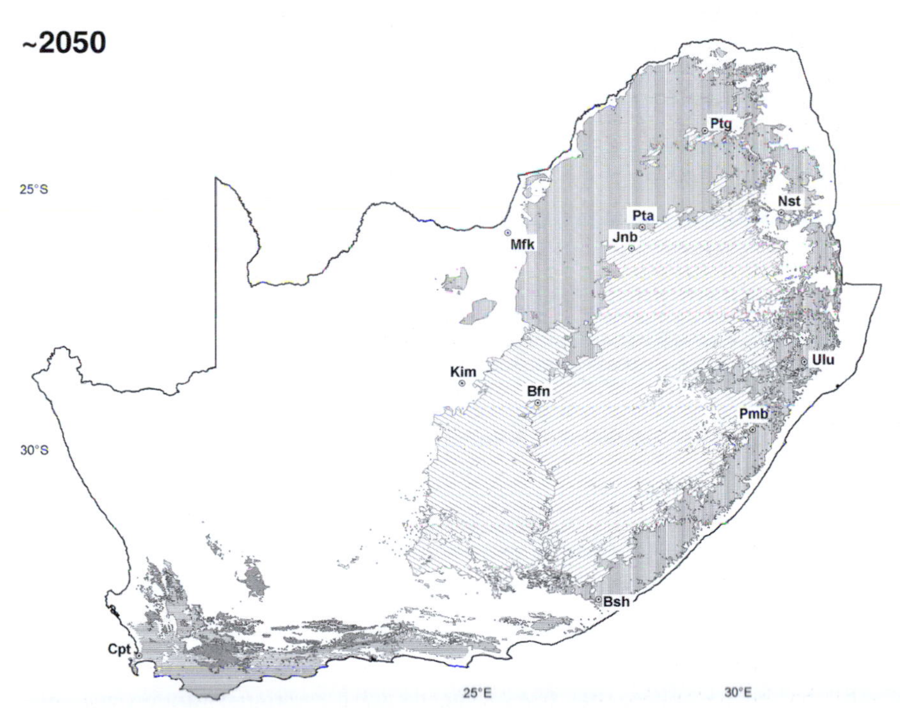

8.3 Biome changes projected for South Africa, using the HadCM2 model with sulphate amelioration, in a scenario of doubled carbon dioxide. Major shifts are predicted for all biomes. The Succulent Karoo and Cape Floristic Region (dark shading along the west coast) are global biodiversity hotspots of high diversity and endemism. Note the southward collapse of the Succulent Karoo hotspot (dark grey), and an overall increase in desertification by 2050 (white). Abbreviations are main cities. (Figure kindly supplied by M. Rutherford and L. Powrie.)

most in need of immediate conservation attention, especially when fine-tuned with quantitative data (Krupnick and Kress, 2003).

8.4 Systematic reserve selection

The approach of Myers *et al.* (2000) additionally requires other inputs for practical planning purposes. At a finer spatial scale, sites need to be weighed against each other not just on their endemic species but their species in general, including their taxonomic relatedness. Such procedures may also include other features besides species, such as communities, structures and processes. The aim is to develop a systematic network of reserve areas (Pressey *et al.*, 1993), using a procedure of complementarity which is a strategy that assesses the content of any existing reserves and then, in a step-wise fashion, select at each step the site or area that is most complementary in the features it contains. But there are many ways to combine sites, and so the outcome, which is a representative network of reserves or reserve areas, must be flexible enough for practical on-the-ground conservation management. But some sites may be common, while others are rare or even unique. So it is crucial to include the concept of irreplaceability which may be defined in two ways: (1) the potential contribution of a site to a conservation goal; and (2) the extent to which the options for meaningful conservation are lost if the site is lost. Often, however, endemic hotspots are located in areas of ecological transition (Araújo, 2002), which can lead to biases in selection procedures.

Other facets must also be built on to Myers *et al.*'s (2000) approach. It is not only centres of diversity and endemism that need to be preserved. Contact zones between species are also of interest, since they are often areas of genetic diversity and active evolution (Prance, 2000), and also need attention. Furthermore, we must be aware that 'areas of endemism' are not hard and fast entities with sharp edges. Additionally, widespread species should be considered on an equal footing with less widespread or endemic ones when determining boundaries (Humphries, 2000). This is underscored by studies on scarab beetles in Bolivia, where ecotonal 'biogeographic crossroads' appear to provide conservation strategists with opportunities to simultaneously conserve high species richness, zones of high beta diversity (species change across geographical areas) and complementarity (areas that complement each other with different suites of species or landscapes), and evolution (Spector, 2002).

Complementarity approaches favour species with restricted range sizes. Furthermore, if extinctions are determined mainly by demographic factors, then selecting areas at the peripheries of species' ranges might be a poor option. But if extinctions are determined mainly by extrinsic factors, then peripheral populations might be important to ensure the long-term persistence of species (Araújo and Williams, 2001). Clearly it is necessary to focus both on hotspots and on associated transitional zones (Smith *et al.*, 2001). It is important however, that the methodologies chosen do not overemphasize transition zones because this may compromise the long-term viability of a reserve network (Gaston *et al.*, 2001).

This is not to say that beta diversity is unimportant. It is important for long-term maintenance of biodiversity (Reyers *et al.*, 2002a). The caveat is that resources for conservation are scarce, making prioritization essential. This means that we must focus on hotspot cores and edges, with an eye all the time on the dictates from global climate change and synergistic impacts. Spatial bet hedging against the future is now becoming a necessity.

In this regard, Williams and Araújo (2000) present an interesting method which considers the probability of persistence of valued features (e.g. species of trees) for selecting networks of conservation areas. The network selected benefits the least-widespread species and results in high connectivity among selected areas. The method also accommodates local differences in viability, vulnerability, threats, costs, or other social and political constraints, and is applicable in principle to any surrogate measure which stands in for and represents 'quality' biodiversity.

One further consideration is optimal reserve design relative to island archipelagos. Boecklen (1997) found that for only about 10% of archipelagos do single islands contain more species than would two or three islands of equal total area. This is principally because of the physical and biological backgrounds of islands. This means that in terms of conservation planning, small islands have their own importance. Indeed they may be refugia from invasive alien insect predators (Kelly and Samways, 2003). Even small habitat islands on mainlands can be important, such as remnant eucalypt patches for beetles (Ward *et al.*, 2002). The habitat quality of such 'mainland archipelagos' then becomes a critical feature, as is maintenance of a variety of such 'islands' to conserve the regional set of species, as shown with prairie butterflies (Swengel and Swengel, 1997).

A final consideration is that systematic reserve selection must, in the case of insects in particular, consider the intrinsic role of topography, which is well-known to play a major role when making decisions with regard to insect conservation (Samways, 1990; Armstrong and van Hensbergen, 1997; Lien and Yuan, 2003). It is a complex issue and depends on spatial scales, ranging from microhabitats to elevational gradients, which in turn, are related to geomorphic and vegetational characteristics (Lawrence and Samways, 2001, 2002).

8.5 Use of surrogates in conservation planning

8.5.1 *Types of surrogates*

Among the systematic approaches developed for reserve selection, some require sound taxonomy and good distributional data, which are often not available for insects. Conservation area selection then depends on using coarse, but not inaccurate, data that are available. These umbrella or surrogate data are then used for making informed conservation decisions. These surrogacy measures may be divided roughly into two groups. The first is species surrogate measures, which include use of higher taxa, species richness (which may be morpho-species), rarity, endemism and alternative taxa. The second group includes environmental surrogate measures. These include vegetation types, land systems or classes and environmental domains. These environmental surrogate classes are derived from information on vegetation types, soil properties, remote-sensing, climatic data and terrain data (Reyers *et al.*, 2002b).

What is not clear at this stage is which individual surrogate measure or combination of measures should be used in practical conservation planning, especially as the goals and choice of methodology will vary from one geographical area to another (see also Chapter 9 on concepts of inventorying relative to surrogacy).

8.5.2 *Species surrogates*

While species surrogate measures are useful, there are risks attached, in that critical aspects of regional diversity may be overlooked. While family richness may be a good predictor of British butterfly species richness (Williams and Gaston, 1994), this approach has the inherent disadvantage of overlooking rare and threatened, endemic species. Similarly, using one taxon as a surrogate for another taxon may not coincide (Prendergast *et al.*, 1993; Lombard, 1995; van Jaarsveld *et al.*, 1998; Reyers *et al.*, 2000b). Ants and carabids, for example, give a complementary and not a coincidental picture of landscape types (Kotze and Samways, 1999a). This means that complementary conservation networks based on one particular taxon will not necessarily represent another one, and almost certainly will not represent all insect diversity. The problem is off-set to some extent when a wide range of data sets is used for each taxon, with the result that a greater range of options becomes available for a reserve network (Hopkinson *et al.*, 2001) (see also Chapter 9).

8.5.3 *Environmental surrogates*

Use of environmental surrogates, when correctly measured, can at times be an umbrella for taxonomic diversity (Faith and Walker, 1996). The argument is that when various environmental classes are conserved, most species will also be protected. This does however, like higher-taxon surrogates, overlook species that are confined to particular habitats (i.e. finer-scale issues) and the needs of migrants (i.e. coarser-scale issues), factors that are often applicable to insects. Furthermore, environmental classes are open to interpretation depending on the measures included. They are also scale-dependent, with heterogeneity increasing with coarseness of scale. Broad classes may therefore miss certain critical habitats.

8.5.4 *Combinations of species and environmental approaches*

As insects are so numerous and scientifically poorly known, the collection of data on a wider range of species and functional types becomes prohibitively expensive. While it may be done for some charismatic taxa, such as butterflies and dragonflies, insect diversity conservationists will have to rely on a combination of species and environmental approaches (van Jaarsveld *et al.*, 1998). Reyers *et al.* (2002b) have done this for several taxa, including butterflies, buprestid beetles, termites and ant lions, as well as mammals, birds and endemic plants. Interestingly, as the insect groups contained many isolated distribution records, it led to insects representing more of the environmental surrogates. A similar pattern emerged with plants. In the case of insects, this may have been partly due to distributional data points having been missed by the original data gatherers. Nevertheless, the results were encouraging for the combined use of both species and environmental surrogates in conservation area selection. This is because together they capture many aspects of biodiversity, missed by one or other approach. The species-based approach, for example, excluded some important vegetation and landtypes, which is a major shortcoming (Maddock and Du Plessis, 1999). Reyers *et al.*, (2002b) conclude that both species and environmental measures should be incorporated into selection procedures to ensure that all facets of biodiversity are represented, including threatened, endemic and rare features (species, assemblages, communities). When this is done however, the outcome is alarming, because about half the land area needs to be conserved to maintain current levels of biodiversity. Of further concern is that this is much higher than the 10% recommended by the IUCN (1993). This has also been found to be the case in Australia (Pressey and Logan, 1997), making the only feasible way forward a combination of protected areas and off-reserve areas. This emphasizes the importance of conservancies, where non-reserve areas are managed in a way to optimize biodiversity and to spatially complement the protected areas. This approach also to some extent builds in safety margins, a fail-safe approach of engineers for complex systems, to increase the probability that the conservation process is effective (Van Jaarsveld, 2003) (Figure 8.4).

It is also important to conserve environmental gradients, not just of species but also of populations (Smith *et al.*, 2001). This is because species are assemblages of populations that are often distributed across a variety of habitat types. These populations have specific adaptations to local environmental conditions. But as populations are being lost at a fast rate, it means that there is loss of variety of adaptations which would otherwise equip the species for future changes. The point is that environmental gradients are important for diversification and speciation. The conservation of such gradients is clearly an important ingredient for ensuring as wide as possible a genetic base for the future.

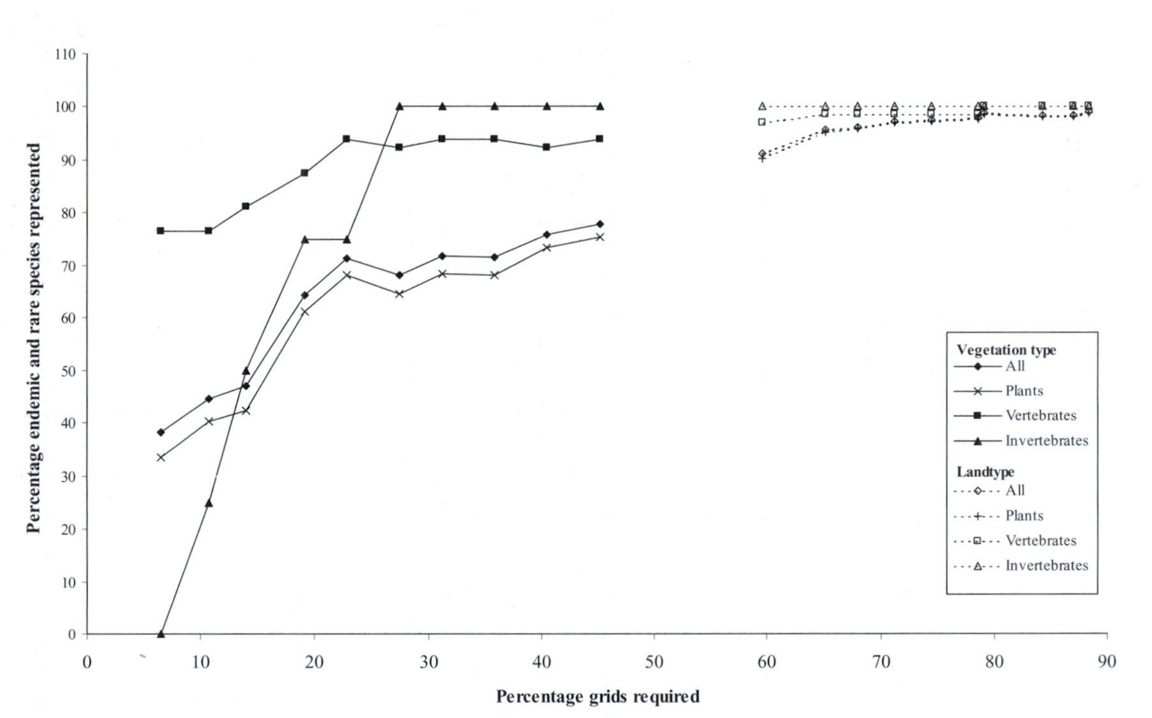

8.4 Percentage of rare and endemic species captured by vegetation type and landtype-based conservation areas in the northern region of South Africa. Points on the lines represent the sliding target vegetation and landtype representation percentages (from 5 to 50%). The points demonstrate the per cent of plant, invertebrate, vertebrate and total endemic species represented by these conservation areas. (From Reyers *et al.*, 2002.)

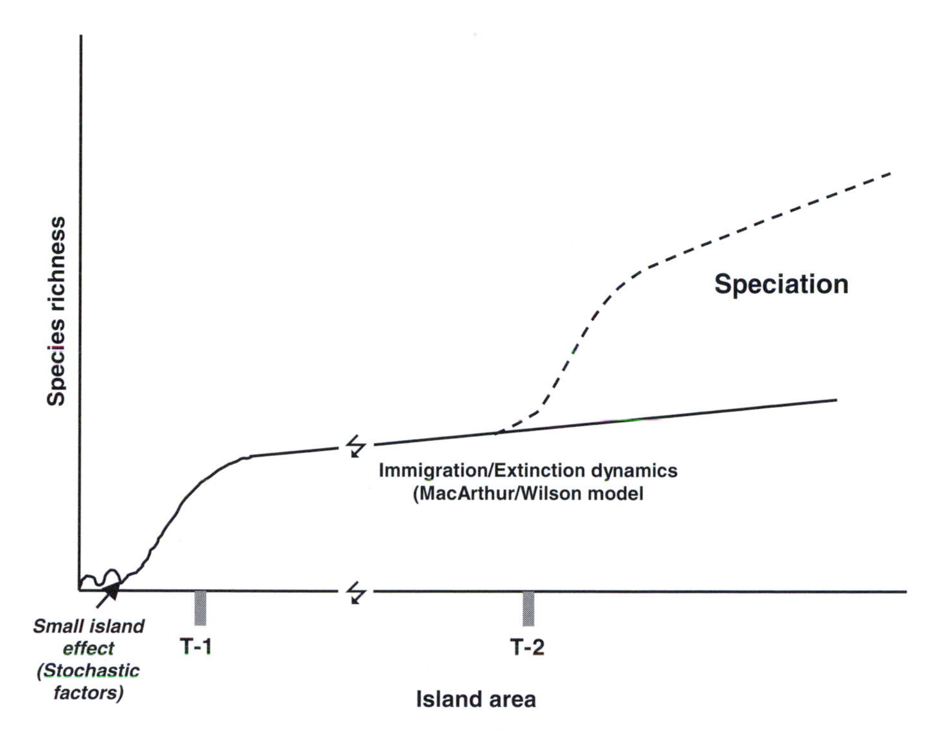

8.5 The scale-dependent nature of the species–area relationship may best be studied by focusing on thresholds which delineate ranges of area where (1) richness seems to be independent of area and is largely determined by stochastic factors (area < Threshold 1 (T-1)); (2) species richness is a function of immigration/extinction dynamics as envisioned by MacArthur and Wilson (1967); range in area between Thresholds 1 and 2 (T-1, T-2); or (3) islands are large enough to allow in situ speciation (i.e. those beyond Threshold 2 (dashed line). (Redrawn from Lomolino, 2000, with kind permission of Blackwell Publishing.)

8.6 Coarse and fine filters

8.6.1 Sampling efficiency

The term 'umbrella species' has been used in the past as a flash of optimism that suggested conservation of the tiger will also conserve all the tiger beetles and other multitudes of organisms in the same area. An important premise is that larger areas have more species than smaller ones. For islands however, this small-versus-large is not a simple relationship, with small islands highly susceptible to stochastic events, both abiotic and biotic, and large islands providing enough heterogeneity for speciation (Lomolino, 2000) (Figure 8.5). This is underscored by the fact that the relative importance of speciation and other factors determining species numbers vary crucially over various spatial scales (Godfray and Lawton, 2001).

The natural heterogeneity of the landscape, which increases with increasing area, means that as more sites are sampled, the greater the number of species.

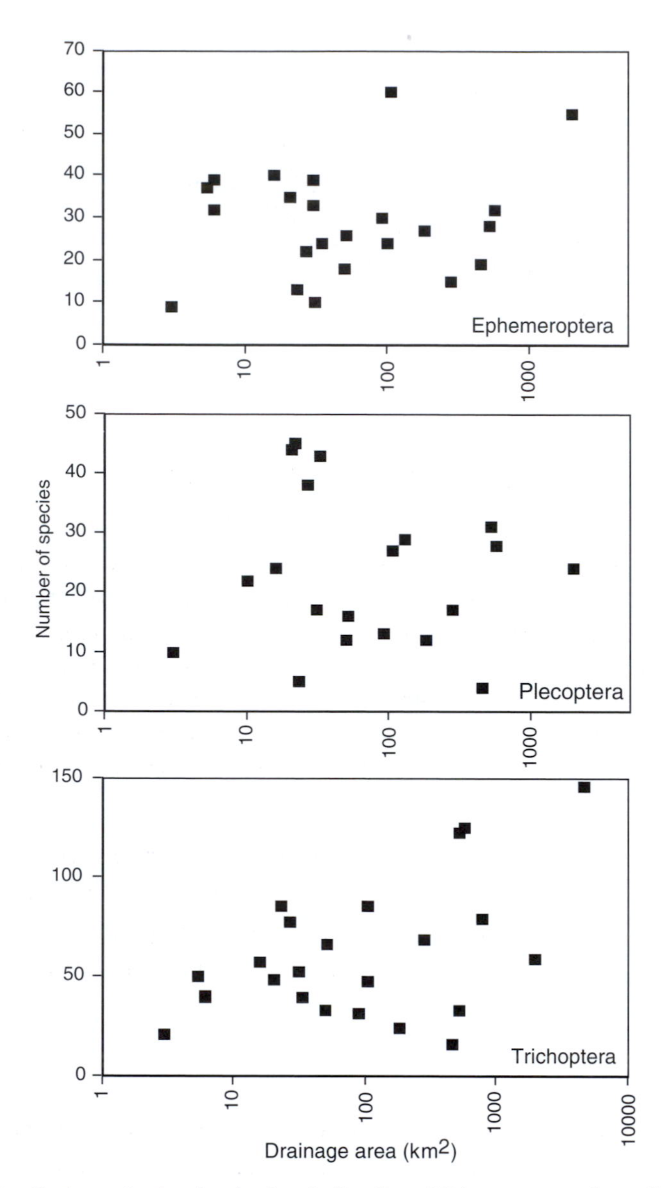

8.6 Drainage basin size in South Carolina, USA versus number of species of mayflies (Ephemeroptera), stoneflies (Plecoptera) and caddisflies (Trichoptera), illustrating that factors such as large sample numbers and drainage area (species–area relationships) are not significant predictors of species richness across streams, nor of much co-variation between taxa. Ecosystem productivity and great habitat heterogeneity appear to contribute most to streams that are species rich. (From Voelz and McArthur, 2000, with kind permission of Kluwer Academic Publishers.)

This is not generally a simple linear relationship, depending on the type and character of the physiographic heterogeneity (Fleishman and MacNally, 2002), and on the taxa being considered. Furthermore it may also vary between different insect taxa in any particular area (Voelz and McArthur, 2000) (Figure 8.6). At any one collecting site, the number of species sampled increases, or accumulates,

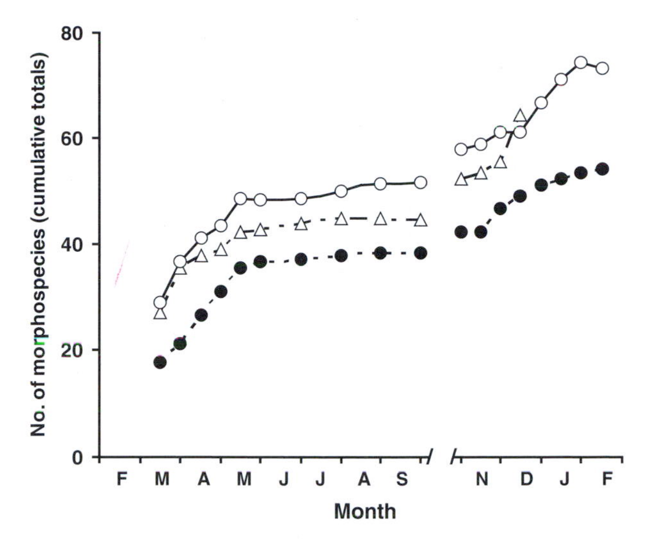

8.7 An example of 'spurious' species accumulation curves, plotted for ant morphospecies caught in pitfall traps in woodland sites in Australia. At the three sites shown here (white circles, white triangles, black circles), an initial trapping period yielded convincing asymptotes (flattening of curves), and suggested that sampling effort was sufficient to define the ant assemblages present. With later resumption of sampling, richness increased considerably. In this case, decreased activity of ants at the onset of winter led to decreased catches and a misleading late autumn asymptote. (Redrawn from New, 2000c.)

until an asymptote is reached. This may not be the actual species richness at the site, with the close-to-real richness only being apparent after a large number of samples have been taken (New, 2000). This is because sampling efficiency is never total, and insect movement is such that newcomers continue to appear (Figure 8.7).

8.6.2 *Modelling approaches*

Species accumulation curves can be used to model multiple species within a biological group as a general surrogate for spatial pattern in that group and other groups (Ferrier *et al.*, 2002a). This form of species modelling can be compared with a wide range of other surrogate mapping approaches, including various types of vegetation mapping, abiotic environmental classification and ordination. Once sites have been selected according to a particular surrogate, the biological survey data for these sites can then be used to assess the number of species actually represented (i.e. included in the hypothetical set of conservation areas) after each selection, and so generating species accumulation curves (Figure 8.8). Ferrier *et al.* (2002a) sampled various community and biological groups. Among the findings (Figure 8.9) was that ground-dwelling arthropods (ants, spiders and beetles) were not served well by any of the surrogate approaches, including species modelling. This returns us to the point made in the last section, that while species approaches may have value

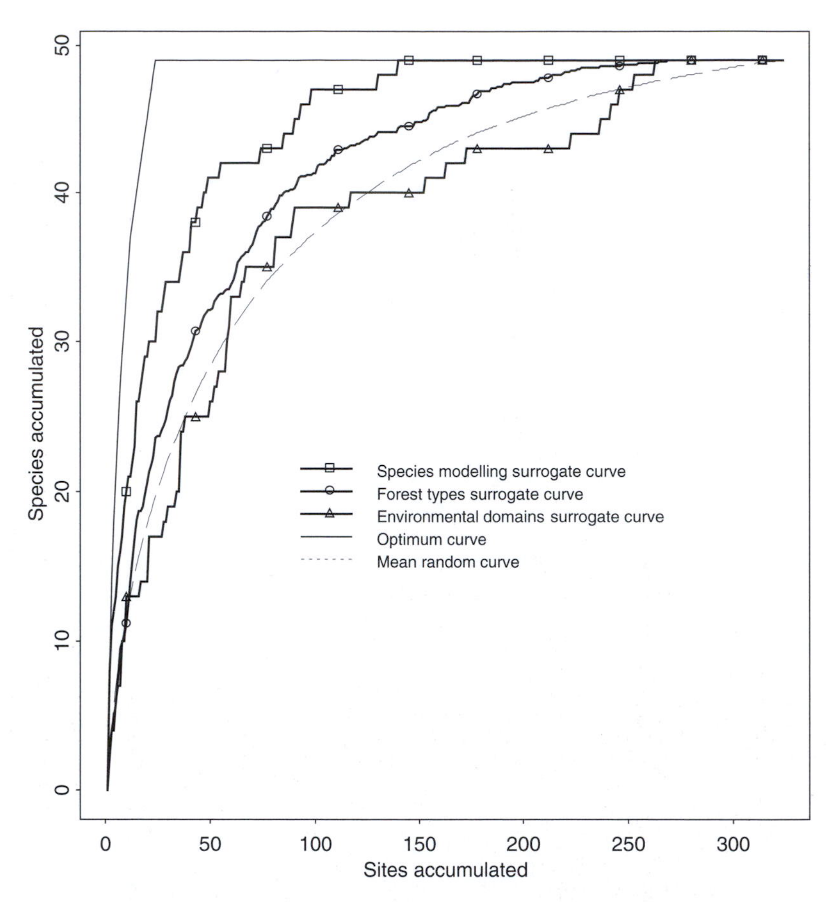

8.8 An example, in this case reptiles in Australia, of modelled species accumulation curves derived to evaluate the performance of alternative biodiversity surrogates. In this case, as the species modelling surrogate curve is closest to the optimum curve and furthest from the random curve, it suggests that species are good surrogates (From Ferrier *et al.*, 2002a, in which more details are given. With kind permission of Kluwer Academic Publishers.) (See also Figure 8.9.)

for certain threatened or particular focal species, regional planning requires an integrated approach with both species and community level approaches (Ferrier *et al.*, 2002b). This is illustrated diagrammatically in Figure 8.10.

Nevertheless, species modelling can indicate potential distributions, including geographical range retractions. This has been shown for the ant *Formica exsecta* in Switzerland, where habitat fragmentation and vegetation transformation have possibly been the main causes of range restriction (Maggini *et al.*, 2002). This underscores the value of combining individual species studies (fine filter) with communities and assemblages (coarse filter), both of which are complementary.

Summerville *et al.* (2003), having studied forest moths, many of which are rare, suggested that a way forward for conservation management is to identify

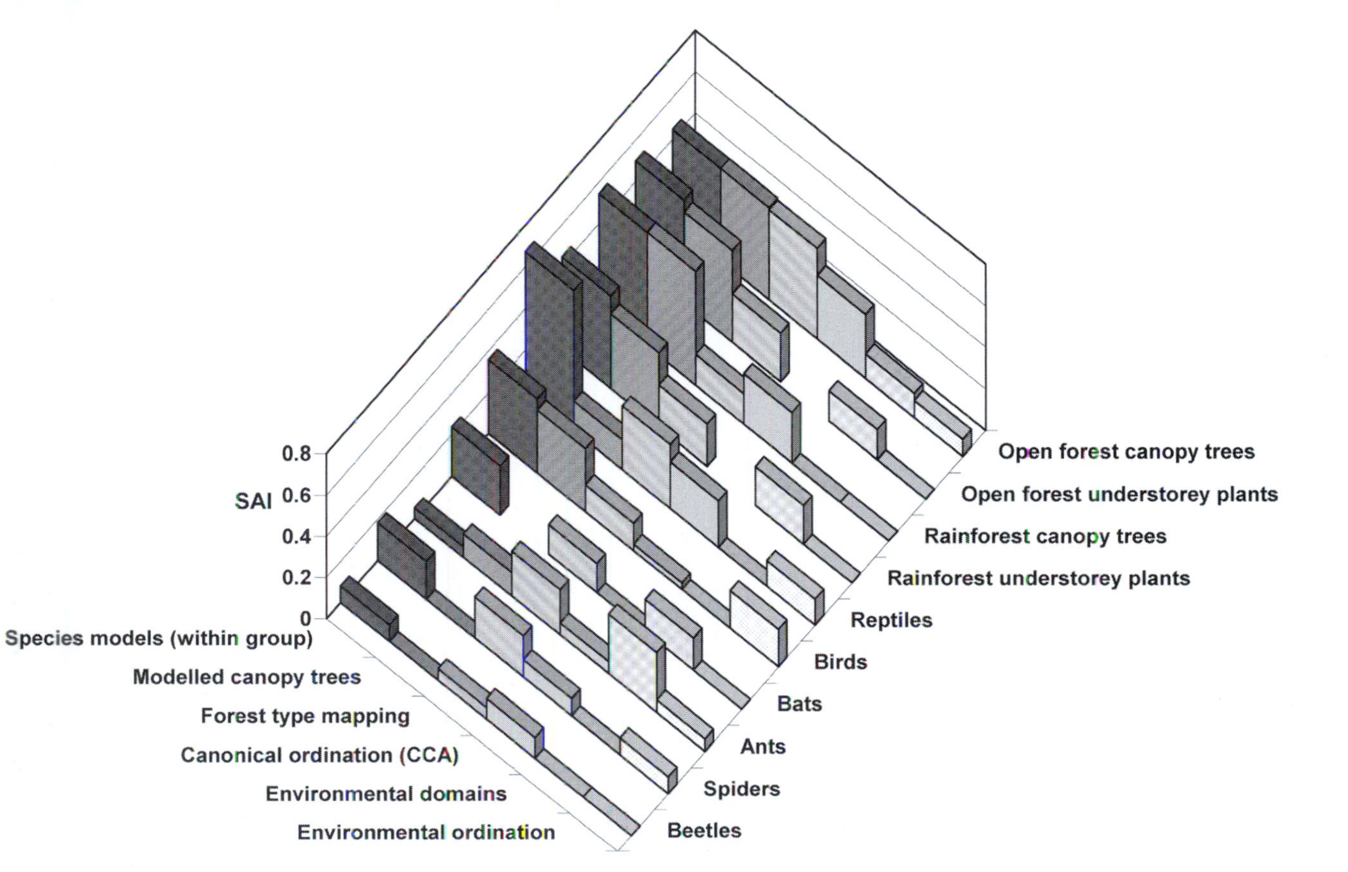

8.9 In Australia, surrogates were evaluated using survey data for ten biological groups. These were then used in combination with various combinations of surrogates. The results indicate that species modelling within each biological data group achieved best overall performance of the evaluated surrogates. Modelling of canopy trees also performed well as a surrogate for understorey plants. Interestingly, none of the ground-dwelling arthropod groups (ants, spiders, beetles) were served well by any of the surrogate approaches, including species modelling, suggesting that these components of the fauna need to be sampled in their own right to obtain a picture of their local geographical distribution and abundance. (From Ferrier *et al.*, 2002a, with kind permission of Kluwer Academic Publishers.)

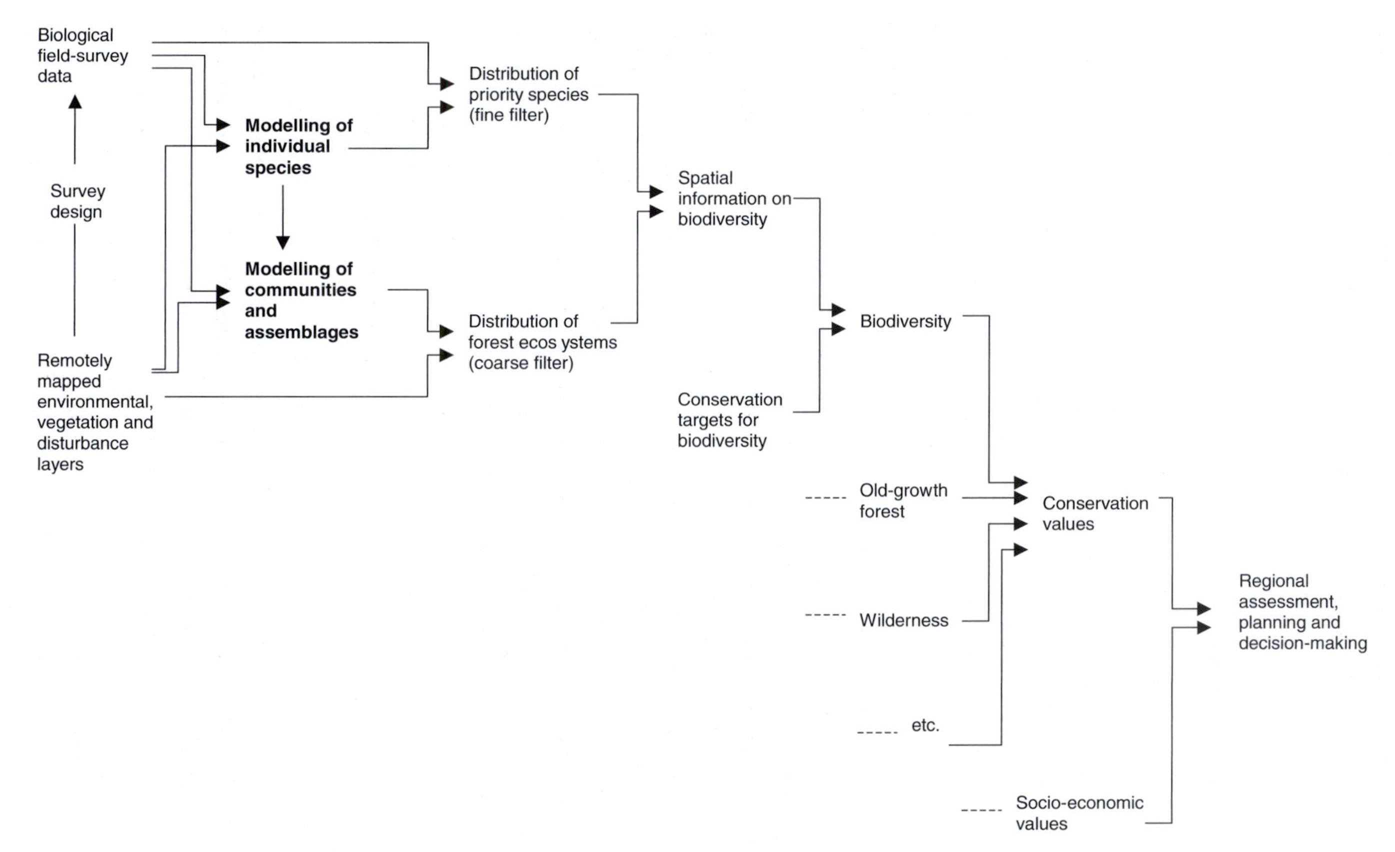

8.10 A summary of the role of modelling in relation to the overall framework for regional conservation planning, in this case, New South Wales, Australia. (From Ferrier *et al.*, 2002a, with kind permission of Kluwer Academic Publishers.)

species that are reasonably abundant within a community (5–10% of all individuals in a sample) and which are unique to particular spatial levels. Such a strategy should produce two desirable outcomes: the conservation of species that make ecoregions distinct and the maintenance of functionally dominant species within the forest.

8.7 Plant surrogates

Arguably, plant assemblages and communities *are* environmental surrogates. Bearing this in mind, it would then be necessary to supplement such environmental surrogates with species surrogates in the case of special insects on particular plants i.e. monophages. The use of plant communities as a surrogate for insect groupings is immensely involved, with biogeographical and population variables, among others, confounding the issue. Furthermore, a particular herbivore rarely has as wide a geographical range as its host plant. Population variables that are important in this context include the fact that rare insects are often less abundant on rare plants than common insects on common plants (Dixon and Kindlmann, 1990). In addition, abundant insects tend to be more polyphagous than rarer insects (Novotny and Basset, 2000).

There may also be other confounding issues. In the case of the Aspen blotch leaf-miner *Phyllonorycter salicifoliella*, it needs two different species of trees, one to feed on and one to overwinter on, and only occurs where these two trees occur together (Auerbach *et al.*, 1995). This means that plants as surrogates for insects is a continuous spectrum from environmental to species surrogates. Nevertheless, plant groups (at various spatial scales, from microhabitat to biome) have considerable appeal as 'umbrellas' for insects. Certainly, at least in one study, plant communities were better predictors of beetle assemblages than were vertebrates (Yen, 1987). A cautionary note on the value of plant surrogates is that certain insect species may also need particular abiotic conditions, such as mud for drinking, rocks for warming, or hills for mate meeting, that plants alone do not provide.

Nevertheless, certain insect assemblages do reflect certain plant communities. Dragonfly species richness in Sweden correlates with vascular plant species richness (Sahlén and Ekestubbe, 2001) (and see Chapter 9). This would seem to be an unlikely relationship with dragonflies being generalist predators. A more expected relationship occurs between butterflies, leafhoppers and certain moths and tallgrass prairie plants (Panzer and Schwartz, 1998). However, there was no coincidence between either prairie plant species richness or common insect species richness, suggesting that the 'hotspot' approach may not include rare insect species at the fine scale.

Although a vegetation-based coarse-filter approach, which emphasizes indigenous plant species and plant community richness, can contribute substantially

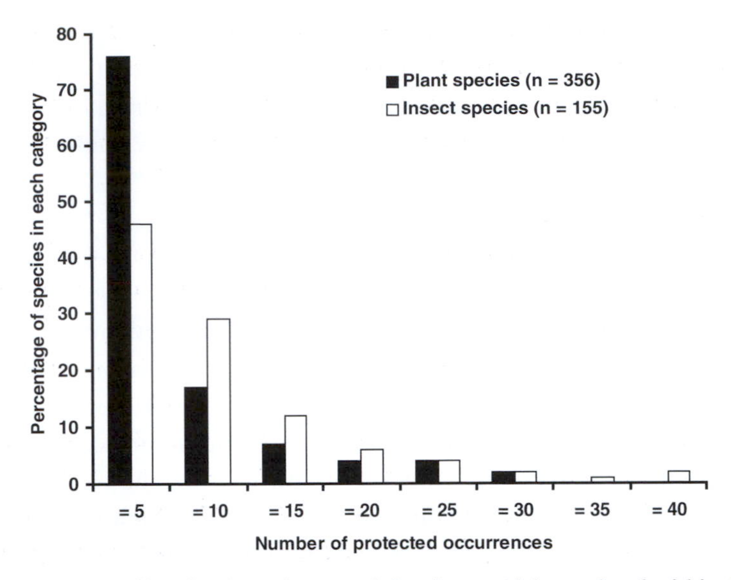

8.11 The coarse-filter (landscape) approach has been widely employed within the Chicago region for selecting reserves, although insects have played little role in the process. Despite this oversight, habitats for at least 93% of the prairie-specialist insect species occur in one or more areas within the reserve system. Also, and illustrated here, is that the median number of sites occupied by prairie-specialist insects (4.5) was nearly double that of plants (2.5), resulting in a significantly lower level of species rarity among insects than among plants. Nevertheless, reserves with rare plants do not necessarily 'capture' the rare insects. (From Panzer and Schwartz, 1998, with kind permission of Blackwell Publishing.)

to insect diversity, it is still necessary to supplement with a fine-filter approach which focuses on particular insect species of concern (Figure 8.11). This was clearly illustrated for saproxylic beetles in Norway where certain Red Listed species had a negative correlation with a group of indicator species (Sverdrup-Thygeson, 2001).

8.8 Animal surrogates

8.8.1 Taxonomic surrogates

Birds have been used as surrogates for insects. At spatial scales of 1–10 km, birds and butterflies show similar trends in species richness and diversity relative to urban disturbance in California (Blair, 1999). Using distributional data on Ugandan woody plants, large moths, butterflies, birds and small mammals, Howard *et al.* (1998) showed that selection of potential forest reserve sites based on butterflies or birds was just as effective at representing all groups as was selection of sites using data on all taxa. However, this was not the case for forest disturbance in Cameroon, where birds and several invertebrate groups, including butterflies, flying beetles, canopy beetles, canopy ants, leaf-litter ants and termites, as indicators of changes in the diversity of other groups, gave a highly misleading picture of overall faunal changes (Lawton *et al.*, 1998). Oliver *et al.* (1998), working in Australia, further consolidate this view, and point out that insects are a vital (and cost-effective) component in biodiversity surveys. In their study, plants, as well as vertebrates, fell short, in the absence of insects, in representing general biodiversity.

This argument may be taken one step further, and emphasis placed on selecting an appropriate complement of insect taxa, such as ants and carabids, which, as diverse groups, can give a very different view of the landscape (Kotze and Samways, 2001). This point is borne out by Rainio and Niemelä (2003), who emphasized that carabids, despite being widely used as indicators or responders of habitat change, are not necessarily a good surrogate for biodiversity in general. They recommended using several species groups with different ecological requirements to indicate biodiversity. Didham *et al.* (1996) take a similar view, and suggest the use of a variety of functional rather than taxonomic groups.

8.8.2 *Morphospecies*

While morphospecies (species with distinctive, recognizable physical features) (Oliver and Beattie, 1996) are a useful tool for conservation, particularly for environmental impact assessment and when inventorying and comparing species richness between similar sites at local or regional levels, their accuracy must be tested for particular taxonomic groups (Derraik *et al.*, 2002). Furthermore, when assessing 'quality' biodiversity, such as identifying endemic hotspots, it is essential to have taxonomic input (New, 1996). Yet the morphospecies and named-species approaches are not necessarily mutually exclusive. In little-known geographical areas and where taxonomic expertise is only available for certain groups, it may be possible, and indeed may be the only option, to combine both approaches and still lead to meaningful conservation protocols, especially in the tropics. This was done for saproxylic invertebrates on a small tropical island (Kelly and Samways, 2003). Andersen *et al.* (2002) used a subset of morphospecies (only large-sized ants as bycatches in vertebrate traps) in an inventory measuring the impact of an industrial plant. Whatever approach is used, it is essential to have some sort of quality control to ensure confidence in the raw data collection (Wilkie *et al.*, 2003) (see also Chapter 9).

8.9 Phylogenetic considerations

A further refinement of the taxonomic approach is to include phylogenetic diversity measures (Faith, 2002). These measure the evolutionary component of biodiversity and can be used to rank areas for biodiversity conservation (Vane-Wright *et al.*, 1991). The point is also that phylogenetic information is one of the most important factors involved in any given level of extinction. This means that loss of evolutionary history depends on the types of diversification processes that gave rise to the species lineage (clade) (Heard and Mooers, 2000). Posadas *et al.* (2001) used a phylogenetic diversity measure which combines taxonomic distinctiveness and endemicity to rank priority areas in southern South America. They used plants and weevils to do this, and identified areas that will ensure the preservation of evolutionary potential and phylogenetically rare species. It provides another complementary approach when assessing and conserving biodiversity in a region.

Conservation planning traditionally has focused on spatial pattern, which is *representation* of biodiversity. Additionally, we should consider the conservation of processes. The inclusion of processes in conservation planning is to consider persistence of species into the future, through maintenance of genetic diversity. Genetic diversity has two dimensions, one concerned with neutral divergence caused by isolation (vicariance) and the other with adaptive radiation. On the one hand, planning for both species and areas should emphasize protection of historically isolated lineages (i.e. evolutionarily significant units), as these are irreplaceable. On the other hand, adaptive features and evolutionary potential must also be conserved. This latter point comes about through maintenance of heterogeneous and viable populations, both of which are the context for natural selection (Moritz, 2002). This view is allied to the point made in Chapter 8, that it is necessary to select a complementary set of areas that represent high species richness and endemism as well as evolutionary potential. Moritz (2002) suggests a strategy where areas are identified that represent both species and (vicariant) genetic diversity. Then the aim is to maximize within these areas the protection of contiguous environmental gradients. These gradients enable selection and migration which maintains population viability and (adaptive) genetic diversity.

8.10 Are 'umbrella' and 'flagship' species of value in conservation planning?

8.10.1 Umbrella species

The surrogacy approaches described above suggest that there is limited value attached to 'umbrella' or 'flagship' species, and they must be used with caution (Andelman and Fagan, 2000). The terms themselves have been poorly defined, are unproved in practice and may detract from wider ecosystem conservation priorities (Simberloff, 1998). An 'umbrella' species is a 'protective umbrella' employed where the conservation goal is to protect a habitat or community of species in a particular area or type of habitat (Caro and O'Doherty, 1999). Usually it is a large species, often a mammal. Although such umbrella species have been widely advocated, there is virtually no proof of their value for over-arching biodiversity in general. Furthermore, the concern is that such an umbrella may be highly vulnerable itself, with its protective umbrella value being patchy and susceptible to dwindling. The term 'umbrella species' seems therefore to have little currency.

The concept of an umbrella species, whose requirements are believed to encapsulate the needs of other species, can be better expanded to include a suite of species, each of which is used to define different spatial and compositional attributes that must be present in the landscape and in their management regimes (Lambeck, 1997). For each relevant landscape parameter, the species with the most demanding requirements for that parameter is used to define its minimum acceptable value. Because the most demanding species are selected, a landscape designed and managed to meet their needs will encompass the requirements of all other species. While this is a useful first approximation to biodiversity conservation, such an approach is taking us down the road again of species surrogacy in general, as discussed above, with necessity to consider complementary landscapes on the one hand (coarse filter) and special species (fine filter) on the other.

It is not that the concept of an umbrella species should be abandoned, it is that it is often considered with little critical research support, and using organisms with different population dynamics and at different spatial scales. Ranius (2002) presents a very interesting case where the European beetle *Osmoderma eremita* serves as a suitable umbrella species for a range of other beetles associated with tree hollows. It is effective in this role as it has similar habitat requirements to the target species it represents. Also, the spatial scale of the habitat units relevant for *O. eremita* and the target species is the same (i.e. hollow trees). The problem, however, with using *O. eremita* as an umbrella species is probably that the occurrence patterns suggest that a few of the target species are more sensitive to habitat fragmentation than *O. eremita*, suggesting that these should also be monitored.

8.10.2 *Threatened organisms as surrogates*

Threatened organisms can be used as a group for conservation planning (Lawler *et al.*, 2003). However, although sites selected with single taxonomic indicator groups provided protection for between 61 and 82% of all other species, no taxonomic group provided protection for more than 58% of all other threatened species. Interestingly, threatened species overall performed well as an indicator group, covering an average of 84% of all other species.

8.10.3 *Flagship species*

'Flagship species' are usually physically large members of taxa that attract attention and garner sympathy. They are sometimes chosen on the basis of their dwindling population size or threat status. Their value for biodiversity conservation lies not so much in their ecological role but in their ability to perform strategic socio-economic roles (Walpole and Leader-Williams, 2002). For example, they attract public visitors to reserves and, in doing so, raise funds and local support for conservation thereby helping conserve wider biodiversity. Butterflies, such as swallowtails (Collins and Morris, 1985), and dragonflies (Moore, 1997; Suh and Samways, 2001) fulfil these roles among insects, with a reserve and site guide developed for viewing both these taxa in Britain (Hill and Twist, 1998).

8.11 Summary

Conservation biologists are aiming to reduce extinction rates and to maintain ecosystem function. To do this we need to be clear on our conservation goals from the global down to the landscape level, and even finer scales for some insects. These goals need to be underpinned by a moral foundation that sincerely provides biodiversity with a long-term, viable future.

One goal has been to identify world biodiversity hotspots, based on levels of endemism and threat, for concerted conservation attention. Hotspot identification must also cater for future global climate change, and it needs to consider the complementary approach, where ecosystems making up ecoregions are also included. At the finer spatial scale, approaches have been developed which select an optimum number of reserve areas to maintain maximum biodiversity. This involves site selection on their uniqueness or irreplaceability. However, reserve area selection needs further to consider biogeographic transition zones, where there is high evolutionary potential, alongside the hotspot evolutionary 'museums'. These networks of sites need also to consider practicalities such as the economic and social context. In the case of actual islands and even 'habitat islands', small-sized areas can be important, so long as habitat quality is high. Complementary sets of such 'quality' small islands can be important for representing regional diversity.

Reserve network selection uses surrogates. These are features, such as higher taxa, alternative species and particular habitats. Results to date suggest that a complementary combination of species features and environmental features produce the most meaningful results. The practical aspects of actual reserve network selection then depend on what can be achieved on the ground, with the alarming conclusion that we are not going to be able to conserve everything. And even if we conserve certain landscapes and species, there must be a sufficient number of individuals to ensure a wide genetic base for the future.

To ensure a future for insect diversity, we need to maintain as much heterogeneity, at all levels, from genetic to landscape, as possible. But as we cannot conserve all, it is necessary to identify a minimum set of ecosystems or landscapes (the coarse filter) that conserves as much as possible with the least number of reserves. Identifying the complementary set of ecosystems or landscapes for insect diversity conservation has been challenged by the fact that vertebrate or alternative insect taxa are not necessarily spatially faithful to each other. Furthermore, the task for insects is vast, and while use of unnamed morphospecies is valuable for comparative disturbance studies, we still need good taxonomy for identifying genetically significant taxa and areas. A refinement on this is to understand the phylogenetic pedigree of focal taxa.

Biodiversity conservation using short-cuts, such as use of umbrella species to afford protection to other biodiversity in their habitat space, have been advocated. However, to conserve the complexity of both historically pedigreed biodiversity and future evolutionary potential, we need to use a variety of complementary approaches involving both the landscape approach (coarse filter), yet additionally cater for particular species in need of special attention (fine filter). Meanwhile, large, charismatic, flagship species, by garnering public sympathy, play a valuable additional role in maintaining reserve areas in which the little, unsung, insect functional heroes also live.

9 Mapping, inventorying and monitoring

Having gone through the upper ranks of Nature, we descend to that of insects; a subject almost inexhaustible, from the number of its tribes, and the variety of their appearance. Those who have professedly written on this subject, seem to consider it as one of the greatest that can occupy the human mind, as the most pleasing in animated nature.

Oliver Goldsmith (1866)

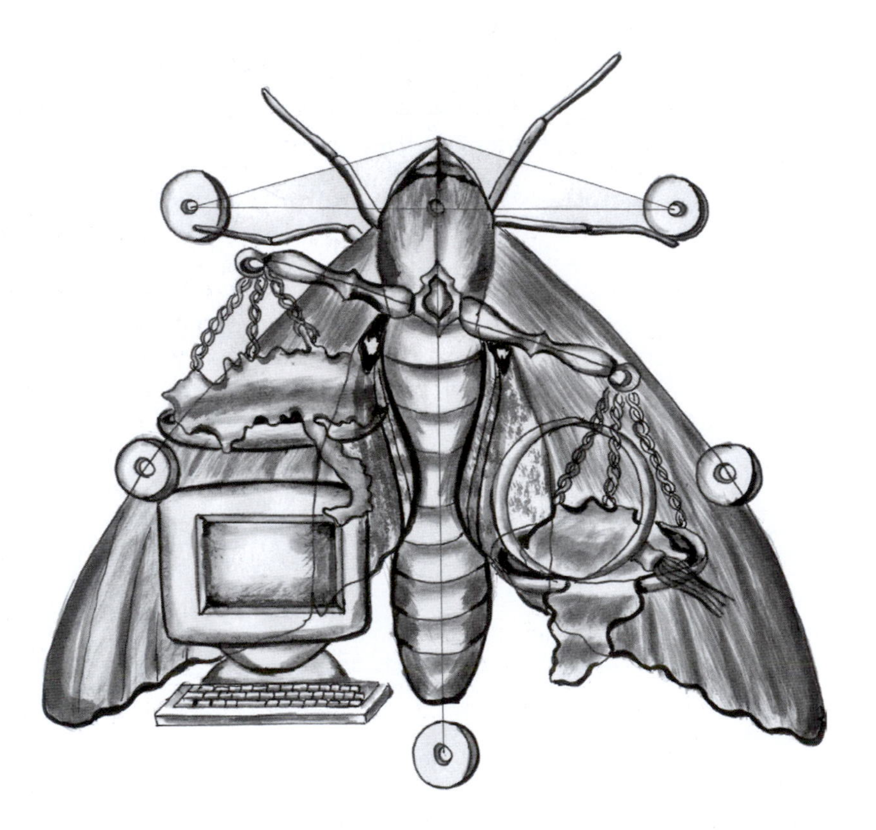

9.1 Introduction

In the last chapter we considered the concept of surrogacy where selected environmental and species variables are used to determine an optimal reserve area network. We concluded that there is no one, absolute solution, and that we need to clearly define the conservation question. Although reserve areas may be selected and function as approximate umbrellas for many species, there are always some species that are not included in those areas and require special attention. Selection of reserve areas is the 'coarse-filter' approach and consideration of individual species (or evolutionarily significant units, which are genetically distinctive subgroups within a species) is the 'fine-filter' approach. These two approaches are not mutually exclusive, with both, ultimately, linked to conservation of ecosystem processes and integrity.

For the fine-filter approach, we need first to know where the species occurs. This involves mapping, and is dependent on sound, underlying taxonomy. We cannot map what we do not know. A useful complementary methodology is to select a location and ascertain the species that occur there. This is inventorying. The profile of species at such a location changes over time, both naturally and from human impacts. Measuring these changes is monitoring, and is an assessment of how a location might be deteriorating or improving.

Mapping and monitoring species relative to threats is Red Listing. It concerns those species whose future is precarious, although in some cases, recovering.

Certain species or groups of species can be used as bioindicators or detectors of overall changes that are taking place in the landscape. This is effectively another form of surrogacy, and arguably the fine-filter, or species approach, at the service of the coarse-filter, or ecosystem approach.

These various, inter-related approaches are considered here with respect to insect diversity conservation.

9.2 Mapping

9.2.1 *Historical, taxonomic and resolution aspects*

During the process of prioritization, species features are used alongside environmental features (Chapter 8). Impacts of various sorts, from fragmentation to climate change, are often measured in terms of effects on particular species. It is important therefore to know geographically where species occur. This though, is not a straightforward matter. Firstly, we need to be clear on whether we are mapping a species whose geographical range has already retreated under human pressure, or whether we are recording its natural range prior to human impact. However, it is not always easy to separate the two. In England, the original extensive forest ('wildwood') was begun to be cleared some 5000 years ago, with only 0.48% of Lincolnshire wildwood remaining. This also emphasizes the importance of complementing coarse-filter, landscape data with fine-filter, species data, to understand how the components of the landscape are changing over time.

Mapping of species also must assume that the taxonomy has been verified. With more widespread use of DNA technology, as well as improved alpha taxonomy using image analysis, some surprises are surfacing. Generally, insect groups are more, not less, speciose than formerly thought. This can render earlier data largely meaningless at the species level.

Having clarified historical and taxonomic aspects, one can proceed to the next step of deciding at what resolution one is going to map. This depends on the conservation question, and on catering for possible questions that might arise in the future. This has been emphasized by Cowley *et al.* (1999), who showed that by mapping at the fine scale, common British butterflies were found to be declining faster than conventional coarse-scale maps suggested.

9.2.2 *National mapping schemes*

Some countries, such as Britain, have national recording schemes. These are immensely valuable for determining how well or not species are doing over time, as well as the extent of the geographical ranges of species. For various taxa, including Lepidoptera, Orthoptera and Odonata in Britain, 10 km squares are used. A network of volunteer recorders send in their records and thus provide relatively comprehensive coverage of the country. The outcome, in the case of dragonflies, has been the production of an altas (e.g. Merritt *et al.*, 1996), which provides an immediate visual overview of present geographical ranges in comparison with the past (Figure 9.1). These types of maps, based on information in 10 km squares, have been used to analyse gross range changes of butterflies, for example, and to predict future ranges (Hill *et al.*, 2002) as well as to determine other landscape effects (Warren *et al.*, 2001).

There are some disadvantages with these so-called 'dot maps'. Firstly, the records are accumulated in an ad hoc way, resulting in taxonomically and geographically biased records (Dennis and Hardy, 1999). Secondly, the coverage is not comprehensive, and the data only record what is present and not what is absent. Thirdly, they do not recognize recorder effort which can bias results (Dennis *et al.*, 1999). Arguably, there is a fourth component which is abundance, which has important survival implications for populations. However, practicalities also play a role, with abundance-recording requiring much greater effort and precision which would not generally be feasible in the context of a volunteer recorder network. The point is that while these approaches are valuable for ascertaining differences in gross geographical range, they can miss finer aspects that are needed for conservation planning at the landscape level. In response, Dennis *et al.* (1999) recommend a supplementary structure to the national recording scheme, which permanently monitors sites. This would provide information on changes in size and timing of butterfly populations, as well as detecting some aspects of changes in distribution.

9.2.3 *'Extent of occurrence' and 'area of occupancy'*

A useful conceptual bridge between coarse and fine geographical scales of recording is use of the terms 'extent of occurrence' and 'area of occupancy'

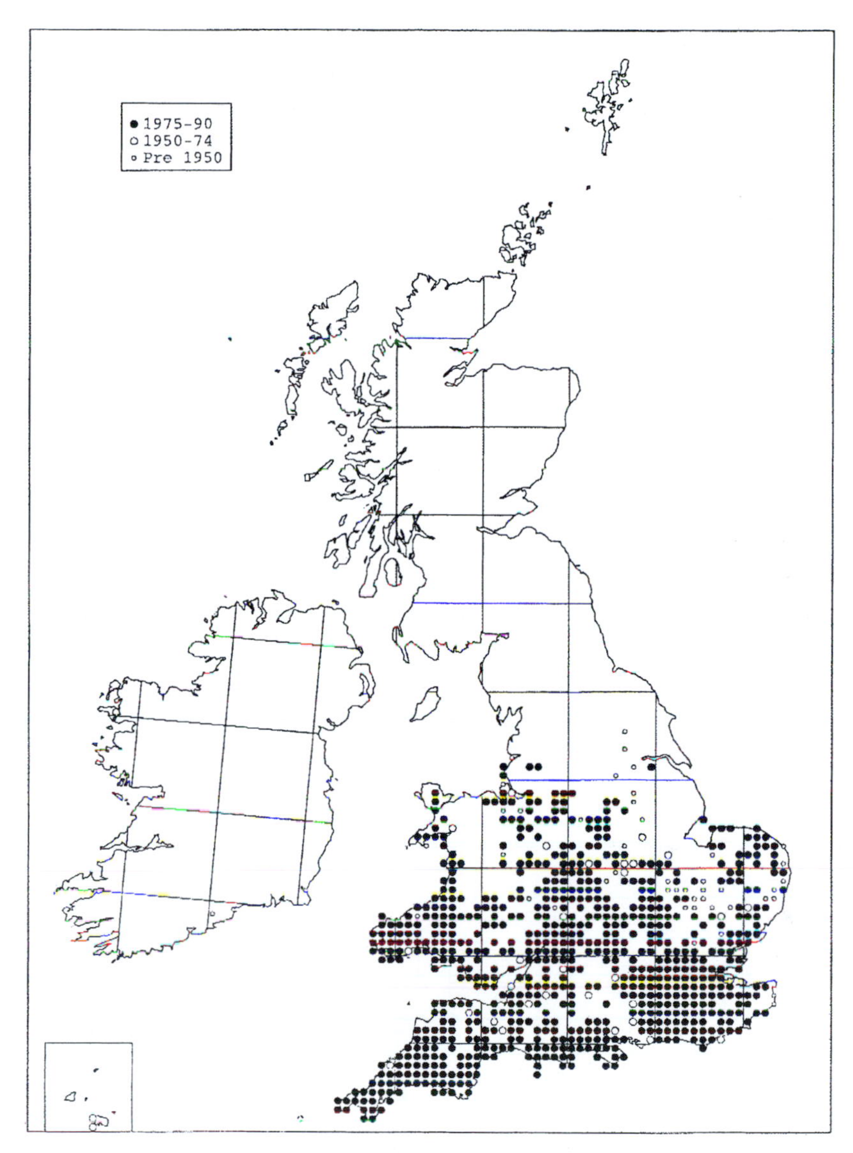

9.1 Geographical range of the Broad-bodied chaser dragonfly (*Libellula depressa*) in Britain, illustrating past and present distribution. As with several other odonate species, there has been a retraction of geographical range in the eastern part of the country. (Map produduced by the Biological Records Centre, CEH Monks Wood, using Dr A. Morton's DMAP software, from records collated by the Dragonfly Recording Scheme.)

(Gaston, 1994) (Figure 9.2). Extent of occurrence equates to gross geographical range and, as biologists know, it does not tell us much about whether the species will actually be found at any one spot within that range. Similarly, it does not guarantee that the species will be breeding across that range, especially at geographical range margins.

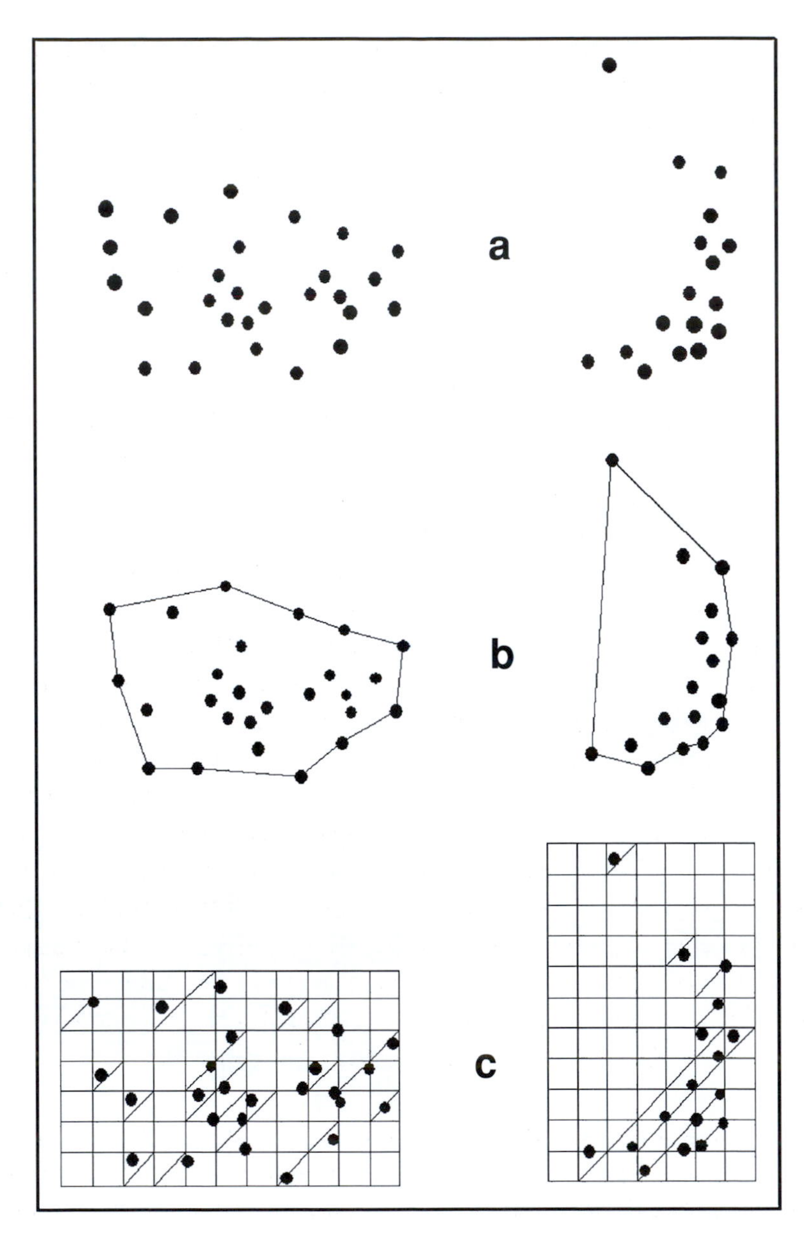

9.2 Two examples of the distinction between 'extent of occurrence' and 'area of occupancy'. (a) The spatial distribution of known, inferred or projected sites of present occurrence. (b) The possible boundary to the extent of occurrence, which is the measured area within this boundary. (c) One measure of the area of occupancy which can be achieved by the sum of the occupied grid. (From IUCN, 2001.)

Area of occupancy subdivides the extent of occurrence and provides a more meaningful picture of actually where the species occurs. This of course depends on the refinement of the mapping. The finer the scale of mapping the more accurate will be the assessments of area of occupancy. The tradeoff is that much greater recorder effort is required at the finer scales. From a conservation planning point of view, the finer and more comprehensive the area of occupancy sampling, the more valuable the information. It may not necessarily alter the extent of occurrence very much, unless new outlier populations are discovered.

9.2.4 *Relational spatial databases*

A way forward for flexible mapping is to develop a relational spatial database (RSD). This is a particularly useful approach in geographical areas without a network of volunteer recorders, and when designing a database to hold a large set of observations from different sources, e.g. past insect collections (Ponder *et al.*, 2001). The database can also accommodate data from a variety of sources, e.g. specimens, sight records and published records. The advantage of RSDs is that the records will have been collected at a variety of spatial resolutions, e.g. from using global positioning systems (GPS) to the nearest 1″ (i.e. about 30 m on the ground), through using a topographic map to the nearest 1′ (about 1.8 km), or finer scale if required, to more generalized data which may simply record a town, farm, reserve or region. Not all historical data are precise, and an RSD can cater for different degrees of precision from day to year. The value of this approach is that the data are entered along with an indication of their level of precision, and data can be filtered and analysed at any of these levels. The RSD can be linked to a geographical information system (GIS), and together they can (1) note the spatial location of any species for a given epoch or range of time periods; (2) show the distribution of a species relative to a range of environmental variables (e.g. elevation, rainfall, vegetation types); (3) show the presence as well as the absence of a species at an epoch for a given season; and (4) be used for spatio-temporal modelling. The fact that absence, as well as presence, data can be recorded is an important refinement for establishing both the extent of occurrence and area of occupancy (Figure 9.3).

As mentioned in Chapter 8, phylogenic considerations can be an important part of the mapping process underpinning conservation planning. Martín-Piera (2001), employing dung beetles in Spain, identified a particularly important irreplaceable area where a rare, endemic species occurs. In a related study, Lobo and Martín-Piera (2002) emphasize that sampling-effort bias has to be reduced to manageable proportions to avoid erroneous predictions of hotspot richness and recommendations for conservation. But Iberian dung beetles, like many insect taxa, cannot be corrected for unequal sampling effort because of the lack of reliable biological information. Lobo and Martín-Piera (2002) propose that this issue be addressed by building comprehensive data bases of bibliographic, museum, and field data and then using forecasting models such as generalized linear models to obtain testable geographic patterns of biodiversity-related variables such as species richness.

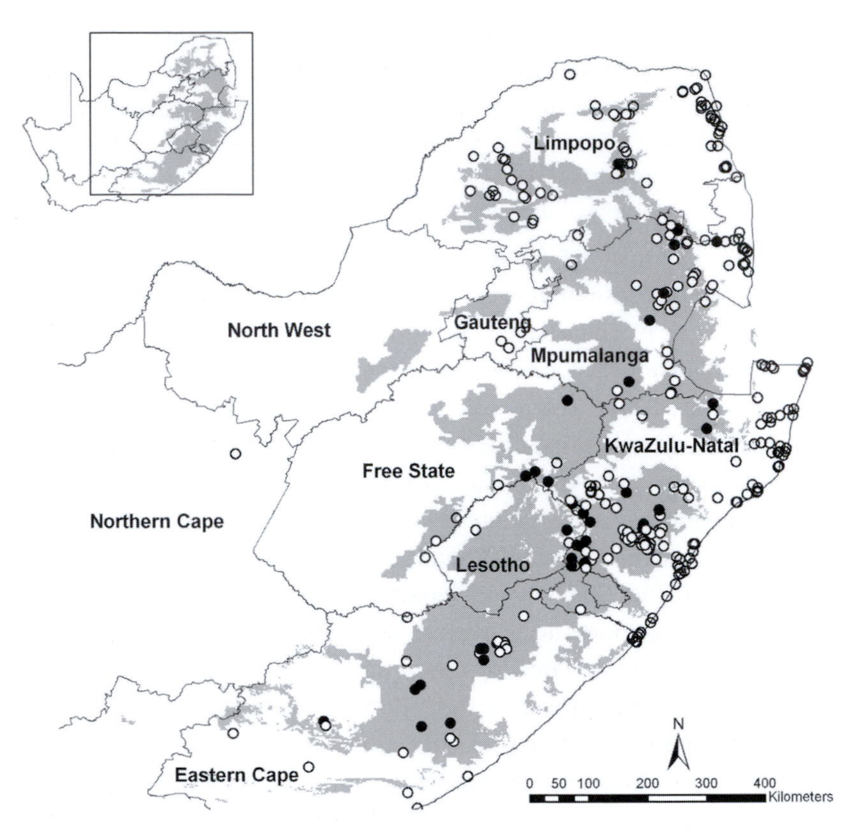

9.3 Actual recorded geographical distribution of the South African endemic Mountain malachite damselfly (*Chlorolestes fasciatus*) based on 'presence' records (black circles) and 'absence' records (open circles). Using a BIOCLIM-type modelling approach, the species' predicted geographical distribution has been drawn (grey area). However, it is essential to be sensitive to nuances of the insect's behaviour and ecology as well as its response to simple climate variables to make these predictions accurate (see Chapter 7). (Map by courtesy of Jemma Finch and Steven Piper.)

9.3 Inventorying

9.3.1 *Snapshot of biodiversity*

Inventorying is the surveying, sorting, cataloguing and quantifying of entities (Stork and Samways, 1995). Generally these are species, but may be other features from genes to landscapes. The spatial scale of an inventory can be from the microscopic to the entire biosphere, while the real value of an inventory lies in its identified entities. This may, for example, be important for determining the uniqueness of a particular geographical location, and is fundamental for planning and prioritization. The irreplaceability of an area depends on the uniqueness of its component features, from genes to landscapes. The variety of genes provides a measure of the viability and adaptive potential of a population, while the variety of species provides a measure, for example, of endemism.

Inventories are never complete as there are always new additions through birth and immigration. Losses also occur through death, emigration or extinction. New variation through mutation may also appear. Inventories are a snapshot of biodiversity at any one time, and form the baseline for measuring change over time, which is monitoring.

9.3.2 *Risks of naïve collecting*

Inventorying insect species has generally been done on single taxa across regions or countries and has led to many identification guides with attendant distribution maps. Conversely, specified areas, particularly nature reserves, compile inventories of a wide range of taxa. Most of these inventories are ad hoc and undertaken by taxonomic experts who happen to collect in the area or are sent a collection. Such inventories can be highly misleading if undertaken by non-experts. This has been very clear from collections of dragonflies we have received in our own laboratory. Non-experts tend to collect diurnal, easily netted individuals at pond margins next to roads, while neglecting species that are crepuscular or confined to specific habitats. The concern is that such lists are then used in biodiversity surveys and eventually into databases used in regional, continental and global planning. Far too little attention has been paid to these fundamental data-gathering flaws in large databases, at least as regards insects. Martikainen and Kouki (2003) have emphasized this point, suggesting that taxonomic specialists and the use of selected, multiple survey methods are essential to assess highly threatened species. Without such a targeted approach, sample sizes have to be prohibitively large.

9.3.3 *Approaches to inventorying*

As inventories are expensive to undertake and can be largely valueless when undertaken casually, they need to be underpinned by effective sampling and estimation procedures. Choice of methods and approaches is best made from a problem-solving perspective: (1) clearly define the challenge, its spatial scale and whether the approach is taxonomic/systematic, ecological or managerial; (2) outline the goals, including the needs of both clients and users; (3) identify the types of biological information required and the scale and intensity of management; (4) determine the statistical precision needed; (5) be clear on the time needed to achieve the goals; (6) determine whether resources (material, financial and human) are sufficient; and (7) use standardized methodologies so that quantitative comparisons can be made at another time or place (see for example, New, 1998; Niemelä *et al.*, 2000; Southwood and Henderson, 2000); (8) maintain a voucher collection (with appropriate permits) for future verification and comparison, including the use of DNA technology. This last point is critical, especially for verifying identifications when preparing a Red List of threatened species (Schlik-Steiner *et al.*, 2003).

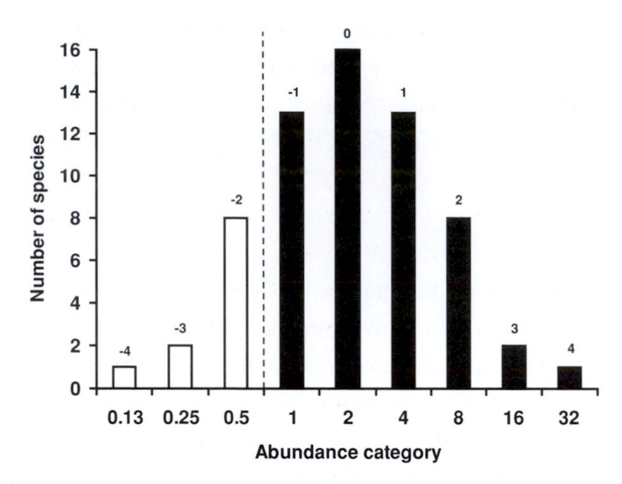

9.4 Graphical depiction for estimating actual species diversity given insufficient sampling effort. This derived approach assumes a log normal distribution of species abundances and calculation of the portion of the curve that is 'missing' from the database (open bars). Abundance categories are on a $\log_2$ scale and are identified by their lower bound. The dotted line represents the level of sampling resolution for the database, and is the boundary between those species represented by at least one specimen in the database (solid bars) and those that are not yet collected but theoretically present in nature (open bars). Values above each abundance bar indicate the 'octave number' of that abundance class, which indicates how many abundance classes a given category is from the mode of distribution. (From Fagan and Kareiva, 1997. Copyright 1997, with permission of Elsevier.)

Fagan and Kareiva (1997) point out that many of the difficulties that arise with inventories is that different areas or regions have been sampled with different intensities. On the one hand, some areas have been sampled to record most of the resident species, while others have been sampled so poorly that only a small fraction of the resident species has been recorded. They suggest two solutions to address this shortcoming. The preferred one is based on the log-normal statistical distribution of numbers of species, which offers substantial improvement over analyses based solely on the raw data, and without risk of over-estimation (Figure 9.4). Ugland *et al.* (2003) point out that extrapolating the traditional species-accumulation curve gives a large underestimate of total species richness (see Figure 8.6). In response, they suggest the use of a 'total species curve' which is based on a combination of species-accumulation curves for various sub-area plots.

Summerville *et al.* (2003) found that for North American forest moths, the assemblage composition was most affected by ecoregional differences, whereas patterns of alpha and beta diversity across spatial scales differed depending on how diversity was measured. Species richness occurred equally across all spatial scales because numerically rare species were continually encountered in response to the varied local conditions. Summerville *et al.* (2003) concluded that because most of these moth species are rare, it will not be possible for conservation biologists to design management plans to account for every species.

The target organisms that are being inventoried may not necessarily be organisms of conservation concern. They may be, for example, high impact species, keystone species, invasive aliens, or even the total fauna, many species of which may be ecologically meek in the sense of playing only a minor role in ecosystem processes. For most insects, we do not know what ecological role they play, and it may be necessary, if the goal is to conserve as much biodiversity as possible, to sample a wide range of functional and taxonomic groups.

Any particular sampling method however, provides only one window on insect diversity. This again emphasizes that the question being addressed needs to be clearly defined before approximate sampling takes place. This in turn may involve sampling a few sites intensively for most or all organisms, and a larger number of sites for a particular subset of focal taxa (Colwell and Coddington, 1994). Decisions then also need to be made on whether presence/absence data or more resource-intensive abundance data are to be recorded.

Dennis *et al.* (1999) suggest, at least for British butterflies, that a representative sampling structure be set up. This would involve stratified selection of sites, for instance based on geology, elevation, land use and other environmental variables, within a systematic framework of a grid. A suitable scale for sampling plots would be 1 ha units, though there may be advantages in aggregating them into larger blocks (e.g. 200 × 200 m squares). As butterflies, and many other insects, have home ranges at this spatial scale, it makes sense to also record host plants and other resources. If these resources are absent, distances to them need to be determined. These sites could be revisited over time as part of a monitoring programme.

Zonneveld *et al.* (2003) have proposed a survey protocol that detects the presence of a particular species which minimizes use of sampling resources. They determined the spacing of a given number of survey days on a transect that minimized the chance of missing a species when it was actually present. While five survey days detected most species with a high probability, they found that rare species, with very small populations, may require many more survey days to be detected.

Another interesting inventorying approach, which has been discussed in the context of studying non-target effects of biocontrol agents, is to construct food webs (Memmott, 2000). These webs describe the feeding relationships between different trophic levels within a community as a picture and, more formally, by mathematical matrices. Summary statistics can be calculated from the web and used to compare different webs. In pictures of food webs, consumer and resource species are shown connected by a line if the former feeds on the latter using 'connectance webs'. Food webs can also represent quantitative aspects, and show the relative sizes of interactions (Figure 9.5).

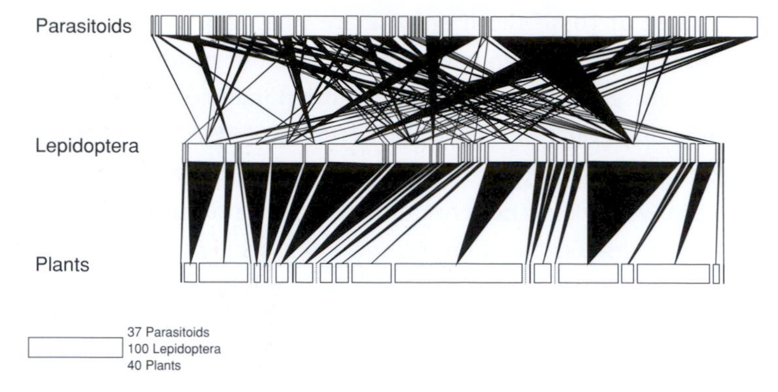

9.5 A quantitative web describing the interactions between plants, Lepidoptera and parasitoids. Each species is represented by a rectangle, with plants at the bottom, the Lepidoptera in the middle, and the parasitoids at the top. The widths of rectangles depicting plants, Lepidoptera and parasitoids are proportional to their abundance, although for clarity the scales for the three trophic levels are different. (From Memmott, 2000, with kind permission of Kluwer Academic Publishers.)

9.3.4 Relating inventory data to concepts of surrogacy

Inventory data may also be used to model, for example, occupancy of habitat patches by particular taxa. Bailey *et al.* (2002) did this for several taxa in England. While they found that particular bird and mammal distributions were correlated to landscape scale measures of forest fragment distribution, this was not generally the case with butterflies, nor plants. The outcome was that different butterfly species showed different responses to different aspects (such as patch isolation and size) of the structure of the woodland landscape. This emphasizes the fine-filter approach over and above the coarse-filter one, as well as cautioning the use of the umbrella concept. This was underscored by Fagan and Kareiva's (1997) study on Oregon butterflies, where the distribution of threatened and rare species rarely coincides with butterfly diversity hotspots, which means that protecting Oregon's butterfly-rich areas would not normally protect most of the rare or threatened butterfly species (Figure 9.6).

How little we know of distribution of rare and threatened species, especially in the tropics, is emphasized by Martikainen and Kouki's (2003) study of boreal forest beetles. To rank ten boreal forest areas to be protected according to the occurrence of threatened species with some reliability may require trapping over 100 000 beetle individuals. This inevitably means that short-cuts need to be identified, such as indicator taxa or umbrella species, but their wider applicability may be limited. Certain stand characteristics such as the volume and diversity of dead wood tend to correlate positively with species richness of beetles, but without knowledge of the previous, continual availability of dead wood, these indicators may not tell us much about the occurrence of threatened species (see also Chapter 8).

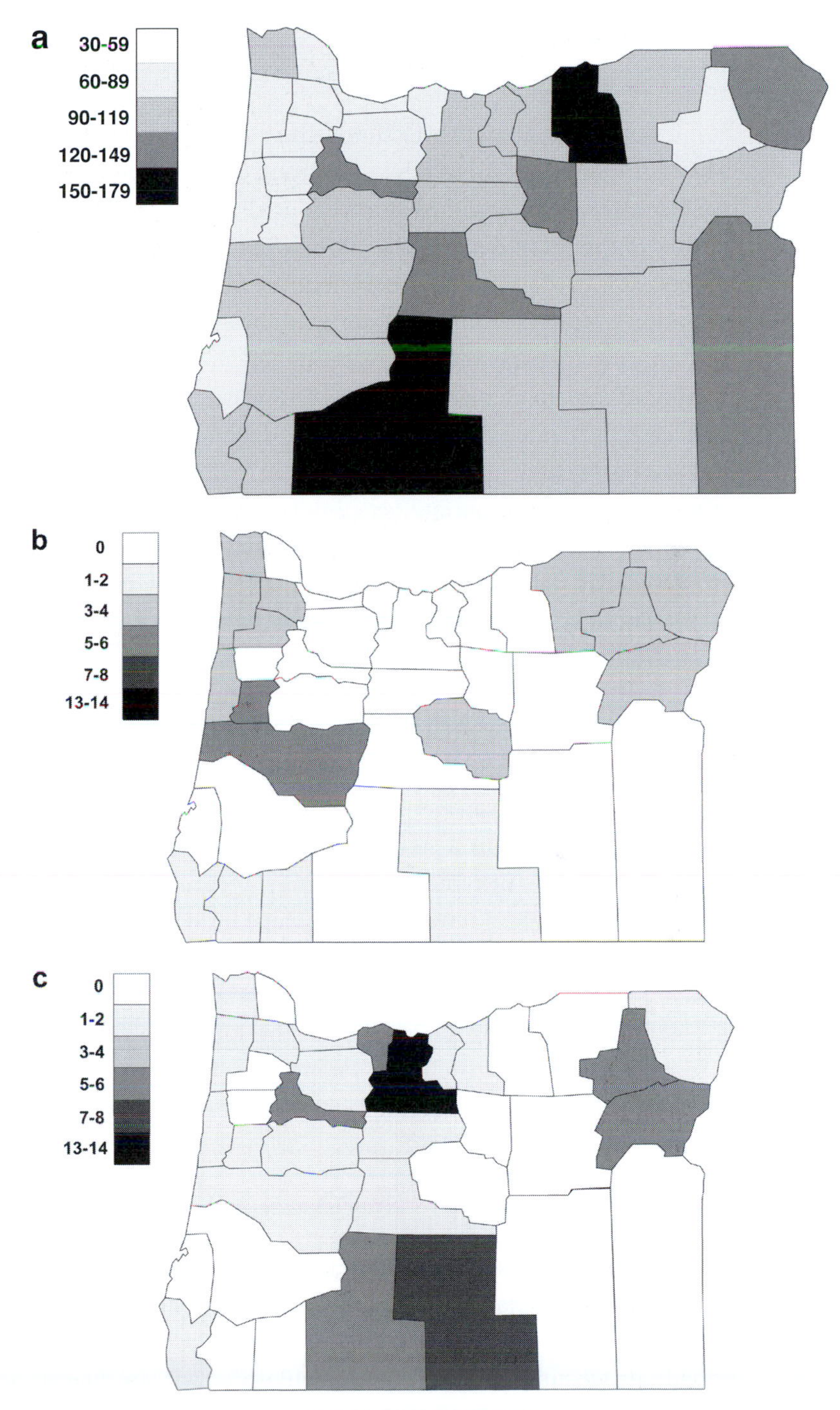

9.6 Geographical distribution of butterfly diversity in Oregon, USA. (a) Estimated butterfly species diversity for Oregon by county, in comparison with (b) 'endangered' and 'threatened' species, and with (c) species that are numerically rare. (From Fagan and Kareiva, 1997. Copyright 1997, with permission of Elsevier.)

9.4 Monitoring

9.4.1 *Aims of monitoring*

This is an activity over time, either regular or irregular, that determines either compliance with, or deviation from, a predetermined norm. It is usually goal-orientated and designed to reveal or illustrate changes in some feature or features of the ecosystem. Data accumulating from monitoring activities can also be a basis for predicting the situation in the future. Monitoring biodiversity aims to develop a strategic framework for predicting the behaviour of key variables so as to improve management, increase management options and provide an early warning of system change (Stork and Samways, 1995).

Success of a monitoring programme depends on having clear goals. These need to be framed around (1) a clear set of spatial and temporal scales of activity; (2) use of appropriate taxa, interrelationships between taxa, and impinging abiotic variables; (3) use of standardized methodologies that can be used across sites, as well as over time; and (4) a feasibility plan which involves catering for financial, material and human resources to see the project through to completion.

It is useful to conceptualize the questions that are being addressed by a monitoring programme using the pressure-state-response model. The state of a system (which can be determined by an inventory) has pressures upon it. Here, we are concerned with anthropogenic pressures. The system then responds to these pressures and moves to a new state (Figure 9.7). Monitoring aims at defining and, if possible, quantifying the pressures, recording the response and changes in state. For example, a butterfly assemblage in forest fragments might be inventoried. Selective logging then takes place. The butterfly assemblage is then monitored at various times after this. But we may suspect ecological relaxation (gradual loss of species from the patches in the long term) may occur. We may also be concerned that global climate change is a factor to be taken into consideration. As a result, we may then decide to monitor frequently at first and then at intermittent intervals in the future.

9.4.2 *Approaches to monitoring*

Monitoring is not just about species, but about all levels of biodiversity from genes to the planet as a whole, and from entities to interactions. From a genetic perspective, there are different levels of monitoring genetic changes: (1) individuals within populations; (2) populations within species; and (3) between-species comparisons. The greatest concern is that there might be a loss of genetic variation and viability, risking the future of a species (Holsinger, 2000; Frankham *et al.*, 2002). But not all species are necessarily automatically doomed. Some may survive because evolution can occur over very short timescales, such as a few generations or years. This is known as 'contemporary evolution'

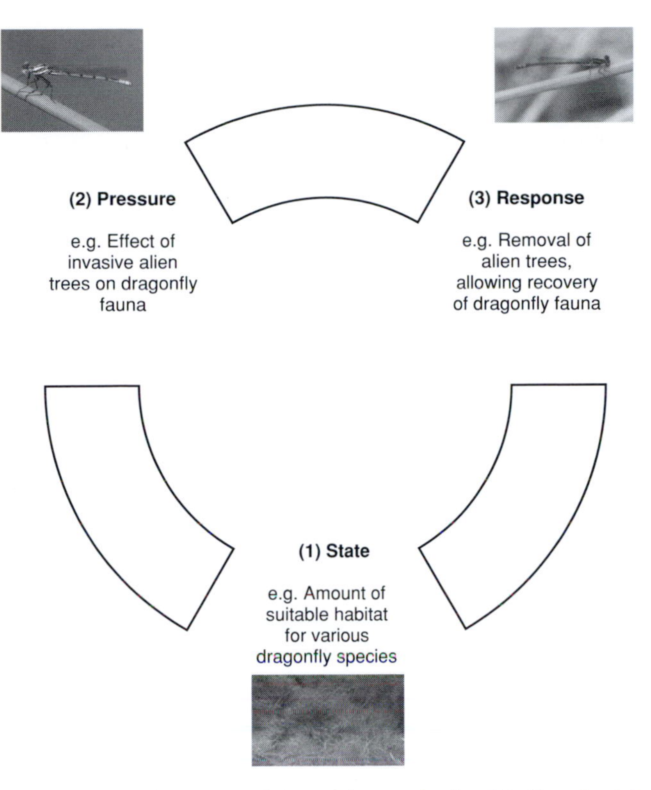

9.7 Pressure-state-response model used in monitoring biodiversity. The *state* of a system may be determined by an inventory, such as numbers and types of species. The *pressure* on the system may then be measured by the changes, such as loss of endemic species. Recording these changes is monitoring. This then leads to a *response*, such as a conservation strategy that reduces the pressure. Monitoring measures the success of the response. In this example, invasive alien bushes and trees have been radically reduced to provide available habitat (bottom picture) of *Metacnemis* spp. damselflies in South Africa. The Kubusi stream damsel (*Metacnemis valida*) is threatened with extinction in the Eastern Cape (top left). Meanwhile, in the Western Cape, the Ceres stream damsel (*Metacnemis angusta*) which was feared extinct, and had not been seen since 1920, reappeared after invasive alien trees were removed, having been surviving in some remote, unknown locality.

(Stockwell *et al*., 2003). It is influenced by complex interactions among population size, genetic variation, the strength of selection and gene flow. This very complexity makes most management scenarios unique, and requiring specific monitoring.

Monitoring of species is still a fundamental cornerstone of insect diversity conservation. While it is desirable, and sometimes essential, to monitor actual, scientifically described species, it may be at times only possible to monitor surrogates. These, for example, may be unnamed representatives of families as indicators of freshwater conditions (Resh and Jackson, 1993). In other cases,

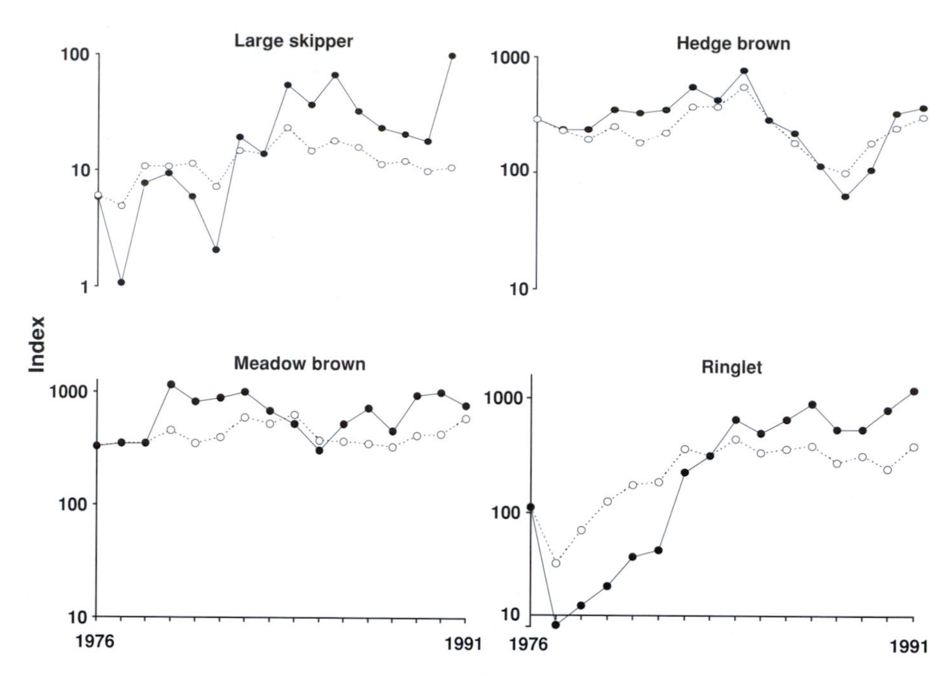

9.8 Fluctuations in abundance index values of four butterfly species at Monks Wood (UK) (solid line), compared with fluctuations at all sites (excluding Monks Wood) (dotted line), 1976–1991. The 1976 all-sites index is given the same value as that at Monks Wood. The general synchrony of fluctuations suggests that a widespread factor, presumably weather, has a major influence on local populations. (Redrawn from Pollard and Yates, 1993, with kind permission of Kluwer Academic Publishers.)

specific species may be monitored both in their own right or as part of a composite picture (Pollard and Yates, 1993) (Figure 9.8). The point is that species monitoring programmes must use standardized methods and use recognized sampling protocols that sample the appropriate species adequately but not excessively. The aim all the time is to have standardized findings from one place or time to the next. Additionally, it is important to select a range of physical and biotic variables that are thought to be influential upon those species, and to monitor them alongside the species.

Monitoring the decline and/or recovery of specific, threatened species is part of the Red Listing process (discussed in the next section) and is the fine-filter in insect diversity conservation. Once a threat has been identified, it should be monitored, as in the case of alien plants impacting on butterfly assemblages in United States National Parks (Simonson *et al.*, 2001).

Coarse-filter monitoring is ongoing assessment of whole ecosystems and landscapes, and is part of general biodiversity monitoring (Stork and Samways, 1995). Insects have played a relatively minor role in coarse-filter approaches, although this is now changing. Methodologies are being refined to incorporate insects and other invertebrates, whether morphospecies or named species. The field has generated considerable debate (summarized by Kitching *et al.*, 2001), and while the goal may be to maximize chances of detecting differences between sites or over time using a suite of taxa, such a multi-pronged approach can readily become resource limited, making it essential to test a range of methodologies as was done by Kitching *et al.* (2001) in South Asian rain forests, and Ranius and Jansson (2002) in Swedish forests. Once such methodologies have been developed and resource availability determined, it is then possible to undertake a monitoring programme tailored to the conservation question being asked.

9.5 Red Listing

9.5.1 Principles

The World Conservation Union's *IUCN Red List of Threatened Species* (IUCN, 2003a) is a global inventory of the worlds' species most threatened with extinction. As the categorization of each species on the list requires revision every few years, it is essentially a monitoring exercise that focuses on the 'fine-filter'. With effective conservation management, some species improve their positions on the Red List (i.e. become less threatened). Despite conservation action, or because of lack of it, other species' positions on the Red List may deteriorate. The point is that the Red List is dynamic and reflects changes in populations, threats and management results. The Red List may, in turn, highlight the plight of a particular species which then requires conservation attention.

In terms of insects, the Red List presents an interesting dilemma. From an ethical standpoint, the Red List takes an 'intrinsic value' approach, with all species being equal, i.e. given equal space on the page, whether wasp or whale. This, however, makes no assumptions on comparative worth or value of a species, nor does it make a statement about the need for, importance or actuality of, conservation measures. For insects, this 'equal space' approach to listing is important, as it raises individual species' profiles considerably. This awareness-raising is offset by the enormous onus this puts on insect conservationists who have the task of assessing threats to several tens of thousands of species, many of which do not even have scientific names.

We must be very clear that the Red List is highly biased in that it principally reflects our state of knowledge and not necessarily the actual status of a particular species' survival potential. This is particularly so in the case of insects (Samways, 2002c), simply because our level of knowledge of all but a few species is insufficient to make really informed assessments. This does not lessen the value or currency of the Red List, which is an enormously influential document.

Another point to consider is that the submission of species names and status assessments to the Red List is precautionary rather than evidentiary. In other words, when a species' conservation status is assessed, it is done in a way that allows time and room to initiate conservation action. Often we cannot wait until there is conclusive and final evidence, because, meanwhile, the species may have slid to the brink of extinction. This does mean that we have to make informed decisions on the extent to which we commit resources to assessments: not too conservative and not too liberal.

A special case is the category 'Extinct'. To list a species as Extinct (i.e. the last individual has died) requires a particularly cautious approach. For the reasons outlined by Harrison and Stiassny (1999), listing as Extinct is a major decision, as this may preclude further searches. This is particularly so for insects which are small, may be cryptic, and easily overlooked. Nevertheless, a charisma surrounds certain 'thought-to-be-extinct' species, with for example, the newsworthy 'extinct' Lord Howe Island stick insect (*Dryococelus australis*) being discovered on a small, neighbouring island (Priddel *et al.*, 2003).

9.5.2 *The Red Listing process*

The Categories and Criteria for Red Listing a species are outlined in the *2003 IUCN Red List of Threatened Species* (IUCN, 2003a) and the booklet *IUCN Red List Categories and Criteria, version 3.1* (IUCN, 2001). Assessments may also be made using the software package RAMAS Red List, which assigns taxa to Red List Categories according to the rules of the IUCN Red List Criteria and has the advantage of being able to explicitly handle uncertainty in the data (Akcakaya and Ferson, 2001).

New global assessments or reassessments of taxa currently on the IUCN Red List may be submitted to the IUCN/SCC (Species Survival Commission) Red List Programme Officer for incorporation into a future edition of the *IUCN Red List of Threatened Species*. Normally this is done via one of the Specialist Groups. There are some insect taxon Specialist Groups (e.g. Odonata Specialist Group) and geographical area groups (e.g. Southern African Invertebrates Specialist Group, South Asian Invertebrates Specialist Group). These Specialist Groups act as Red List Authorities through which submissions on species' statuses can be made, vetted and submitted to the central IUCN database, the Species Information Service. Records can be updated anytime but must be so every 10 years. This poses a problem for many insect species because there are simply not the human and financial resources to undertake the task. There is a minimum set of information which should accompany every assessment submitted for incorporation into the *IUCN Red List of Threatened Species*, which are outlined in IUCN (2001).

With any assessment there is always some degree of uncertainty. With regard to Red Listing, uncertainties may be identified as (1) natural variability resulting from the fact that species' life histories and the environments in which they live change over space and time; (2) semantic uncertainty arising from vagueness in

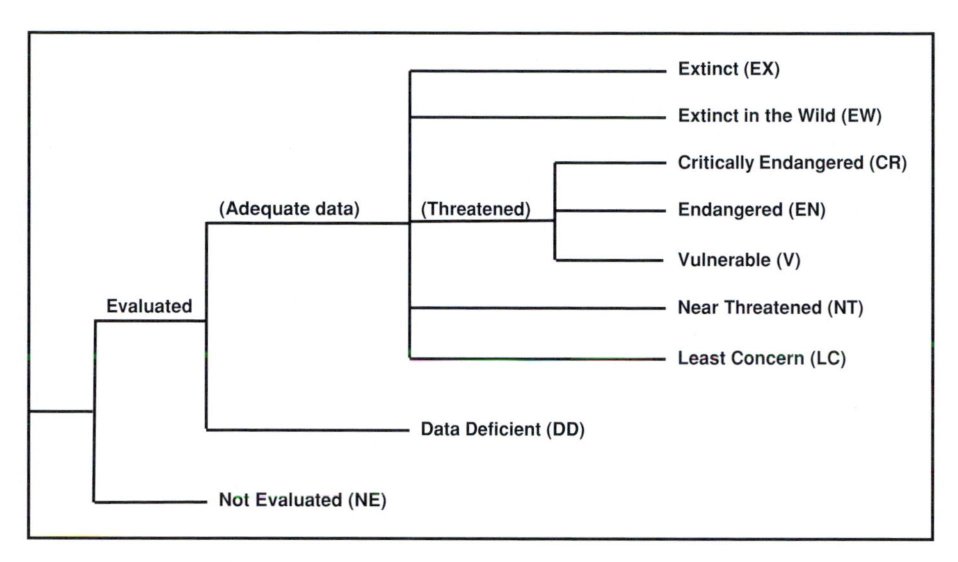

9.9 Structure of the categories used for evaluating Red Listed species according to IUCN criteria. (From IUCN, 2001.)

definition of terms, or lack of consistency among different assessors; and (3) measurement error which arises from the lack of precise information about the parameters used in the assessments (Akçakaya *et al.*, 2000). Measurement error is by far the greatest factor in insect assessments, not because entomologists are inaccurate, but because it is difficult to assess the population parameters of any one insect, let alone a multitude of them.

The aim of an assessment is to assign a single Red List category to a particular species or evolutionarily significant unit (ESU). This is done by following the dendrogram from left to right (Figure 9.9). A species may not yet be evaluated at all, and even if it is, there may be insufficient data to allocate a category. This Data Deficient (DD) category applies to many insect species. Where there are adequate data, the species may be assessed, into one of three groups: (1) Least Concern (LC), Near Threatened (NT); (2) Threatened (Vulnerable (VU), Endangered (EN) or Critically Endangered (CR)); or (3) Extinct (Extinct in the Wild (EW) or Extinct (EX)). Allocation of a category to a species is done using a particular set of criteria. Each species is evaluated against all the criteria, which are known broadly as A–E, and are detailed in IUCN (2001). They refer in general terms to: (A) reduction in population size based on four types of measurement; (B) geographic range in the form of either extent of occurrence or area of occupancy, or both; (C) population size estimates linked to estimated or actual declines or fluctuations; (D) population size very small; and (E) probability of extinction in the wild. Criteria A and B are the most useful ones for insect species assessments, followed by D and E for some species where detailed demographic information is available.

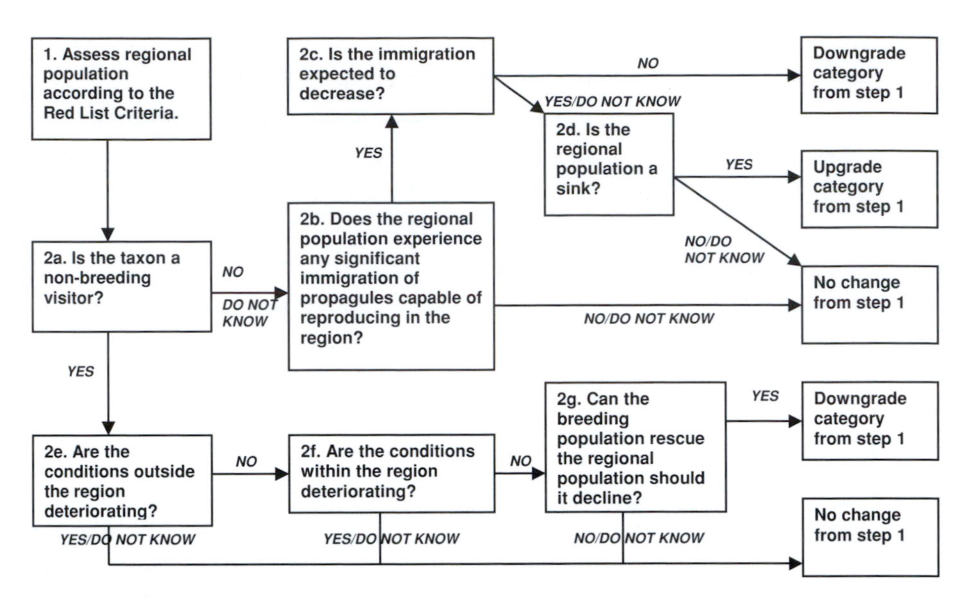

9.10 Conceptual scheme of the procedure for assigning an IUCN Red List category at a regional level. In step 1 all data used should be from the regional population – not the global population. The exception is when evaluating a projected reduction or continued decline of a non-breeding population – in such cases conditions outside the region must be taken into account in step 1. Likewise, breeding populations may be affected by events in, for example, wintering areas, which must be considered in step 1. (From IUCN, 2003b.)

9.6 Application of IUCN Red List Criteria at regional or national levels

The boundaries delimited for conservation management are largely political rather than biogeographical. This has important implications for regional or national Red Listing of taxa (Gärdenfors, 2001; Gärdenfors *et al.*, 2001; IUCN, 2003b) (Figure 9.10). Most important is that a global category may not necessarily be the same as a national or regional category for a particular taxon. For example, a species listed as Least Concern globally might be Critically Endangered within a particular country or region. Conversely, a species that may be classified as Vulnerable on the basis of its global declines in numbers or range, might be of Least Concern within a particular region where its population is stable.

Marginality is another pivotal factor that needs to be considered when undertaking national Red Listing (Samways, 2003c). Marginality may be defined as the state of populations at a specific location and time when environmental and landscape factors decrease the probability of population survival and persistence (Shreeve *et al.*, 1996). The degree of marginality may vary with the intensity of those factors which influence survival and persistence, and it will vary in both space and time. As such it is usually associated with populations at or close to the edge of their geographical ranges. Typically these edge populations are small, sparse and isolated from one another (Lawton, 1993). Also, they may differ from conspecifics towards the centre of the range, in morphology, biochemistry, physiology, life history, and in genetic composition (Shreeve *et al.*, 1996). This may be important for future survival of species. In a global warming study on four British insect species, Thomas *et al.* (2001) showed that marginal species have recently colonized a wider range of habitat types. This has led to an increased potential for dispersal. One of these species, the bush cricket *Metrioptera roeseli*, has either short or long wings, and the long-winged individuals are now in higher proportions in recently established populations than in old established ones.

Marginality and the process of national Red Listing must also take into account vagaries of weather conditions. This is seen in the eastern part of South Africa where certain marginal, nationally Red Listed odonate species have become locally extinct in recent years, having succumbed to adverse El Niño weather conditions (Samways, 2005). But as they are common farther north, they are not globally threatened.

9.7 Insects as bioindicators

As insects are so speciose, abundant and often easy to sample, it is not surprising that they, and other arthropods, have been widely used for indicating changing ecological conditions (Kremen *et al.*, 1993). Often, however, it is not clear whether we are really using indicators (i.e. heralds of some change in environmental conditions) or responders (i.e. seeing how a particular group of organisms responds to certain changing environmental conditions). Many studies have shown how a particular group of insects responds to a changing set of environmental conditions, and then the members of this group are described as ‘good indicators’. This has led McGeoch (1998) to suggest that there needs to be clear definition of the objectives of bioindicator studies, formal testing of the robustness of insect indicators and recommendations on how they might be used. However, she also points out that advocating rigorous, long-term protocols to identify indicators may presently be questionable given the urgency with which conservation decisions have to be made. Andersen (1999) has also pointed out that the use of indicators is most effective when supported by a predictive understanding of the responses of target taxa to environmental stress and disturbance, at multiple spatial and temporal scales. He also suggests that in addition to demonstrating that insects really are reliable indicators, we need robust short-cuts for revealing the responses in question and user-friendly protocols that can be readily adopted in the land management process.

McGeoch (1998) provides a conceptual framework to clarify ‘when’ and ‘for what’ we should use bioindicators (Figure 9.11). There are three indicator

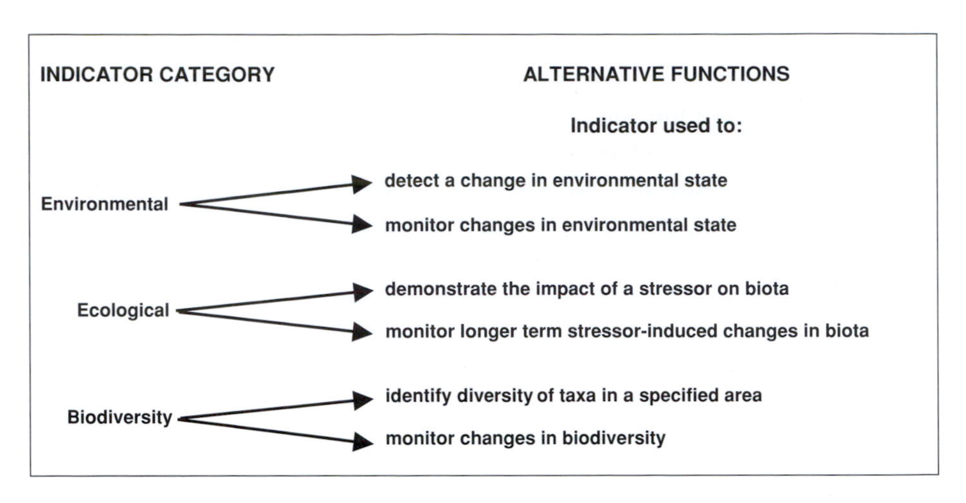

9.11 The functions of bioindicators in each category of bioindication. (From McGeoch, 1998, Copyright Cambridge Philosophical Society 1998, reproduced with permission.)

categories: (1) environmental (a species or group of species that responds predictably, in ways that are readily observed and quantified, to environmental disturbance or change in environmental state); (2) ecological (taxa that demonstrate the effects of environmental change (e.g. habitat alteration and fragmentation, climate change) on biotic systems, rather than functioning merely as gauges of changes in environmental state); and (3) biodiversity (i.e. surrogates of biodiversity) (discussed in Sections 8.5–8.10). These categories are then used to detect, demonstrate, identify or monitor some sort of change. While doing this, one needs to be acutely aware of risks associated with changing spatial scales. Indicators at a large spatial scale may not be so at finer scales (Wiens, 1989). This can carry considerable risks in terms of conservation. Cowley *et al.* (1999) demonstrated that conventional coarse-scale distribution maps widely used by conservation biologists, grossly overestimate the areas occupied by British butterflies and grossly underestimate decline. This emphasizes too, the importance of clearly determining area of occupancy and not simply whether a species occurs or not in a large square.

Exploratory ways are being found to undertake rapid assessments using conspicuous insect indicator groups as surrogates for lesser known groups. Hughes *et al.* (2000) found that dipteran species richness was marginally correlated with hymenopteran species richness and was significantly correlated with the total number of insect orders, suggesting that insect taxa may be reasonable surrogates for one another when sampling is done across habitat types. Similarly, Kerr *et al.* (2000) found that butterflies and skippers can be used to predict species richness and morphospecies richness among Hymenoptera at the landscape scale, providing a rapid assessment method, as butterflies are easily recognized and monitored (Figure 9.12) (see also Section 8.8).

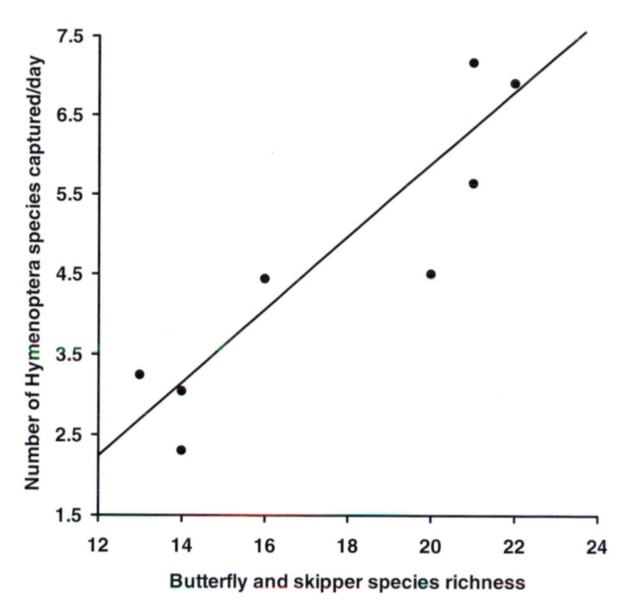

9.12 Hymenoptera species richness shows a positive correlation with butterfly and skipper species richness among oak savanna fragments in Canada. (Redrawn from Kerr *et al.*, 2000, with kind permission of Blackwell Publishing.)

Along similar lines, butterflies can be valuable for indicating vascular plant richness (Simonson *et al.* 2001) and fragmentation (Zschokke *et al.*, 2000) (see also Section 8.7 for contrary views). This is not necessarily to argue the case for butterflies, but to choose an appropriate suite of taxa or functional groups for the task at hand (Kotze and Samways, 1999a). A further point is that a particular set of indicators may or may not be effective across wide geographical areas or different types of ecosystems. Ephemeroptera, Plecoptera and Trichoptera appear to be valuable worldwide for monitoring stream water quality (Resh and Jackson, 1993). Yet grasshoppers are more valuable for monitoring grassland condition in Africa (Samways, 1997a) than in Central Europe (Zschokke *et al.*, 2000). The reasons for this are complex and depend on historical and regional processes as well as immediate, local ones. Taking a functional perspective, frugivorous Costa Rican butterflies, across a landscape of different levels of disturbance, were relatively good indicators of other butterfly guilds, although rare species in both groups may fall outside this relationship and require special attention (Horner-Devine *et al.*, 2003) (Figure 9.13).

Spatial scale can also play a role when comparing across taxa. In Australia, ants performed poorly as indicators of Victorian grasslands because of the fine-scale heterogeneity of the landscape and the relative habitat-tolerance of the ants (New, 2000b).

One of the shortcomings with indicator work involving insects, with the exception of freshwater species, is that there is rarely a follow up from the

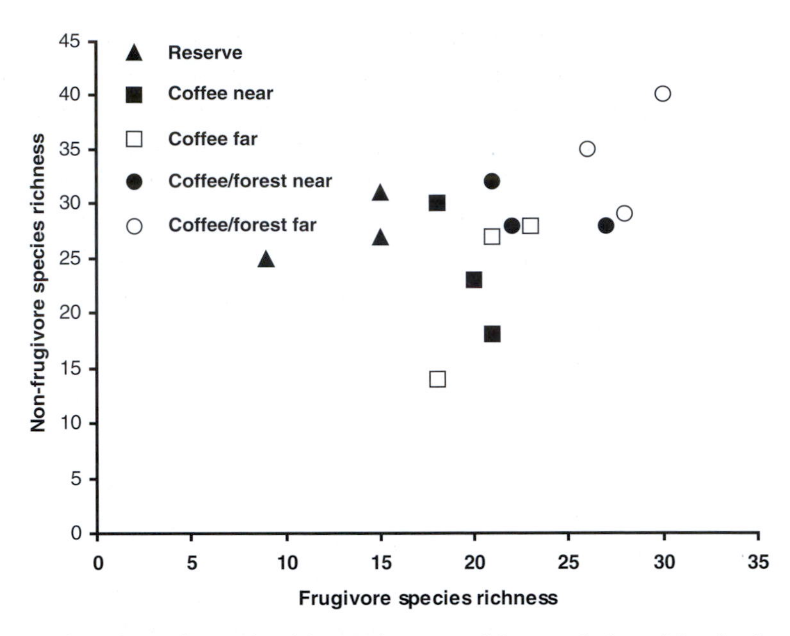

9.13 Frugivore butterfly species richness shows a positive correlation with other butterfly functional groups (non-frugivore) across a range of disturbed and undisturbed habitat types in Costa Rica. (Redrawn from Horner-Devine *et al.*, 2003, with kind permission of Blackwell Publishing.)

original research to verify the initial findings. Bioindicators of habitat quality and environmental change must be identified quantitatively and tested independently to confirm their usefulness. McGeoch *et al.* (2002) did this for dung beetles. They used the indicator value (*IndVal*) method (Dufrêne and Legendre, 1997), which combines measures of habitat fidelity (abundance and distribution within sites) and specificity (unique to particular sites), and showed that species with strong habitat specificity (characteristic species) are unlikely to provide information on the direction of ecological change despite high vulnerability. Rather, detector species that span a range of ecological states are likely to be better in this role. McGeoch *et al.* (2002) point out that because of *IndVal*'s resilience to changes in abundance, it is a particularly effective tool for ecological bioindication.

There needs to be a two-stage process when selecting bioindicators (quantitative identification and verification) to establish the degree of confidence with which they can be applied. This not only improves the efficiency of bioindication systems but is likely to increase their successful adoption as management tools. Furthermore, employment of a range of species with different combinations of specificity and fidelity values will maximize the information on habitat quality extracted from bioindicator assemblages.

The *IndVal* approach is particularly useful for insect diversity conservation because it provides a simple method for identifying the value of indicator species that is robust to differences in the numbers of sites between site groups, to differences in abundance between sites within a particular group, and to differences in the absolute abundance of very different taxa which may show similar trends. This is because each *IndVal* measure is absolute (expressed as a percentage), and is calculated independently of other species in the assemblage. Direct comparisons of indicator value can be made between different taxonomic groups, different functional groups and even those in different communities. Furthermore, where taxa show very similar specificity and fidelity trends, but differ in overall abundance, their *IndVal* remains the same, making direct comparisons possible (McGeoch and Chown, 1998).

An interesting corollary is that once a comprehensive baseline study has been done which has ascertained the full species richness for the focal taxon or taxa in a particular area, it is then possible to select indicator species and simply to sample them at the height of seasonal abundance at various sites. Interestingly, Landau *et al.* (1999), using moths in the USA, also indicated the value of short-term studies for indicating the actual long-term values of species richness for the same area. While there are obvious shortcomings with this method, as discussed by Landau *et al.* (1999), it again provides a practical and expedient methodology for assessing the vastness of insect diversity.

9.8 Reference sites

Colwell and Coddington (1994) illustrated the importance of having reference sites to assess the true richness and composition of species assemblages, to measure ecologically significant ratios between unrelated taxa, to measure taxon ratios and to calibrate standardized sampling methods. Other writers have also emphasized the importance of reference sites, whether as a complement to standardized, national mapping (Dennis *et al.*, 1999), so as to include biology to assist overcoming mapping biases (Dennis, 2001), to verify long-term changes (Woiwod, 1991), as well as for determining bioindicators (McGeoch *et al.*, 2002).

9.9 Summary

Much of our understanding of how insect diversity is responding to anthropogenic environmental change is coming from studies of single species or assemblages of known species. This approach depends on sound underlying taxonomy. It also depends on a clear understanding of the conservation goal, and addressing it using the appropriate spatial scale. Some countries have national recorder and mapping schemes which have led to valuable insight into geographical range changes over time when measured at a coarse scale. These mapping schemes are largely ad hoc, but could be complemented by certain sites that are permanently monitored and which provide more detail on populations.

The overall geographic range of a species is its 'extent of occurrence'. Within that area, any particular species only occurs in particular habitats or sub-areas. These finer-scale areas make up the 'area of occupancy'. Overcoming some of the problems associated with data collected in many different ways by different recorders at different spatial and time scales is to use a 'relational spatial database' which has considerable inherent flexibility.

Mapping is part of the wider process of inventorying, which is the surveying, sorting, cataloguing and quantifying of entities, from genes to landscapes and from the microscopic to the entire biosphere. Inventorying only provides a temporal snapshot, and its real value at the species level, is when the species are scientifically named. It is essential that the species inventory is done by, or at least receives advice from, taxonomic experts. Sampling too, must be well-planned and take cognisance of resources available to undertake it. Depending on the aims of the project, a sampling protocol may involve sampling a few sites intensively for most or all organisms and a larger number of sites for a particular subset of focal taxa. It is critical that appropriate spatial scales are chosen to address the question being posed, which may, for example, be on functional ecology, and involve sampling food webs. Inventorying may also be used to verify models, such as actual versus predicted occupancy of remnant patches.

Monitoring ascertains compliance with or deviation from a predetermined norm, and aids management as well as providing early warning of system change. Again, it is vital to have clear goals before monitoring begins. Monitoring, like inventorying, can take place at various functional scales from genes to landscapes, although monitoring of species abundance is the cornerstone. Monitoring often employs surrogates, such as species or higher taxa, to indicate system change, as in streams or rivers. On the other hand, monitoring may focus on particular species of interest. This may be a threatened species, whose population viability and vulnerability are assessed, along with the threats upon it. This is termed 'Red Listing'. Such a species is assigned a threat category, which is revised over time. Insects and other invertebrates pose difficulties in that there are so many species, and an assessment of their populations has many uncertainties. Although Red Listing aims to determine the global status of a species, the process may also be carried out at national or regional levels to assist in local conservation planning.

Insects, being speciose and abundant, have often been used as bioindicators. However, much more rigour is required to ascertain their value and reliability in this role. They may be categorized as environmental, ecological or biodiversity indicators. Their usefulness in these roles depends on their sensitivity at various spatial scales. It also again depends on the conservation goal and on resources available. Progress is being made on developing bioindicator insect species protocols. Much of the rigour in development of these methods is coming about through having well-researched reference sites against which other sites can be calibrated.

10 Managing for insect diversity

Yet maybe we naturalists and insect mongers are too prone to thus philosophise over our delectable spots and socalities . . . A great forest will, if ransacked prove to contain more compressed information than our trim shrubberies and hanging rosaries; a neighbouring heath or marsh land tells more tales than rich Levantine orchards and cattle-cropped parks.

A. H. Swinton (*circa* 1880)

10.1 Introduction

Wilderness is the ultimate natural value. Large areas of land with no or minimal human interference have become increasingly important repositories of quality biodiversity. As insects are small in size and often have limited home ranges, tracts of land that are, say, too small for large, wandering vertebrates, can nevertheless be gems for insect diversity. The value of parks and reserves lies in maintaining their natural ecological processes, although for the smaller areas in particular, some management may be necessary to maintain these processes.

As insects are so varied and largely unknown, we need to find ways of maintaining as much natural landscape variation as possible. Such a wide range of ecological conditions maximizes opportunities for insects, with their huge array of biologies. To maintain this range of conditions, it may be necessary to apply management processes and procedures that maximize opportunities for most, if not all, the species. However, some particular specialist and rare species

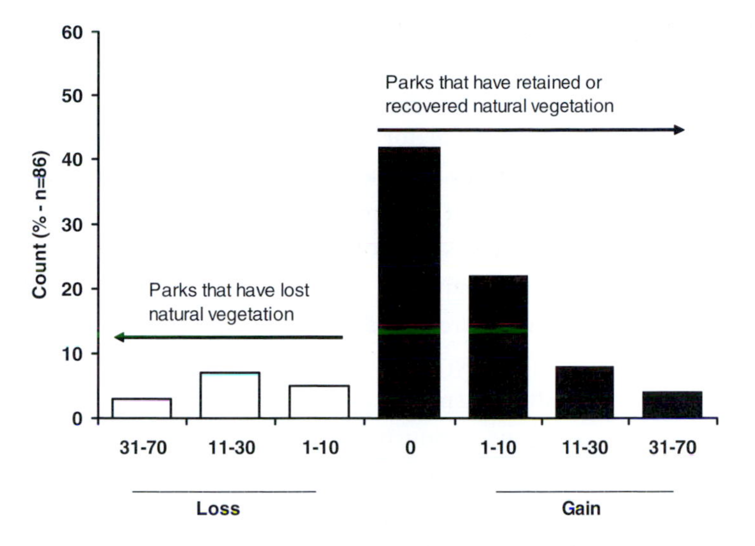

10.1 Change in the area of natural vegetation since establishment for 86 tropical parks. The majority of parks have either experienced no net clearing or have actually increased natural vegetation cover. Median park age is 23 years. (Reprinted with permission from Bruner *et al.*, 2001. Copyright 2001 AAAS.)

may require particular attention. So we need to consider the 'fine-filter' species approach in addition to the 'coarse-filter' landscape one.

There is still much we do not know about how we should manage natural and transformed landscapes for insect diversity, but some principles are beginning to emerge. In this chapter, we explore some of the management activities and opportunities that are currently being researched.

10.2 Importance of parks, reserves and remnant patches

10.2.1 Value of wilderness

Arguably, there is no wilderness left unsullied by human impact (McKibben, 1990). Nevertheless, wildlife parks in the tropics, where most biodiversity is situated, have been surprisingly effective in protecting the ecosystems and species within their borders (Bruner *et al.*, 2001). Despite being chronically underfunded, these parks have been especially effective in preventing land clearing, and remain a central component of conservation strategies. Some parks have actually increased natural vegetative cover (Figure 10.1). Caution is required, because there may be long-term ecological relaxation and loss of species not yet detected (the 'extinction debt' (Tilman *et al.*, 1994)), as has already occurred in plants in urban parks (Drayton and Primack, 1996).

The importance of wild areas is emphasized in the case of birdwing butterflies. The natural area network in Indonesia may be sufficient to maintain the full range of swallowtails, although this may not necessarily be the case in Papua New Guinea (Collins and Smith, 1995). Many other insect species, particularly in tropical forests, are being maintained in natural areas. As many of these are pollinators, it is mutually beneficial for both plants and insects to maintain these reserve areas. In the case of Japanese wood-boring longicorn beetles, it is necessary, for example, to maintain old-growth forests (Maeto *et al.*, 2002).

But current trends worldwide suggest that sustainability of a large wildland will only be achieved by bestowing what Janzen (1998) calls 'garden status to it, with all the planning, care, investment and harvest that implies'. This means combining biodiversity aspects with others, such as social, economic and spiritual, to develop a 'multiple aspects approach' to landscape planning (Fries *et al.*, 1998). This has been done in South Africa where networks of indigenous grassland linkages maintain indigenous butterfly diversity. When the indigenous linkages are about 250 m or more wide, they are effectively reserve areas within the managed landscape (Pryke and Samways, 2001) (see Figure 1.2).

Even though Vietnam butterfly species richness is high in mixed habitats of agriculture, scrub and cleared lands, it is clearly the undisturbed or very moderately disturbed forest habitats that are a top priority for conservation of rare, endemic butterflies (Spitzer *et al.*, 1997; Lien and Yuan, 2003). Fabricius *et al.* (2003), working in a fairly arid area, emphasized that reserve areas are key to conserving those species that decrease under heavily grazed and disturbed conditions. They found, for example, six new weevil species but these were confined to the reserve area. Clausnitzer (2003) emphasized the same point for dragonflies, which are habitat specialists in Kenyan forests. This underscores again, the importance of maintaining remnant areas that have suffered little human impact. This assumes of course that global climate change will not have too great an impact on insects that are confined to reserve areas. Yet there are no guarantees, as reserves subject to El Niño climatic events cannot necessarily maintain their insect fauna with the changing conditions (Samways, 1997c).

10.2.2 Maintaining natural disturbance

The edges of parks are often sharp ecotones, at least to the human eye. In the African context, and through the eyes of arthropods, the boundary is not necessarily so sharp, so long as the impact of domestic cattle outside the reserve is of a similar intensity to that inside from the impact of indigenous megaherbivores (Rivers-Moore and Samways, 1996). Indeed, it is essential, particularly for grasshoppers, to have some grazing and trampling outside the reserve to simulate conditions inside. Waterholes, as congregation points, provide a range of natural disturbances in the case of wild game inside the reserve and of simulated disturbance by cattle outside it (Samways and Kreuzinger, 2001) (Figure 10.2).

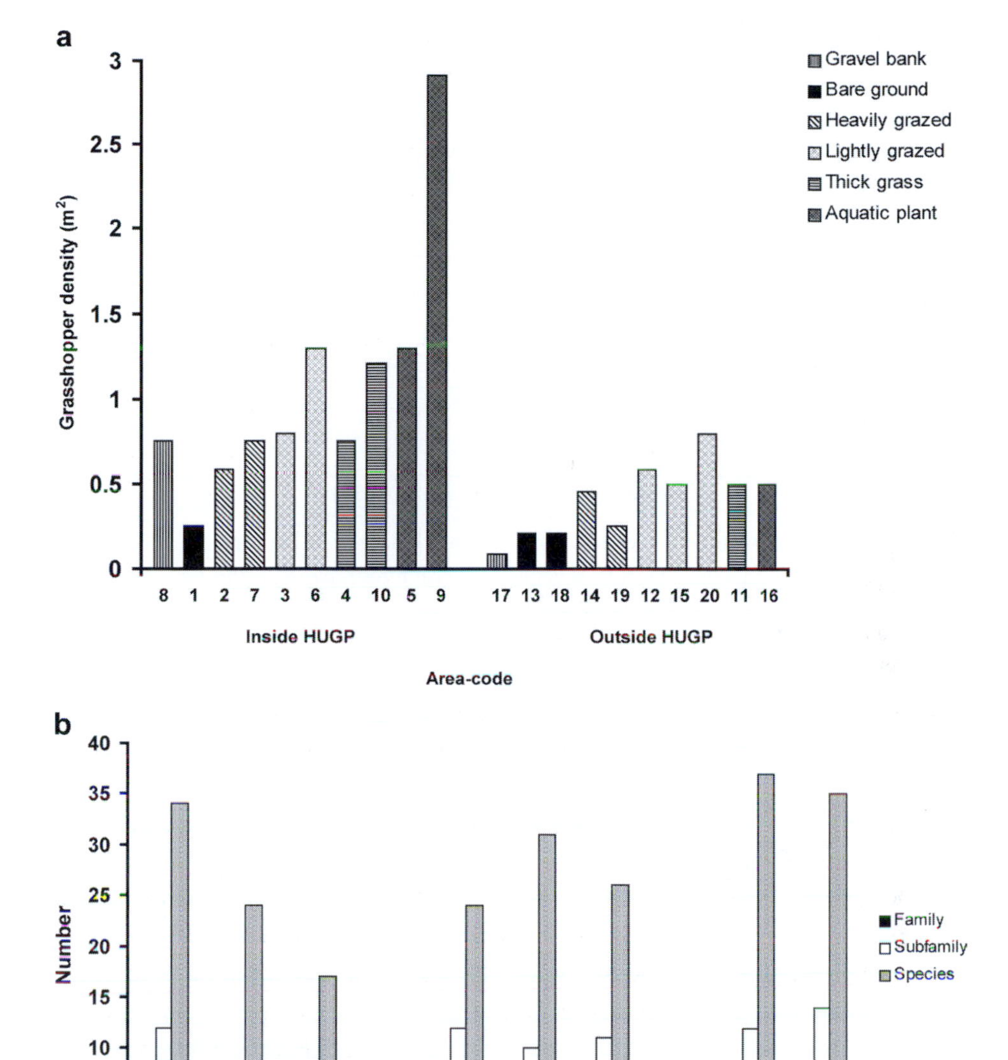

10.2 (a) Grasshopper densities inside/outside Hluhluwe–Umfolozi Game Park (HUGP) at 20 different sites associated with waterholes. Densities are lower outside where cattle replace indigenous game animals. (b) Abundance of families, subfamilies and species of grasshoppers at three waterholes inside and three outside HUGP show that species richness was overall the same inside versus outside the reserve. (Redrawn from Samways and Kreuzinger, 2001, with kind permission of Kluwer Academic Publishers.)

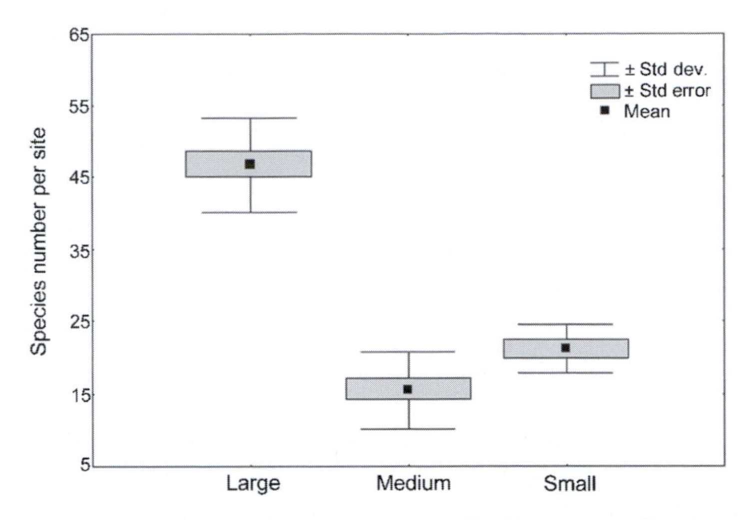

10.3 Number of leaf litter ant species per site (10 m^2) in large (438 km^2), medium (*c.* 43 km^2) and small (*c.* 1.5 km^2) plots in Malaysian tropical forests. (From Brühl *et al.*, 2003, with kind permission of Kluwer Academic Publishers.)

10.2.3 *Significance of remnant patch size*

Size of remnant patch is important, with butterfly species increasing proportionately with size of the remnant (Brown and Hutchings, 1997). Similar findings have come from Malaysian ants, with the alarming discovery that plots 43 km^2 in size are not large enough to retain the original ant diversity of extensive forest (Brühl *et al.*, 2003) (Figure 10.3). However, it is apparent that small remnant forest patches can play a significant role, at least for some neotropical butterflies (Horner-Divine *et al.*, 2003). Similarly, small fragments of Australian heath and woodland do not necessarily support a depauperate arthropod fauna of larger fragments, suggesting that habitat quality and not just size is important (Gibb and Hochuli, 1999). This emphasizes the importance of maintaining large areas of patch heterogeneity for maintaining a wide range of insect diversity. Indeed, insect species turnover across sites can be very high, as in wood-inhabiting insects in Sweden (Wikars, 2002). Insect diversity turnover across the landscape can be much higher than that of birds (French, 1999), indicating that a reserve network must be comprehensive enough to encompass all taxa.

This need for maintaining large areas with sufficient heterogeneity means that management must be sensitive to the different survival requirements of different species, as Swengel and Swengel (2001) have shown for prairie butterflies, and Lockwood and Sergeev (2000) for temperate grasshoppers. In the case of insects in small prairie remnants, this means burning each fourth year (Panzer, 2002), or for butterflies in grassland remnants in Japan by 'low frequency' of mowing (Kitahara and Sei, 2001).

Other forms of management might be necessary to keep indigenous sites 'natural'. Preservation or restoration of appropriate and constant hydrological conditions to prevent peat bogs being overwhelmed by closed forest is essential for some specialized Lepidoptera and Coleoptera (Spitzer *et al.*, 1999).

10.2.4 Landscape context

Edges of reserves and patches can influence the insect assemblages in the surrounding areas. Indigenous remnant forest patches, for example, can encourage butterflies 50–100 m into the disturbed matrix (Horner-Divine *et al.*, 2003), and conversely, alien pines can inhibit grasshoppers 30–50 m into natural grassland matrix (Samways and Moore, 1991). Moth assemblages, however, appear to expand from the remnant forest patch into the disturbed matrix by 1–1.5 km, possibly because nocturnal moths are more vagile than diurnal butterflies (Ricketts *et al.*, 2001). Shady field margins in El Salvador extend the period during which dung is suitable for colonization by dung beetles (Horgan, 2002).

As natural habitat becomes scarce and/or too disturbed, populations are lost from the system, with, in cases, some species colonizing secondary anthropogenically disturbed habitats. The Violet copper butterfly *Lycaena helle*, for example, now colonizes almost exclusively anthropogenic ephemeral habitats (abandoned moist meadows) in Central Europe (Fischer *et al.*, 1999). Indeed, rare species close to the extinction threshold in a particular landscape can show the greatest variance in the value of particular fragments that they inhabit (Lewis and Bryant, 2002). Terblanche *et al.* (2003) further emphasize this point by clearly showing the subpopulations of the threatened lycaenid *Chrysoritis aureus* are adapted to the slightly different vegetation habitat conditions.

Caution is needed when making generalizations about local extinction risks for different species and even for different higher taxa. Donaldson *et al.* (2002) for example, show that species richness of bees, flies and butterflies did not vary significantly between different-sized remnant South African shrubland patches, while the abundance of particular species of bees and scarab beetles did differ. Of concern is that these differences had an adverse effect on the seed set among certain plants.

Whether a species can colonize a disturbed habitat depends on the extent to which it has resources and conditions suitable to all aspects of its survival. At the level of community and at a spatial scale of only a few metres, this can be highly significant with substrate moisture and organic matter shaping invertebrate communities under rocks (Ferreira and Silva, 2001).

From the foregoing, it is clear that the value of reserves as source habitats depends not only on size but also on the quality of the habitat they contain, their context (i.e. relationship with surrounding area) (Wiens, 1995), their colonization history and their disturbance history (York, 1999), among other factors. Quality old-growth forest remnants can act as a source area for insect species, many of which may be undescribed (Winchester, 1997). On balance, it leads us to the self-evident First Premise of landscape management, that is, to maintain as much quality, natural, near-pristine land as possible. Let us now consider further principles.

10.3 Importance of landscape heterogeneity

10.3.1 Large-scale spatial considerations

The relationship between ecological productivity and species diversity changes with spatial scale (Chase and Leibold, 2002). Animal species diversity is highest in ponds with intermediate algae productivity at the local scale. At the regional scale however, it was the most productive groups of ponds (in different watersheds) that were the ones with highest diversity. This is because in productive watersheds, although ponds were relatively low in species richness, they shared few species. This emphasizes the importance of maintaining large-scale spatial heterogeneity.

Maintenance of a heterogeneous landscape, where a variety of habitats is conserved, is essential for the full array of local species, from bumblebees (Kells and Goulson, 2003) to dragonflies (Steytler and Samways, 1995). British bumblebees need a variety of field and forest boundary types, while South African dragonflies need a variety of lakeside vegetational structural types.

British (Greatorex-Davis *et al.*, 1994) and Irish (Mullen *et al.*, 2003) studies have emphasized that it is essential to maintain sunlight levels in pine plantation herbaceous linkages ('rides') to encourage both plant and invertebrate species richness. Mullen *et al.* (2003) suggest that, ideally, a forest road edge should contain a variety of plant structures including bare ground, low vegetation, tall vegetation, scrub and trees, thereby providing a range of microhabitats, emphasizing again the importance of maintaining landscape heterogeneity. Interestingly, Weibull *et al.* (2003), working on butterflies, carabids, rove beetles and spiders, concluded that species richness was highest on Swedish farms with a heterogeneous landscape, such as a mixture of arable fields, pastures and forests. The individual landscape patch therefore becomes an important player in biodiversity conservation. These findings contrast with those on carabid beetles in Germany where the species richness of comparable grassland areas was relatively low compared with the regional potential (Irmler and Hoernes, 2003). Clearly there is much we still need to find out as regards such spatially explicit situations, and while there may be some generalizations, there seem to be important differences with regard to taxa, type of landscape elements and regional history.

10.3.2 Temporal considerations

Overlaying such spatial concepts are temporal ones. At sites in Germany, butterfly species richness did not change during plant succession, although the species composition changed substantially (Steffan-Dewenter and Tscharntke, 1997). Life-history features of butterflies changed significantly from pioneer to early- and mid-successional fields. There was also decreasing body size and migrational ability, decreasing numbers of species hibernating as adults, decreasing numbers of generations and increasing larval stage duration, but, surprisingly, no increase in plant specialization. This leads to the conclusion that it is essential to maintain a range of successional habitats, as Paquin and Coderre (1997) showed for forest soil macroarthopods in Canada. This does require, though, adequate migration between like seral stages to avoid local extinction (Pagel and Payne, 1996). These local migrants tend to be species with intermediate dispersal ability and localized dispersal. However, in the case of the rare, Red Listed Swedish butterfly *Lopinga achine*, less than 0.1% of the eggs are dispersed beyond 500 m (Bergman and Landin, 2002). Such dispersal may not be random because, at least in the Meadow brown butterfly *Maniola jurtina*, when released at larger distances from their habitat, individuals used a non-random systematic search strategy in which they flew in loops around the release point and returned periodically to it (Conradt *et al.*, 2000) (Figure 10.4).

The importance of seral stages was also emphasized by Butterfield (1997) who showed the peak in carabid density and diversity coincided with the successional stage at which the ground flora in conifer forests was most diverse (in both species and structure) and the densities of other soil surface macro-invertebrates were also highest. This mirrors to some extent the situation in deciduous forests where an increase in size and structural diversity of plants led to an increase in insect diversity at all trophic levels (Southwood *et al.*, 1979).

Adequacy of connectance and hence migration between seral patches should be determined by monitoring the species with the lowest dispersal capacity (Nordén and Appelqvist, 2001) (Figure 10.5). Without migration and with increasing attrition of the size of remnant patches, it is the weaker dispersers that succumb first, as illustrated by insects of fragmented European heathland (Webb and Thomas, 1993; Assmann and Janssen, 1999).

10.3.3 *Maintaining 'stepping stone' habitat*

Behavioural activities have important implications for metapopulation models. A metapopulation (Hanski and Gilpin, 1997; Hanski, 1998) is a complex of connected populations, each of which would not persist on its own but is viable as a dynamic whole. They exist in various forms (Harrison and Hastings, 1996) (Figure 10.6), but metapopulation models assume random dispersal. Yet the results on the Meadow brown butterfly *Maniola jurtina* suggest that systemic searching behaviour could lead to a relatively large number of long-distance dispersers because the resulting 'stepping-stone dispersal' and the likely higher search efficiency should reduce the losses of individuals over longer net dispersal distances (Conradt *et al.*, 2000). This being the case, it is essential to maintain

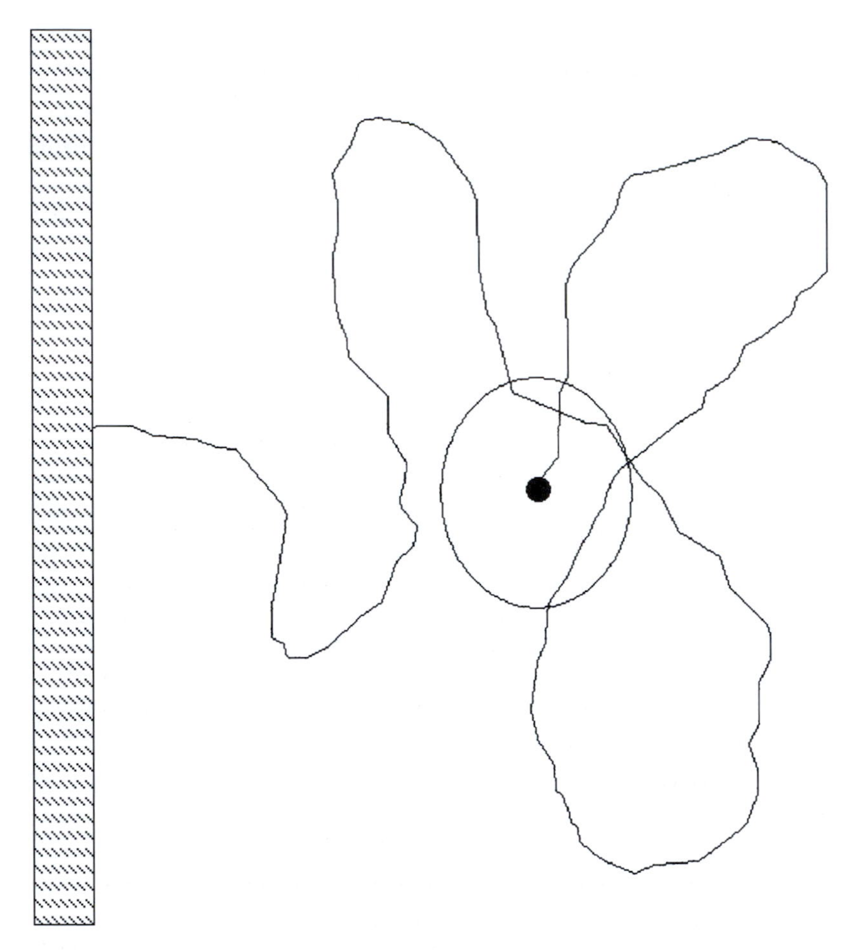

10.4 A type of searching flight behaviour in the Meadow brown butterfly *Maniola jurtina* when released (black spot). They flew in a series of loops returning close to their release site before finally heading for their breeding habitat, a grassy bank (shaded bar). (From Conradt *et al.*, 2000.)

reachable 'stepping stones' of quality habitat. For carabid beetles in Finland this has been emphasized by Koivula *et al.* (2002) who recommended that while harvesting timber, connectivity between mature forest stands should be maintained close to each other (a few tens of metres) and the matrix quality be improved. In Britain, Purse *et al.* (2003) suggest that management efforts to conserve the rare Southern damselfly *Coenagrion mercuriale* should maximize the likelihood that individuals recolonize sites naturally within 1–3 km of other populations. This can be aided by removing scrub boundaries to facilitate stepping-stone dispersal movements.

10.3.4 *Catering for extreme weather conditions*

When creating suitable patch habitats, one challenge is to be aware that habitat preference can vary according to extreme weather conditions. The bush

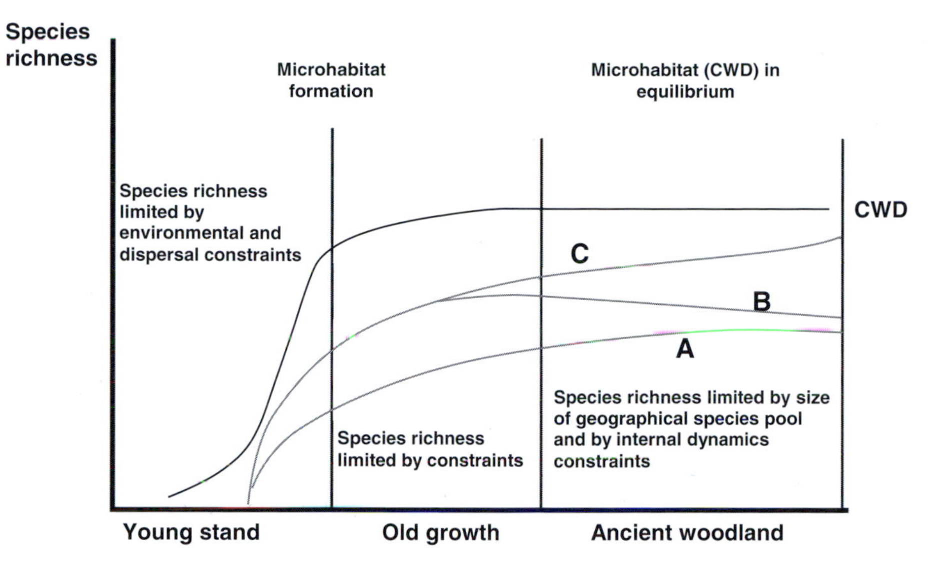

10.5 Alternative possible changes in species richness of saproxylic biota dependent on Coarse Woody Debris (CWD) in a forest stand after initial colonization of tree-less ground. Large amounts of dead wood are characteristic of old-growth forests (Kirby *et al.*, 1991). Microhabitat formation here means the time needed to create 'equilibrium levels', at which input rate and decay rate of CWD (bold line) are in balance in the stand. Alternative scenarios are possible. (A) Here the geographical species pool is severely constrained by isolation or other factors, and the forest stand will have relatively few species. (B) This alternative is a forest stand that is not isolated, and where biodiversity decreases in late succession due to competition, and local extinction. (C) In this situation ('Ecological continuity'), biodiversity increases slowly and continuously over a long time, but implies a large geographical species pool. Disturbance and patchy successional events may alter these curves. (From Nordén and Appelqvist, 2001, with kind permission of Kluwer Academic Publishers.)

cricket *Metrioptera bicolor*, for example, normally prefers low grassland in Sweden, but during extremely dry conditions moves to tall grassland (Kindvall, 1995). This returns us again to the importance of conserving as much heterogeneity in and between patches and landscapes as possible. How much exactly is difficult to answer, but it does reflect Leopold's (1949) 'Precautionary Principle', of keeping all the parts. For our purposes here, this maintenance of as much heterogeneity as possible, we may call the Second Premise of ecological landscaping for insect diversity conservation. Let us now consider the Third Premise.

10.4 Countryside-wide management

10.4.1 *The landscape mosaic*

Classical metapopulation theory views patches as discrete entities surrounded by a contrasting but homogenous matrix. In reality, this is not the case,

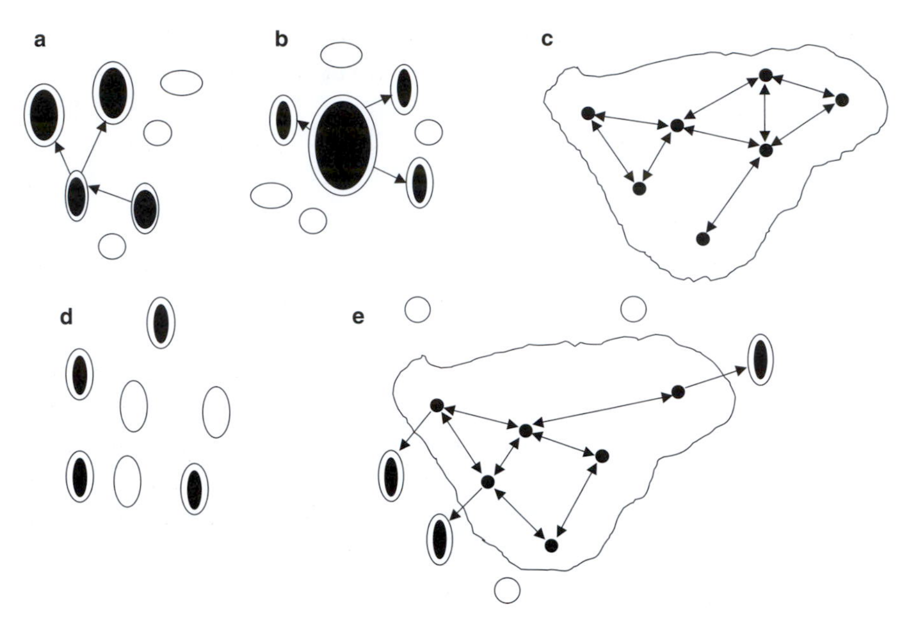

10.6 Different types of metapopulations. (a) 'Classical'; (b) mainland-island; (c) patchy population; (d) non-equilibrium; (e) intermediate case combining features of (a)–(d). Filled symbols, occupied habitat patches; unfilled symbols, vacant habitat patches; arrows, dispersal; outer black lines encompassing filled symbols (either groups or individual symbols), boundaries of local populations. (Redrawn from Harrison and Hastings, 1996.)

with most landscapes being a mosaic of adjacent patches with differing features (Wiens, 1995) (see Figure 5.4). The mosaic has various measurable parameters, such as context, contrast, patch orientation, perimeter: area ratio, boundary form, connectivity and size distribution of patches (Wiens *et al.*, 1993). Inevitably, this leads to an almost infinite number of landscape types, each with different and varying biotic communities which are determined by a combination of history and the characteristics of the organisms involved. Furthermore, there is an interactive process between landscape and organisms, each affecting the other over time. Succession is one example. Additionally, as organisms, even plants, move, the boundaries are permeable and dynamic.

As insects are small and plants larger, insect populations are generally affected by the boundaries at distances beyond what we as humans perceive as the vegetation boundary. This inevitably leads to an almost infinite number of landscape management options. From among these we seek some generalizations. One is that management activities should focus on the wider countryside rather than on single patches. Reserve areas, whether patches of boreal forests (Niemelä, 1997), temperate forests (Usher and Keiller, 1998), Afromontane forests (Horner-Divine *et al.*, 2003), or prairie (Panzer, 2003) can play an important role as source habitats, having a positive effect on the surrounding areas. And this applies to plants as well as insects (Smart *et al.*, 2002). However, a rider here is that small fragments are subject to edge impacts and attrition, and that tropical forest fragments of < 5000 ha are in serious and immediate danger of suffering receding edges (Gascon *et al.*, 2000). This is partly dependent on the harshness of the matrix, with 'high-quality' matrices contributing more, for example to ant diversity in the tropics (Perfecto and Vandermeer, 2002) than 'poor-quality' matrices. This follows Webb's (1989) findings that heathland remnants with 'softer edges', where the matrix was similar to the heathland patch, had richer invertebrate assemblages. These results and those of Magura *et al.* (2001) suggest a Third Premise of ecological landscaping and that is to reduce as far as possible the contrast between remnant patches and the disturbed adjacent patches or matrix. Dover *et al.* (1997) have emphasized this point and indicated that shelter from wind, which maximizes time for activities such as feeding, movement and basking, are an important component for butterfly survival in the wider countryside. Let us now move on to the Fourth Premise.

10.5 Importance of patch size relative to habitat quality

10.5.1 Patch size per se

Patch quality and size are interrelated in the sense that as remnant patches become much smaller, the impacts on the edges become proportionately great. Small size, however, may not always be negative, with small calcareous grassland patches being important for a range of butterflies, even threatened ones (Tscharntke *et al.*, 2002). Nevertheless, patch size per se does appear to have a negative effect on some species. The Silver-spotted skipper *Hesperia comma*, for example, has higher emigration and immigration rates (proportionate to patch size) when patches are small (< 0.07 ha) than when they are larger (0.33–5.66 ha) (Hill *et al.*, 1996). Similarly, the bush cricket *Metrioptera bicolor* regularly becomes locally extinct when habitat patches are less than 1 ha, and recolonization is infrequent when patches are more than about 100 m from each other (Kindvall and Ahlén, 1992) (Figure 5.6). Similarly, Biedermann (2000) produced some similarly striking results with the froghopper *Neophilaenus albipennis* in Germany, where patch area had a dramatic effect on population extinction rate (Figure 10.7). These results are partially predicted from Island Biography Theory (IBT) when the matrix between patches is essentially totally unsuitable as it would be for oceanic islands. Indeed, when the small oceanic island of Cousine in the Seychelles was provided with open water habitat, the establishment of dragonflies was that predicted by IBT (Samways, 1998a) (Figure 10.8). This is an extreme case, but illustrates that movement and establishment are major factors for pioneering populations, as suggested by metapopulation ecological theory (Hanski 1999).

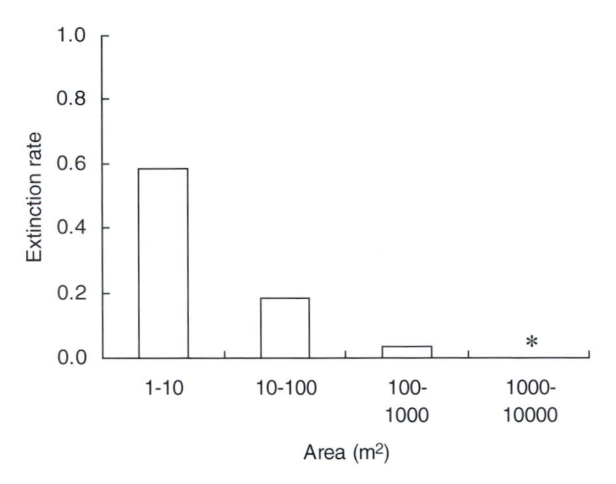

10.7 Effect of patch area on extinction rate of 90 occupied froghopper *Neophilaenus albipennis* habitat patches between 1994 and 1995. (Asterisk means no population turnover observed.) (From Biedermann, 2000, with kind permission of Kluwer Academic Publishers.)

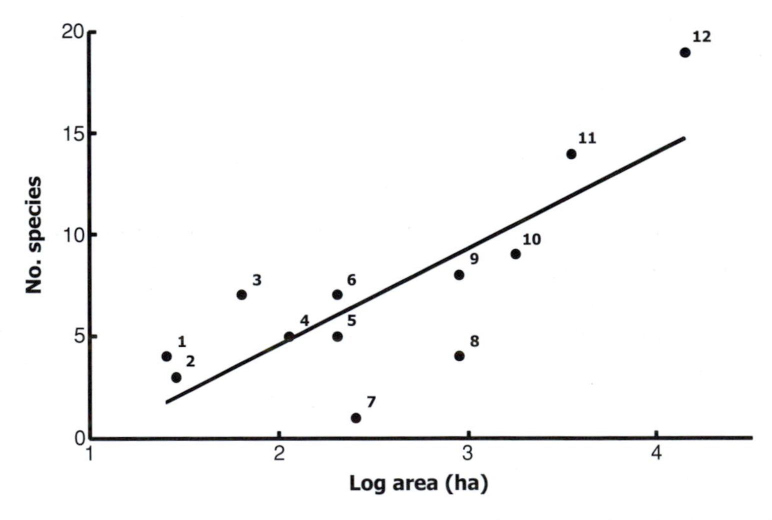

10.8 Dragonfly (Odonata) species richness on 12 Seychelles islands. Island No. 1 (Cousine Island) was formerly waterless, but with the creation of drinking and bathing pools for birds and tortoises, dragonfly species began to arrive and reach an asymptote predicted by Island Biogeography Theory. (From Samways, 1998a.)

Across terrestrial landscapes, the matrix may not be completely hostile, and that metapopulation dynamics and island effects are not necessarily exclusive (Clarke *et al.*, 1997; Thomas *et al.*, 1998b). Nevertheless, the level of unsuitability of a matrix depends on the focal species. For the Glanville fritillary *Melitaea cinxia* in Finland, conservation of an isolated population was more successful in

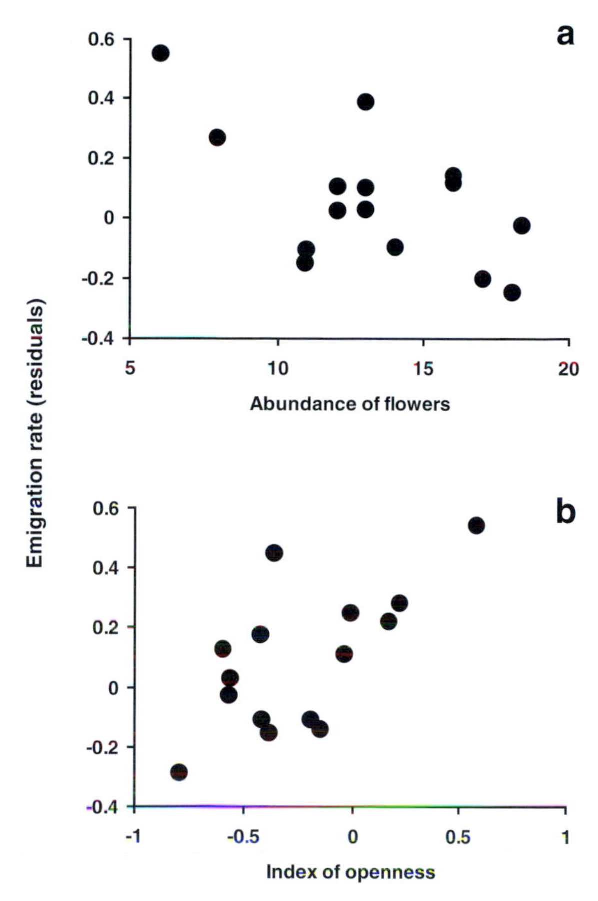

10.9 The fraction of emigrants of the Glanville fritillary butterfly *Melitaea cinxia* varied much among the 16 patches into which the butterflies were released. In males (shown here), emigration rate decreased with increasing abundance of flowers (as a food, nectar source) illustrating the importance of habitat quality for maintaining the population (a), and emigration increased with the fraction of open field (which is windy and not preferred) as patch boundary (b). (From Kuussaari *et al.*, 1996, with kind permission of Blackwell Publishing.)

an area with physical barriers to migration than in an open landscape (Kuussaari *et al.*, 1996) (Figure 10.9). This contrasts with some of the findings outlined in Section 10.4.

10.5.2 *Patch quality per se*

Habitat patches are rarely uniform in character, and can be variable in quality. Large patches may act as metapopulation units in their own right and small patches simply as locations with aggregations (units of patchy populations) (Sutcliffe *et al.*, 1997a). Patches, in turn, are subject to various

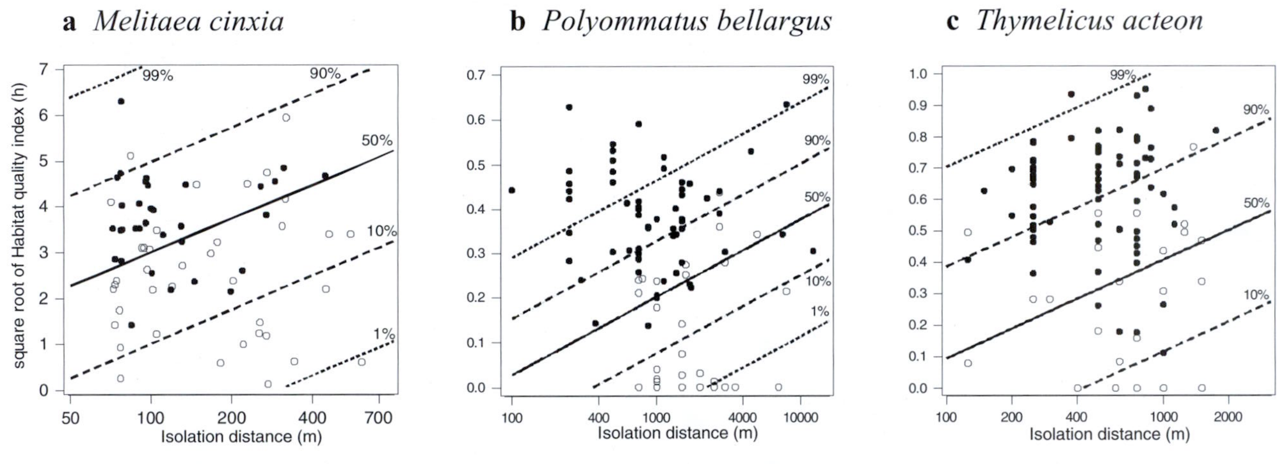

10.10 The distribution of occupied (black circles) and unoccupied (white circles) habitat patches for three species of British grassland butterflies (Glanville fritillary, Adonis blue and Lulworth skipper) in relation to the patch quality (h) and isolation of the patch. High quality, nearby patches are particularly important for the butterflies. (From J. A. Thomas *et al.*, 2001.)

stochastic environmental effects (Sutcliffe *et al.*, 1997b), and so quality refers not just to when environmental conditions are optimal, but also when they are adverse.

For three species of British butterfly, variation in habitat quality explained which patches supported a species' population two to three times better than site isolation, emphasizing that habitat quality is the third parameter in metapopulation dynamics (in addition to habitat or patch area and isolation) (J. A. Thomas *et al.*, 2001) (Figure 10.10). These three factors (habitat quality, patch size and isolation) produced a correct classification of > 88% of the Large heath butterfly *Coenonympha tullia* sites (Dennis and Eales, 1997). However, small patches of high habitat quality are not necessarily without value and, as mentioned earlier, can act as stepping-stones, as with the Ringlet butterfly *Aphantopus hyperantus* (Sutcliffe *et al.*, 1997a,b).

In certain landscapes, such as eucalypt woodland remnants on farms in Western Australia, small, high-quality habitat patches can play an important role for sustaining an ant, a scorpion and two termite species (Abensperg-Traun and Smith, 1999). Single species (fine-filter) management is then dependent on a knowledge of the habitat requirements of particular species, which may include some disturbance, such as maintaining short and open grassland for the threatened lycaenid *Aloeides dentatis dentatis* in South Africa (Deutschlander and Bredenkamp, 1999).

Management of a particular landscape normally has to consider a range of species (the coarse filter). This has been emphasized by Collinge *et al.* (2003) who showed that grassland type was the primary determinant of species richness of grassland butterflies in Colorado, USA, and that habitat quality was a secondary factor (Figure 10.11). However, in the case of prairie butterflies, living in indigenous remnants, 40% depended entirely on habitats unmodified by humans (Panzer *et al.*, 1995).

This brings us to the Fourth Premise of ecological management for maintenance of insect diversity, which is interrelated with the other premises. Outside formal reserves, it is necessary to maintain as much undisturbed habitat as possible. However, this implies building in management practices that simulate natural disturbance without imposing anthropogenic disturbance that decreases habitat quality. Particular species may require appropriate management actions, but these can be chosen in a way so that both threatened specialists and more abundant generalists can benefit. Such management may involve multiple approaches (Swengel and Swengel, 2001), which also need to be sensitive to other non-insect taxa. We see this with prairie management, where the fine-filter specialist butterfly considerations, as well as coarse-filter total butterfly assemblage approaches, need also to cater for fire-sensitive snails (Nekola, 2002).

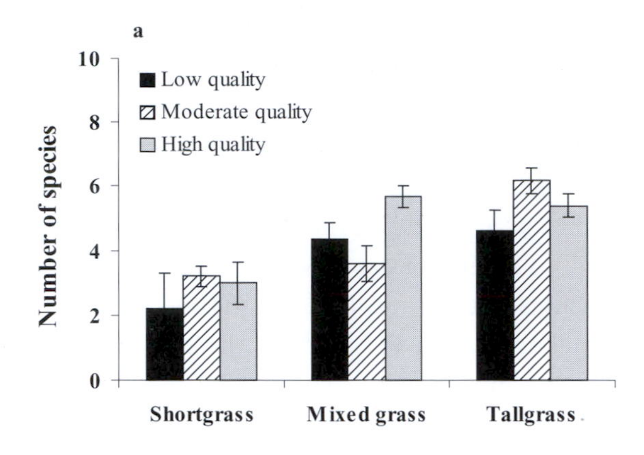

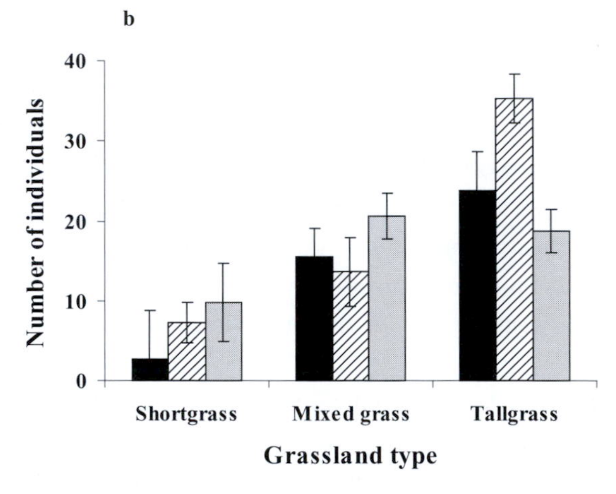

10.11 North American prairie butterfly species richness (a) and abundance (b) in relation to three different levels of habitat quality in three grassland types, and illustrating the importance of conserving a range of habitat types, preferably high in quality, for maintaining the full butterfly assemblage. (From Collinge *et al.*, 2003, with kind permission of Blackwell Publishing.)

10.6 Simulating natural conditions and traditional practices

10.6.1 Natural succession

Large remnant patches (as quality reserves or only moderately disturbed landscapes) often, but not always, require some form of management. This is because natural disturbance factors may be excluded. The reason for having continuing management of a site is because it changes, usually because of succession (Morris, 1991). This change may not always be adverse, at least for insects that live in senile trees and fallen logs. Indeed, saproxylic species form a high proportion of the tropical forest insect fauna, and are often specialized and with poor powers of dispersal (Stork, 1987), pointing to the importance of having sites containing mature trees for maintaining the diversity of saproxylic organisms

(Kaila *et al.*, 1997; Nilsson and Baranowski, 1997). Similarly, diverse rare species such as moths of *Calluna vulgaris* (Haysom and Coulson, 1998) and stem borers of fynbos plants (Wright and Samways, 1998, 1999) require long periods without any intrusive management or major natural disturbance events.

10.6.2 *Managing to simulate natural conditions*

Sometimes it is difficult to ascertain exactly what are the 'natural' conditions, especially where there has been a long history of human disturbance. In central Europe, extensively grazed or mown calcareous grasslands and their successional stages ('old fallows') have declined drastically over the last few decades, and yet are an important reservoir for threatened butterflies (Balmer and Erhardt, 2000). Also, in central Europe the Chestnut heath butterfly *Coenonympha glycerion* inhabits different types of relatively intensively grazed pastures, while in the Carpathians it prefers moderately or sporadically grazed areas in the vicinity of woodland edges. The conclusion is that this butterfly occupies intensively grazed pastures in central Europe not because it prefers such habitats, but because more suitable habitats have long since vanished (Elligsen *et al.*, 1998). This sort of ecological plasticity greatly aids management decisions. But not all species are necessarily so tolerant, with rare butterflies often with narrow tolerances (New *et al.*, 1995).

Where these specialists occur in fragments, often some appropriate management is required. This is illustrated in prairie butterflies, where most specialists showed significantly increased numbers associated with less frequent and/or less intrusive management. However, leaving habitat entirely unmanaged was rarely optimal. Single occasional wildfires were typically more favourable for specialist abundance than rotational burning, which often produced very low numbers (Swengel, 1998). This is interesting, because strict rotational burning, based on equally spaced times of burns or mowing, may not always be the best strategy. Similar conclusions have been reached for the conservation of rare antelope in the Kruger National Park, where small patch fires that provide green grazing over extended periods have been recommended (Grant and Van der Walt, 2000), a method which benefits African grasshoppers (Chambers and Samways, 1998). To maintain the full array of arthropods in mixed-wood forest, simulating natural fires was found to be preferable even to sensitive forest harvesting, as fires lead to stands of different physical structure from those characteristic of a landscape dominated by forestry practices (Spence *et al.*, 1997).

This adaptive management approach appears also to be appropriate for prairie butterflies because no single management type was clearly favourable for all specialists of a given habitat. Swengel (1998) concluded that for conserving specialist butterflies, both consistency of management type *within* site and deliberate differences in management type *among* sites of *like* habitat is desirable. This returns us again to the importance of maintaining landscape heterogeneity at various spatial scales without allowing degradation of the landscape, as has been demonstrated for butterflies in Borneo (Hamer *et al.*, 2003). Constant monitoring of effectiveness of such adaptive management is essential. Current bracken (*Pteridium aquilinum*) management regimes in Britain, for example, may not benefit all fritillary species at the site (Joy, 1997).

10.6.3 Adaptive and rotational management

Under African conditions, the landscape is naturally variously disturbed as megaherbivores move across the savanna and congregate at waterholes resulting in a rich orthopteran assemblage (Samways and Kreuzinger, 2001). Domestic livestock to some extent mimic this impact and maintain the orthopteran assemblage, albeit at a lower level of general abundance (Rivers-Moore and Samways, 1996). While adaptive management would seem to be the ideal situation, it is not so feasible with livestock as with game or burning, and so rotational grazing, at appropriate stocking rates, is the best compromise for creating a heterogeneous, near-natural vegetational landscape and grasshopper assemblage (Gebeyehu and Samways, 2003). Such a rotational approach produces a mosaic of patches ranging from bare ground to tall grasses, provides a structurally and microclimatically complex vegetational pattern and ensures a diversity of successional stages available at one time suitable to a range of species from rare specialists to widespread generalists (Chambers and Samways, 1998).

This has strong parallels with the situation for prairie butterflies (Swengel and Swengel, 2001) and with other prairie insects which benefit from patchy burns and having refugia from which to disperse (Panzer, 2003). It also emphasizes the point made by Morris (1981) that no single management regime will suit all species. This has an evolutionary basis which cannot be ignored, and, as Swengel (2001) concludes in her overview, insect responses to fire and other management activities can be interpreted on the basis of biological mechanisms and traits that do not assort by ecosystem type. This may not be the situation for all trophic levels. Steffan-Dewenter and Leschke (2003) found that the higher trophic groups of bees, eumenid wasps and sphecid wasps did not differ between grazed, mown or abandoned management types. In summary, the evidence on balance nevertheless suggests the importance of retaining considerable spatiotemporal variation among sites of the same ecosystem type in the frequency of fire, other management activities and natural events, without which leads to a paucity of countryside-wide insect diversity.

10.6.4 Regulation of management actions

These principles are well and good but it still leaves the question of where, how and how often one should apply management activities. It also asks: for whom? Arguably, in this situation, the fine-filter approach here has priority, on a local basis, over the coarse-filter approach, with focal rarities, such as Red Listed lycaenid butterflies, receiving particular focus at their population nodes (e.g. the very rare *Orachrysops* species in South Africa (Edge, 2002; Lu and Samways, 2002)), without which they may well go extinct. Once such highly threatened species have been considered, then a whole landscape approach can be taken. This has received considerable attention in recent years with emphasis, for example, on less intensive grazing or a return to traditional methods to conserve species richness and endemicity of Spanish (Verdú *et al.*, 2000) and Italian (Barbero *et al.*, 1999) dung beetles, Ukrainian (Elligsen *et al.*, 1997) and German (Dolek and Geyer, 1997) butterflies, as well as Australian spiders (Zulka *et al.*, 1997).

For a variety of German insects, Kruess and Tscharntke (2002) concluded that a mosaic of extensively grazed and ungrazed grasslands, with resumption of grazing after a few years to prevent succession into woody habitats, may be the best strategy with which to maximize biological diversity and the strength of trophic interactions, which appear to be disrupted by disturbance through regular grazing. Similarly, in Britain, maximizing hemipteran species richness may be achieved by using a grazing regime that maintains a mosaic of dwarf shrub and grass cover (Hartley *et al.*, 2003).

Management activities have an impact that is more than simply maintaining the ecological status quo. Inevitably they play a role in 'contemporary evolution' (Stockwell *et al.*, 2003). This potential for management-influenced evolution is emphasized by the effect that burning management has on ants in Australia (Vanderwoude *et al.*, 1997). Not only does burning have a differential effect on different ant functional groups (Figure 10.12(a)) but, interestingly, also has a differential effect depending on which biogeographic groups the ants are in. This effect, however, was on proportional abundance rather than on proportional species (Figure 10.12(b)).

Landscapes across the world have been variously disturbed from recent times to many millennia, and an appreciation of the type, intensity and extent of those disturbances underpins conservation decisions. This leads to the Fifth Premise of landscape management. Wherever possible, insect diversity conservation should simulate the natural condition and disturbance. On the one hand this recognizes natural impacts over deep time, as with the megaherbivores of the African savanna. On the other hand, there needs to be recognition of traditional agricultural practices prior to intensive mechanization and agrochemical input. This provides a conservation platform that may not conserve the original biota but nevertheless maintains a cross section of insect diversity which otherwise would decrease with intensive agriculture (Firbank *et al.*, 1994; Feber *et al.*, 1997). Various spatial methods are available to improve conservation in cereal fields for example. Conservation headlands, where the outermost 6 m of the cereal crop is left untreated with pesticides, were originally proposed to enhance gamebirds (Sotherton, 1992) but are now known to have a positive effect on butterflies

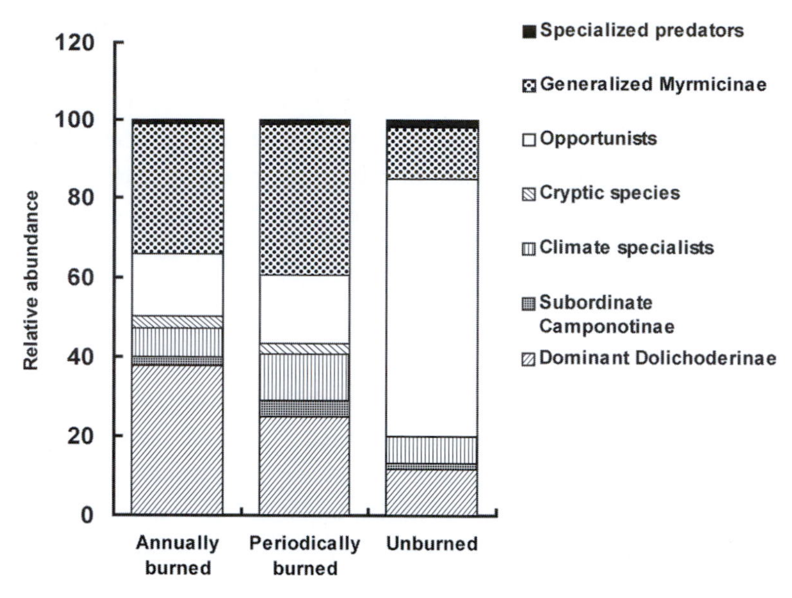

10.12(a) The way in which the Australian landscape is managed by burns affects the abundance of ant functional groups. Specialists, in particular, differentially benefit from patches that are periodically burned. (Redrawn from Vanderwoude *et al.*, 1997, with permission of CSIRO Publishing, Melbourne, Australia.)

(Dover *et al.*, 1997). An allied approach, where borders of farmland were composed of more dicotyledonous plants, rather than being grassy, encouraged bumblebees (Bäckman and Tiainen, 2002).

Kirby (1992) provides an excellent overview and practical recommendations for insect conservation through practical modification of the landscape.

10.7 Corridors

10.7.1 Perceptions of corridors

Corridors or linkages are continuous linear strips of habitat that are meant to connect and therefore improve the chances of survival of otherwise isolated populations. There has been considerable debate as to their merits and disadvantages (summarized by Bennett, 1999), but also misunderstandings arising from different perspectives (debates in Beier and Noss, 1998; Hess and Fischer, 2001). The roles of corridors derive from six ecological functions: habitat, conduit, filter, barrier, source and sink (Figure 10.13). When viewing these roles from an insect diversity viewpoint, there are two additional considerations. Insects are generally small, and they are speciose. This means corridors for insects may not be suitable for other, larger animals, and even for other insects with completely different ecologies and behaviours. Generalizations thus become difficult.

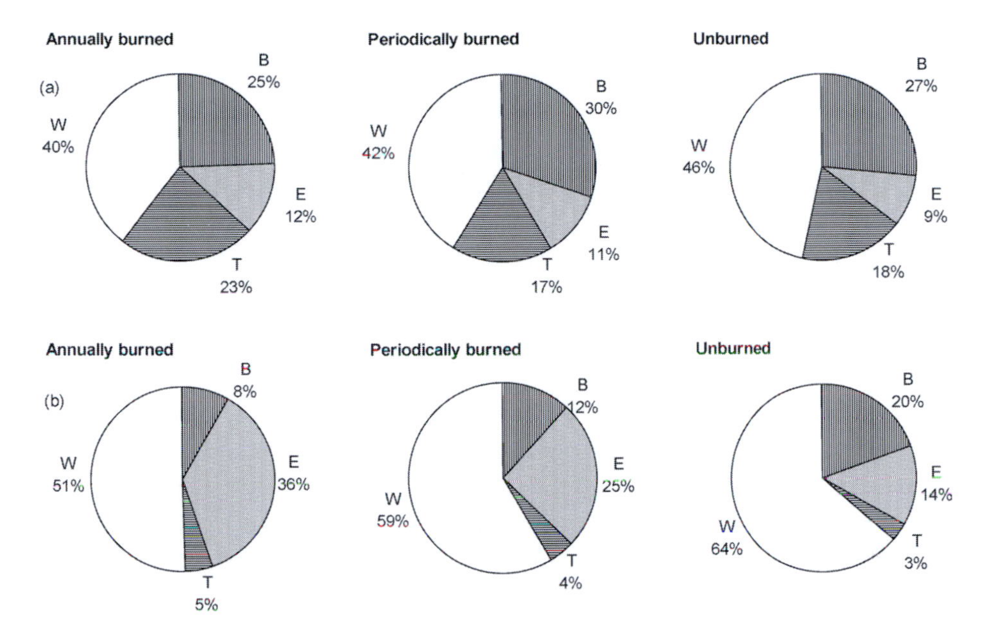

10.12(b) Type of burning management of the Australian landscape has little effect on the proportions of species of ants from different biogeographical groups (a) but it does on the proportional abundance of these groups (b) W, widespread; B, Bassian; T, Torresian; E, Eyrean. (Reproduced from Vanderwoude *et al.*, 1997, with permission of CSIRO Publishing, Melbourne, Australia.)

10.7.2 Corridors as habitats

Corridors can be residual, i.e. remnants of natural conditions and can function as such as habitats, as in the case of butterflies in South African grasslands (Pryke and Samways, 2001). Corridors alternatively may be introduced habitats. Planted as well as preserved rare fig trees can promote fig–fig wasp mutualisms (Mawdsley *et al.*, 1998). This is a stepping-stone effect which benefits, for example, the North American butterfly *Junonia coenia* (Haddad, 2000). Amenity areas such as rides in English woodland can also benefit species, such as the Wood white butterfly *Leptidea sinapsis* (Warren, 1985). Quality of habitat for butterflies in corridors is essential if they are to be source habitats. For African butterflies this includes nectar, oviposition, drinking and sunbasking sites (Pryke and Samways, 2001) (see Figure 1.2).

Corridors in the broadest sense may not necessarily be continuous habitat. Dennis (1997) suggests that a closely spaced network of small Scottish woods, in particular in a matrix of hedgerows or shelterbelts, could generate mutual shelter, so that under varied and seasonal weather conditions, some sites will always mitigate severe and adverse wind-mediated edge effects. In this way, summer canopy or winter litter stages of arthropod populations will be maintained in at least a fraction of the available woods. In contrast, when trees are isolated and only colonized (by a microlepidopteran leaf miner) from the leaf litter, local population extinction occurred (Connor *et al.*, 1994). The important point as regards management of phytophagous insects on beech trees is that several conditions must be met, such as reduced draughtiness, dense leaf litter and connectivity through hedgerows and lines of trees, to suit the life history strategies of the local species, so as to avoid local extinction (Dennis, 1997).

10.7.3 *Movement corridors*

Corridors can act as conduits for insect movement (Sutcliffe and Thomas, 1996; Haddad, 1999). Evidence from European carabid beetles (Vermeulen, 1994) and African butterflies (Pryke and Samways, 2003), however, suggests that corridors can be conduits even when the habitat quality is low, with individuals actively searching, and forced to do so linearly, for suitable habitat. The African butterflies fly fastest along narrow, highly disturbed corridors (Pryke and Samways, 2001). Generally however, for the community in general, a corridor is a differential filter, more suitable (either as habitat or conduit) for certain species over others. This may be due to various structural features of the landscape to which different butterfly species are variously sensitive (Wood and Samways, 1991; Dover and Fry, 2001). This point has also been emphasized by Rosenberg *et al.* (1997) in terms of streamside riparian areas which, as wildlife habitat, are not necessarily movement corridors.

Movement in some insects, such as the bush cricket *Metrioptera roeseli*, can be fairly straight along corridors to find suitable habitat (Berggren *et al.*, 2002). Nevertheless, when they walk and disperse like this they do so more slowly than they would over a hostile matrix. The model of Travis and Dytham (1999) also predicts individuals in corridors evolving much lower dispersal rates than those in the mainland populations, especially within long, narrow corridors.

When corridors are both suitable habitat and conduits, as with moths in Finland, the results have a positive effect on dispersal and population persistence (Mönkkönen and Mutanen, 2003).

For the highly threatened Swedish moth *Dysauxes ancilla*, it was not the centre of the corridors that were the suitable habitat but the edges (Betzholtz, 2002) (Figure 10.14). Similarly, Norwegian butterfly movement between patches is more likely over suitable habitat than linear distance per se, i.e. the 'least-cost path' model was a better predictor of butterfly movement than Euclidean distance (Sutcliffe *et al.*, 2003). Similar findings come from the Speckled wood butterfly *Pararge aegeria* in Belgium (Chardon *et al.*, 2003). Thus the concept of a differential filter (Ingham and Samways, 1996) runs across the corridor as well as along it. This begs the question of how wide should a corridor be? This depends on the conservation goal and the focal species. For an African butterfly assemblage this is about 250 m when the corridor is for movement as well as being a habitat source (Pryke and Samways, 2001). Interestingly, Hill (1995) found a similar

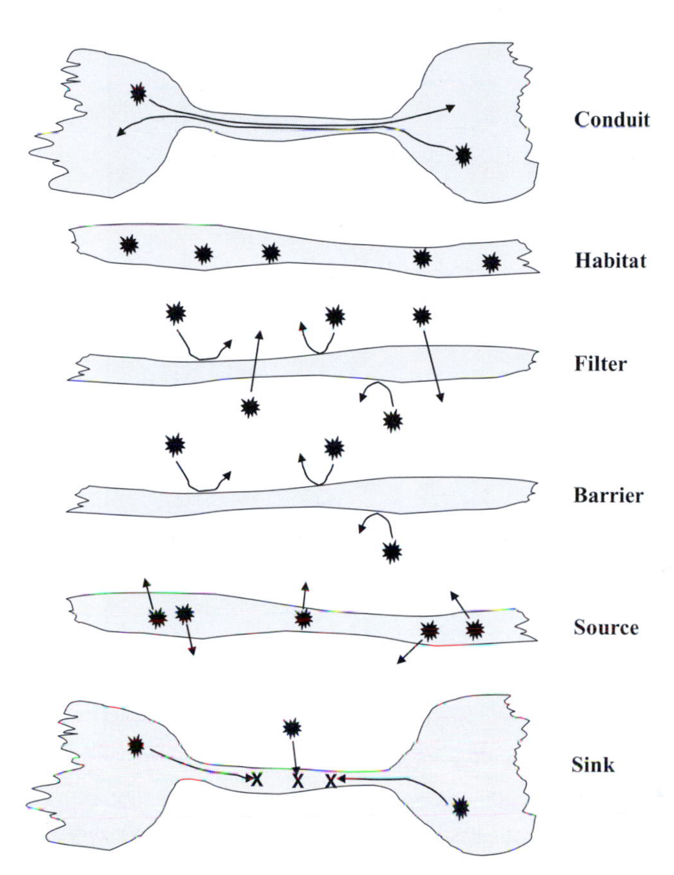

10.13 Corridor functions. Conduit: organisms pass from one place to another, but do not reside within the corridor. Habitat: organisms can survive and reproduce in the corridor. Filter: some organisms or material can pass through the corridor; others cannot. Barrier: organisms or material cannot cross the corridor. Source: organisms or material emanate from the corridor. Sink: organisms or material enter the corridor and are destroyed. (From Hess and Fischer, 2001.)

figure of 200 m for dung beetles in tropical Australian forest. In the agricultural context, and at least for some common insects, even small corridors can play a valuable role. Conservation headlands (the selectively and sensitively sprayed outer margins of crop fields) (Dover, 1997) and field margins (Feber *et al.*, 1996) can improve conditions for butterflies, for example, across the wider countryside (Feber and Smith, 1995).

Johannesen *et al.* (1999) caution that there may be discrepancies between short-term movement events as are normally observed (i.e. over ecological time), when in fact for two European grasshoppers what may be important in real conservation terms in the long term are medium and particularly long-distance dispersal events. This emphasizes strongly the importance of a large enough

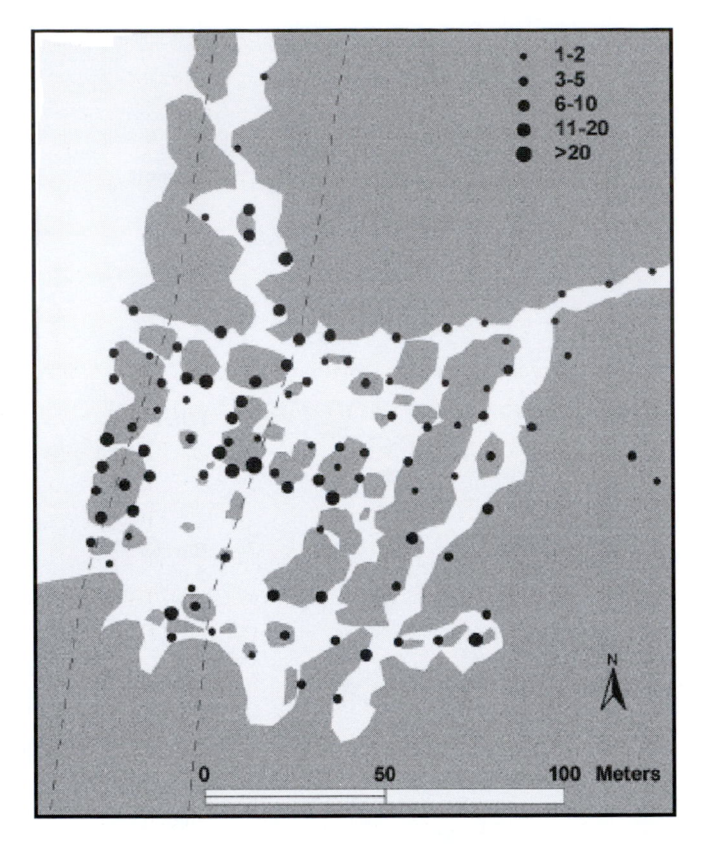

10.14 Pattern of positions of individuals of the threatened moth *Dysauxes ancilla*, on the Baltic island of Öland, illustrating a preference for edges of dry meadow corridors (white) and deciduous woodland (grey). Dot size indicates number of observations. (From Betzholtz, 2002, with kind permission of Kluwer Academic Publishers.)

network of suitably connected patches for long-term survival of a species, such as the threatened European butterfly *Proclossiana eunomia* (Néve *et al.*, 1996).

10.7.4 *Spatial scale of corridors*

Consideration of spatial scale (and proportionately great edge effects in very narrow corridors) may be why Collinge's (2000) experiments which involved minute, experimental corridors, 1 m wide, did not significantly influence overall rate of patch colonization, but did slightly increase the probability of colonization by less vagile species. Her results clearly showed that the function of small corridors may depend upon species characteristics, landscape context, patch size and environmental variation. These findings also underscore the debate between Haddad *et al.* (2000) and Noss and Beier (2000) that not only do we need to uncover general principles that predict behavioural and population responses to corridors across species and landscapes, as through experiment, but we also need to explore real-life conservation priorities across whole landscapes and even regions (Erwin, 1991), so that ecological processes and evolutionary lineages can survive into the future (Fagan *et al.*, 2001). Our Sixth Premise of management action then becomes: connect like habitat patches as much as possible with corridors as wide as can be, and of the highest quality. But as Mönkkönen and Mutanen (2003) emphasize, the positive effects of corridors should not be used to justify more habitat destruction.

10.8 Landscape management in urban areas

There is generally a decrease in species richness of insect assemblages closest to city centres (Davis, 1978; McGeoch and Chown, 1997). Nevertheless, urban commons, parks, leisure gardens, private gardens and cemeteries provide refugia for wildlife (Gilbert, 1989), especially when linked by corridors or greenways (Smith and Hellmund, 1993; Forman, 1995). These refugia may or may not be left to undergo succession. The 'brownfield' sites in Britain (sites fundamentally altered by humans) are usually at various stages of natural change, where the vegetation changes reflect on the insect assemblages, such as carabid beetles. As the bare ground becomes herbaceous and then grassy with age, the beetle species tend, among other things, to be larger and with more carnivores, yet not necessarily rare species (Small *et al.*, 2003). This emphasizes the point made earlier with rural habitats, that to maintain as much insect diversity as possible, we should maximize habitat heterogeneity.

It is the patches of semi-natural habitats that can lessen the impact of urban cover on species richness. For British butterflies in and around Manchester, adult butterflies are opportunistic nectar users, and nectar sources are more widely spread and thus less influenced by urban development than are specific butterfly hostplants. This may be why some species are able to locate and breed on tiny areas of host plant within extensively built-up areas (Hardy and Dennis, 1999). Indeed, 'green areas', particularly of relatively undisturbed natural vegetation, can be important for maintaining local insect diversity in various countries (Frankie and Ehler, 1978; Samways, 1989b; Owen, 1991; Clark and Samways, 1997; Rösch *et al.*, 2001). A cautionary note however, is that for a South African lepidopteran assemblage abundances and/or densities were lower in the city centre (Rösch *et al.*, 2001), supporting Californian results (Blair, 1999).

In terms of management, the task in the urban context is similar to that in the intensive agricultural one. However, landscape contrast, particularly where roads are severe borders to habitat patches (Samways *et al.*, 1997), is particularly great in the urban context. This makes urban patches often relatively discrete islands. However, Collinge *et al.* (2003) showed that it was not the urban context that influenced remnant urban patches in Colorado, but the type and quality of the vegetation within the patches.

Metropolitan Open Space Systems are coming into being, as in Durban, South Africa (D'MOSS), which is an ecologically enhanced network of linkages and patches. However, using pitfall traps to assess the D'MOSS project, Whitmore *et al.*

(2002) concluded that movement of invertebrates was not particularly enhanced along corridors of semi-natural vegetation nor that large areas of urban open space increased species richness. These findings contrast with those of Brown and Freitas (2002), working on butterflies in São Paulo, Brazil, and concluded that connectivity of urban forest remnants, as well as presence of water and mixed vegetation including flowers, were important determinants of butterfly diversity.

Beneš *et al.* (2003) found that in central Europe, limestone quarries can play an important role as refuges for xerophilous butterflies, provided that there is maximum habitat heterogeneity within the quarries and that xerophilous sites are preserved and adequately managed.

Clearly, our knowledge of urban insect diversity management is rudimentary, and among the first challenges is to establish appropriate indigenous vegetation (Cilliers *et al.*, 1999). Social factors are also important in the urban context (Bradshaw *et al.*, 1986). Ponds in particular have huge educational value and have been promoted both in Japan (Primack *et al.*, 2000) and in South Africa (Suh and Samways, 2001), while in Britain there is a call to cluster ponds and protect them from intensive land use so as to mimic the natural processes of pond creation (Vines, 1999).

A Seventh Premise of landscape management for insect diversity conservation is that the urban environment has some similarities to the intensive agricultural one in terms of landscape ecology principles (patch size, quality, isolation etc.) and conservation planning. However, a difference between the agricultural and urban situations is that immediate human social and educational aspects become particularly important in urban planning. A challenging corollary is that in urban green patches, children are being exposed in general terms to remnant populations of largely 'weedy', habitat-tolerant or vagile species, without regard to more specialist species of extensive wild areas. Nevertheless, such exposure is better than none at all.

10.9 Summary

Wildlife parks play an important role in maintaining biodiversity, particularly rare and specialized species of later vegetational successional stages. Some areas, such as the African savanna, depend on large indigenous mammals to maintain the natural dynamic landscape. Although large reserves with a variety of landscapes are the ideal, some small reserves and patches can nevertheless be important for maintaining certain insect assemblages. However, we need to be sensitive to the range of species that these remnant wildlands contain. We need to apply appropriate management to these remnants, especially as they may be quality source habitats supplying the surrounding disturbed matrix.

As insect diversity changes with spatial and temporal scale, it is essential to maintain heterogeneity of habitats, including a range of successional habitats. Ideally, insects should be able to move from similar habitats and contemporary seral stages to avoid local extinction, particularly as weaker dispersers succumb first. Habitats may not be ideal, but may be 'stepping stones' between the most favourable habitats. This connection between suitable habitats is particularly important during times of prolonged, adverse weather.

As most landscapes are disturbed by humans, it is often important to consider countryside-wide management. When we do this, the emphasis again is on maintaining as much natural habitat variety as possible, with similar patches being as close as is practical. This can be assisted by reducing disturbance in the areas surrounding reserves. In other words, the matrix around reserve patches can be important for insect diversity conservation in its own right, and intensity of disturbance to it should be reduced as much as possible. All this relates back to habitat quality, which can be as important as reserve patch size or isolation of patch, and is extremely important outside reserve areas. In such areas, agricultural practices should be least disturbing, and done in a way that mimics, as far as possible, natural conditions. Usually this involves management for compositional and structural variation in vegetation. This often means some sort of rotational or spot management of the landscape. Nevertheless, it may not always be possible to maintain all species equally, and certain species may need to be targeted for special attention. A return to traditional, low-intensity farming practices can be highly beneficial in this regard, so long as they maintain as much heterogeneity, at all spatial scales, as possible.

Where linear strips of land differ substantially from their surrounding matrix, they may be corridors which can act as habitats, conduits and source habitats for sink habitats. But they may also be filters along which not all organisms can move all the way. Sometimes they are even barriers. Yet the evidence on balance suggests that movement and habitat corridors can play an important role in insect diversity conservation. Like reserves however, their beneficial role depends on the quality of the habitat that they contain.

Generally, insect diversity decreases towards city centres. In terms of management, the urban context is similar to the intensive agricultural one, although much more difficult to think of in terms of countryside-wide management as a result of the mosaic of fixed structures. The main difference is that urban reserves and networks need to be acutely sensitive to social issues. In a positive sense, urban reserves are exposing young conservationists to wildlife issues, albeit that they are being exposed to the more habitat-tolerant 'weedy species'.

11 Restoration of insect diversity

> Modern agricultural methods are unfortunately lethal both to wildflowers and butterflies. Cowslips, buttercups, and blue and copper wings have been cultivated, drained, and bulldozed out of our fields . . . But with time and trouble and experimentation one can get wildflowers to grow in profusion in the grass or mixed in with the good old cultivated varieties. Thus we can entice a few butterflies back into our daily lives and hope they will dawdle and dally round the *Buddleia*.
>
> Miriam Rothschild (1998)

11.1 Introduction

There is no substitute for wild and unspoilt original landscapes. Whatever natural areas we still have left must be maintained. It is these wild areas where many of the rare, specialist endemics occur. They are also areas of natural selection, with no or minimal human agency.

At the other extreme, there are areas, particularly where the human population density is high, where biodiversity is reduced to widespread generalists. In between, are agricultural and suburban areas where some intervention management can make an enormous difference and restore at least some of the former biodiversity.

It is essential before any restoration activity takes place to be very clear on the conservational goals. To date, the restoration of insect diversity has been little explored. Nevertheless, some guidelines are beginning to emerge. These relate closely to other aspects of landscape management, and, in particular, emphasize the importance of maintaining landscape heterogeneity and quality. We explore here the restoration of general insect variety as well as the special needs of particular species. We are also cognisant of the fact that restoration is an expensive exercise and rarely a real substitute for avoidance of habitat damage in the first place.

11.2 Principle of restoration triage

A useful starting point for restoration ecology is to conceptualize a spectrum of activity. In response to loss of biodiversity worldwide, a first option for a particular physical area may be to do 'nothing', or alternatively, 'something'. On the positive side, where we do 'nothing', it may be because biodiversity is so intact that urgent attention will make little difference. This is the case with certain wildlife reserves that are principally set-aside areas of land where naturalness is largely intact, and at most, there is minimal management to address a particular conservation goal. On the other hand, we may do 'nothing' because the biodiversity is so degraded, as in a city harbour. In this second case, to do something truly meaningful in terms of naturalness or biodiversity recovery would be a monumental task. This does not mean, however, that some sort of ecological landscaping cannot be done, we are simply conceptualizing extremes. Between these two extremes of 'doing nothing' is an area where 'doing something' can have a major, positive effect on local naturalness and on quality biodiversity. This is a three-pronged approach of two 'doing nothings' and one 'doing something' which we may call 'restoration triage' (Samways, 2000b) (Figure 11.1).

The central activity prong of restoration triage, where doing something meaningful in terms of restoring ecological integrity, has two important riders. Firstly, and inevitably, our starting point for action is somewhere along an interrupted ecological succession. In other words, the level of restoration required depends on how disturbed the ecosystem was at the time that the restoration activity began. Secondly, we can never truly restore because we can never be sure exactly what the original state was.

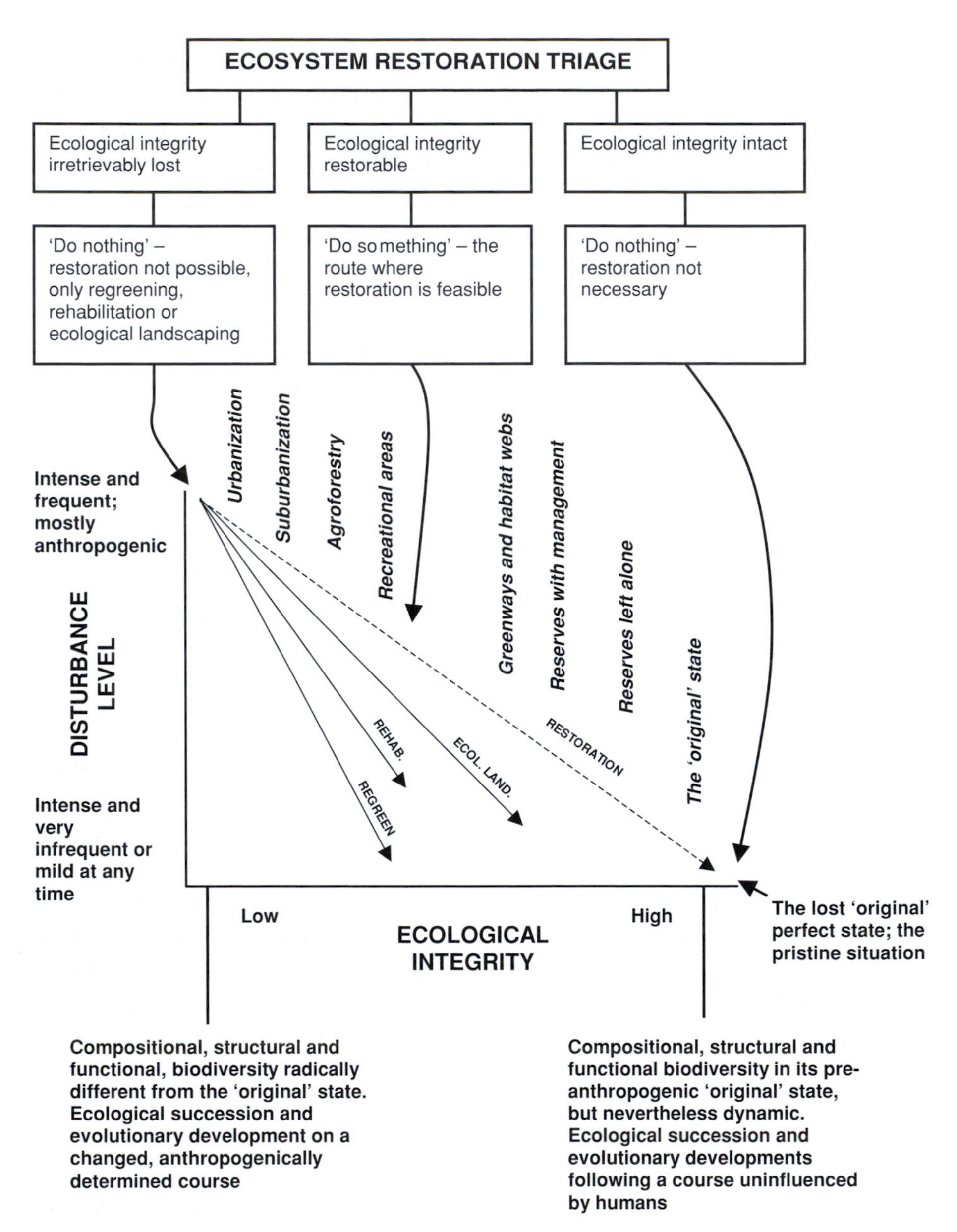

11.1 A conceptual model of ecosystem restoration triage. There is a gradient from ecosystems with intense and frequent disturbance to those with infrequent and less severe disturbance. Urbanization is one end of this gradient, and 'original' is in quotes because (a) it depends how far we go back in time, and (b) because we can never know what all the compositional, structural and functional diversity was at the time. The lost 'original' state is the pristine state, which no longer exists, as anthropogenic impacts have reached all parts of the world. Restoration here is a biocentric, deep-ecology view, where there is a genuine aim to bring back all aspects of ecological integrity.

The central area of restoration triage involves taking positive action. But if the original ecological integrity is irretrievably lost, only regreening, rehabilitation or ecological landscaping is truly possible. (*Continued on next page.*)

Arguably, there are other polarizations attached to the exercise of restoration. On the one side, restoration may be considered as a biocentric, deep-ecology exercise. On the other side, it may be seen as an anthropocentric, aesthetic, vegetation-orientated exercise, which is the realm of ecological landscaping (landscaping with an ecologically reasoned approach, but with aesthetics and human cultural considerations not a top priority). Rehabilitation (ecological recovery with an underlying social context, and with an ecological component appropriate to the situation [e.g. grass covering old mine dumps]) also comes into this category, and does not necessarily recreate the 'original' ecological integrity. These categories nevertheless blend into each other and with vocabularies that depend on the ethical and cultural conceptions involved (Hendersen, 1992; Higgs, 1997).

Perrow and Davy (2002) provide a comprehensive overview of the rapidly emerging field of restoration ecology, with Majer *et al.* (2002) giving particular attention to invertebrates, both as a subject per se and as indicators of the restoration process.

11.3 Restoration of species or processes?

Defining conservation goals relative to time frames is particularly important for ecological restoration. In Sundaland, although vegetation had not

---

11.1 *cont.* Regreening is simply putting back a vegetation cover with more consideration for aesthetics and engineering value than for ecological integrity (e.g. grass cover along road cuttings). The maximal ecological integrity value for regreening is roughly at the level of recreational areas, with disturbance ranging from intense and frequent (e.g. mowing), to infrequent or not at all, when succession takes place. Rehabilitation aims at recovering some ecological integrity but has a major aesthetic and/or human cultural component combined with ecological considerations (e.g. mine dump rehabilitation, removal of pollutants from a stream). Like regreening, the maximal ecological integrity achievable through rehabilitation, in the short term at least, is low. This contrasts with ecological landscaping that deliberately aims to recreate what we believe to be a 'natural' ecosystem. Such a recreated 'natural' ecosystem may or may not be motivated by aesthetic values, over and above deep-ecological ones. Carefully planned planting of indigenous trees and other 'natural' vegetation is an example of ecological landscaping. Researched well, ecological landscaping can have great ecological integrity value, at least over time after indigenous biodiversity returns in most of its entirety. Finally, and arguably, true restoration can only be done on minimally degraded ecosystems (hence the dashed line in the central area of the figure). It aims for the 'original' state, but this is rarely actually achievable (because of, for example, invasive aliens) (hence the dashed line in the lower right of the figure). ECOL. LAND. = Ecological landscaping; REHAB. = Rehabilitation; REGREEN = Regreening. (From Samways, 2000b, with kind permission of Kluwer Academic Publishers.)

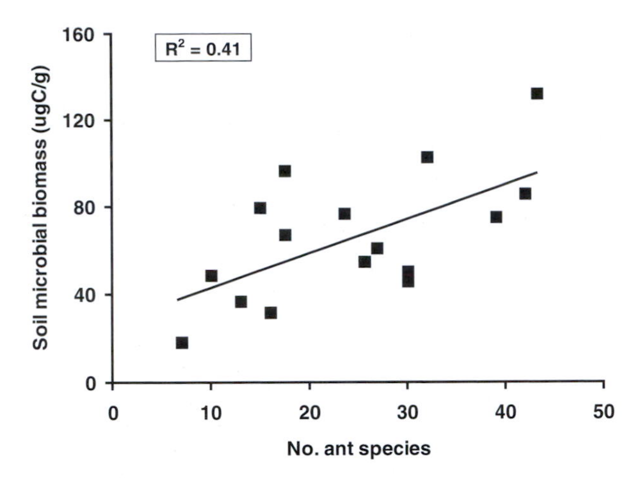

11.2 There is a correlative link between above-ground ant activity and below-ground decomposition processes at disturbed sites in Australia, thereby providing support for the use of ants as indicators of restoration success following disturbance. (From Andersen and Sparling, 1997.)

completely recovered after 50 years following forest removal, the termite assemblage did restore itself and associated ecosystem services when near a source area of primary forest (Gathorne-Hardy *et al.*, 2002). There is a parallel with the Amazonian ant assemblage which recovers at different rates where there are different land-management practices. Recovery of the ground-foraging ant fauna appears to be faster than regeneration of the woody plant community (Vasconcelos, 1999), the latter of which can take as long as 200 years. In the USA, however, the ant assemblage recovered in a restored riparian woodland in comparison with a reference site in just 3 years (Williams, 1993), although in Brazil the full complement of ants on rehabilitated mines had not recovered after 11 years (Majer, 1996). But both Crist and Wiens (1996) and Andersen (1997a) have emphasized that while monitoring using ants has value, particularly with respect to below-ground restoration success (Andersen and Sparling, 1997) (Figure 11.2), it is essential to be sensitive to the spatial scale at which one is measuring recovery success.

These various ant studies have underlined the fact that ecological restoration, as opposed to simply regreening, is an immensely complex process. Andersen (1997b) has suggested, again using ants, that the use of functional groups can simplify apparently complex patterns of species composition, and provide an ecological context to these patterns within a general framework of stress and disturbance. This has also been confirmed by Gómez *et al.* (2003). Indeed, there seems to be merit in rehabilitating mine sites in Germany in a way to maintain successional processes and habitat diversity for maximum beetle species diversity, including rare ones (Brändle *et al.*, 2000). Returning to ants, Andersen *et al.* (2003) emphasize that rehabilitation of mine sites rarely reconstitutes the

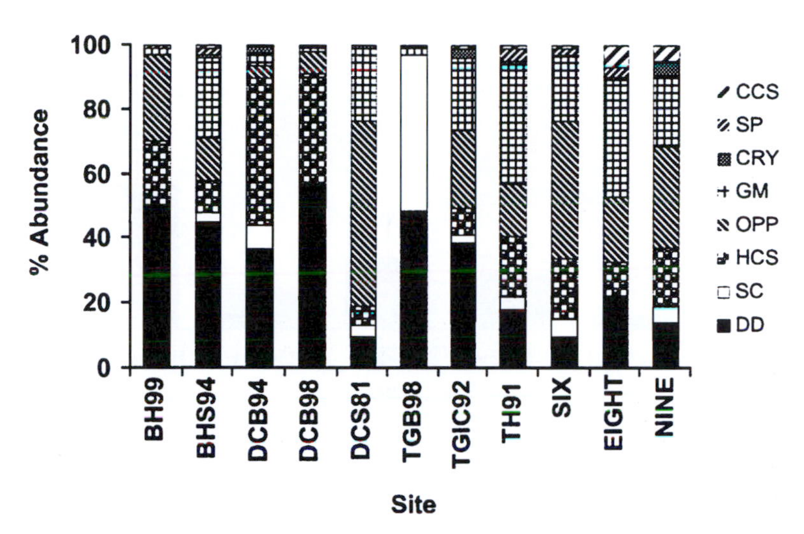

11.3 Relative abundance of Australian ants at various mine sites. The important point is that some sites (e.g. EIGHT and NINE) have leaves, grass, twigs and logs, and, by having these, are able to encourage a full range of ant functional groups. CCS, cold climate specialists; SP, specialist predators; CRY, cryptic species; GM, generalized Myrmicinae; OPP, opportunists; HCS, hot climate specialists; SC = subordinate Componotini; DD, dominant Dolichoderinae. (From Andersen *et al.*, 2003.)

original fauna, with functional groups such as cryptic species, cold climate specialists and specialist predators generally underrepresented, while there is a high relative abundance of dominant Dolichoderinae, hot climate specialists and/or opportunists (Figure 11.3). It seems to be that litter development and logs need to be available and suitable for the specialists at rehabilitated sites. Andersen *et al.* (2003) also emphasize that it is essential to establish locally indigenous vegetation and appropriate ground cover to truly restore the ant fauna.

Restoration may be considered effective only when the original balance of predators, grazers and plants is established. This is necessary because important predators (i.e. higher trophic interactors) can have a major effect on the diversity of an ecosystem and how it functions (Duffy, 2003).

One of the reasons why there are various outcomes to restoration efforts is that ecological dynamics in degraded systems can be very different from dynamics in less-impacted systems. Although traditional restoration has focused on re-establishing or enabling natural successional processes, strong feedbacks between biotic factors and the physical environment can alter the efficacy of succession-based management efforts. It appears that some degraded systems are difficult to encourage through a successional process. This is because of constraints such as changes in landscape connectivity and organization, loss

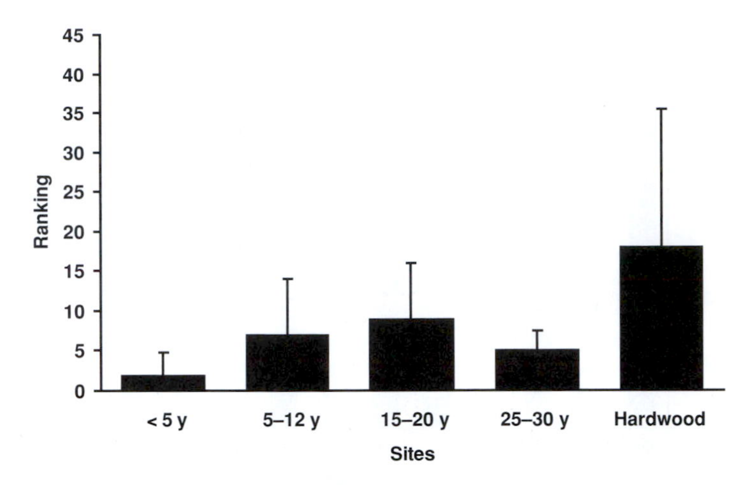

11.4 Mean (± 1 SD) values for rarer diurnal Lepidoptera at reclaimed coal surface-mined sites and at natural hardwood sites in North America. Most lepidopteran species commonly occurring in hardwoods were also in later-successional reclaimed sites (25–30 years), although several rarer species did not re-establish, giving the low bar. (Redrawn from Holl, 1996, with kind permission of Blackwell Publishing.)

of indigenous species pools, shifts in species dominance, trophic interactions and/or invasion by aliens and even changes in biogeochemical processes. The result is that alternative system states are possible, and restoration efforts can sometimes send systems along unintended trajectories (Sudling *et al.*, 2004).

11.4 Coarse-filter and fine-filter approaches to restoration

Mature habitats may be very difficult to fully reassemble, as with rare lepidopterans of North American hardwoods (Holl, 1996) (Figure 11.4). Holl's (1996) study, like Brändle *et al.*'s. (2000) beetle study, indicates that the restoration process is remarkably dynamic, with different species and population abundance being favoured at different times during the recovery process (Figure 11.5). This has led to a call to use a broad taxonomic base (i.e. functional variety) when assessing restoration projects (Mattoni *et al.*, 2000). Furthermore, Maina and Howe (2000) emphasize that as many species are as rare in restored, as they are in natural communities. These rare ones are least likely to appear in the restoration process in favour of vagile, common and widespread species. Again this emphasizes the role of the fine-filter, species approach as a supplement to the coarse-filter, community approach when attempts are being made to restore the original ecological integrity.

Restoration may depend on having the appropriate food plants, such as Bird's foot trefoil (*Lotus corniculatus*) in restoration plots for the Common blue butterfly *Polyommatus icarus* (Davis, 1989), or a range of appropriate pollinators for

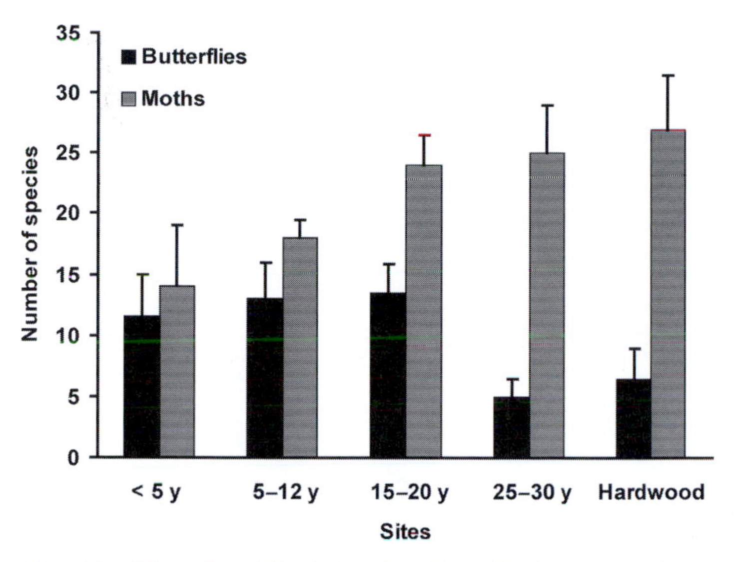

11.5 Mean (± 1 SD) number of North American diurnal Lepidoptera species in restored sites of different ages in comparison with a natural hardwood. Not only does the restoration process take a long time to establish the equivalent of the natural complement of species, but the process is remarkably dynamic. (Redrawn from Holl, 1996, with kind permission of Blackwell Publishing.)

the plant community being restored, as in the Everglades National Park, USA (Pascarella *et al.*, 2001). Nevertheless, studies on the froghopper *Neophilaenus albipennis* are cautionary, in that restoration of existing habitat patches was not entirely effective when there was a prior history of long and severe disturbance to the habitat patches being restored (Biedermann, 2000). In other words, as emphasized in the last section, we need to question not just what we are restoring to, but also from what we are restoring. This applies particularly to rehabilitation of already modified landscapes. We may, for example, rehabilitate to maintain an ecological status quo that was not necessarily the original, pre-human one. This has been illustrated in Germany, where sheep are necessary not only for maintaining the plant species richness of calcareous grassland but also play a substantial role in distributing orthopteran species on their fleece (Fischer *et al.*, 1996).

During the process of restoration, it is necessary to supply all spatial and biological supports for particular species or for assemblages. While the planting of species in restored areas provides abundant nectar sources for adult butterflies, there are other habitat variables, such as food plants for larvae, that are an essential part of a successful restoration process (Holl, 1995). Even with predatory functional groups, such as carabids, plant type can be important, with establishment of sown wildflower swards being richer than grass or clover cover, but still lacking carabid species of unmanaged, natural habitats in the same area (Blake *et al.*, 1996). Nevertheless, this should not discourage the planting of seed

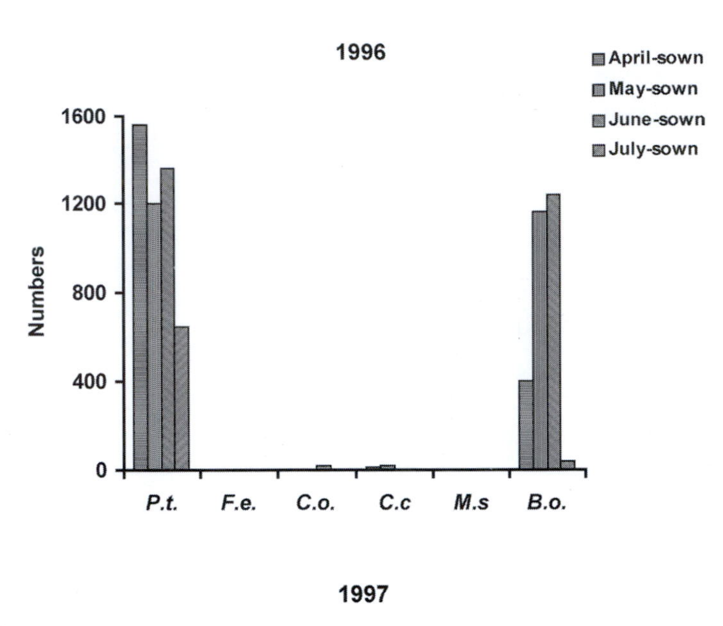

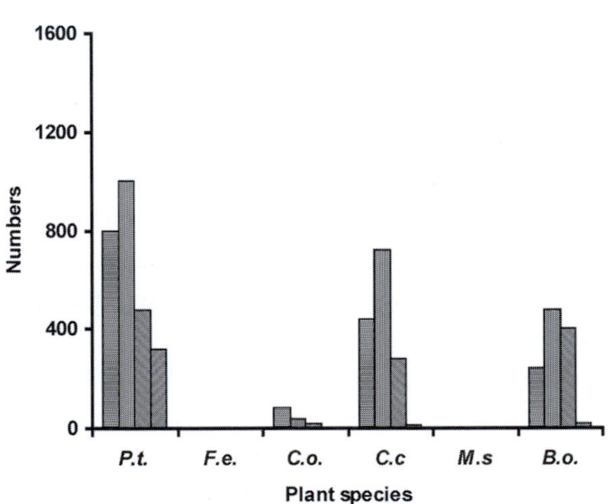

11.6 Numbers (accumulated totals) of bees observed on six plants species when sown in April, May, June or July (bar shading from left to right) indicating that time of sowing can be adjusted to attract a particular suite of species. *P.t., Phacelia tanacetifolia; F.e, Fagopyrum esculentum; C.o., Calendula officinialis; C.c., Centaurea cyanea; M.s., Malva sylvestris; B.o., Borago officinalis.* (Redrawn from Carreck and Williams, 2002, with kind permission of Kluwer Academic Publishers.)

mixtures which can be proportionately adjusted to attract a particular suite of insects (Carreck and Williams, 2002) (Figure 11.6). The optimum, besides retaining natural patches for particular rare species, is both to deliberately establish and to allow natural establishment of certain bushes, so as to create a taxonomically as well as functionally diverse arthropod community (Burger *et al.*, 2003) (Figure 11.7).

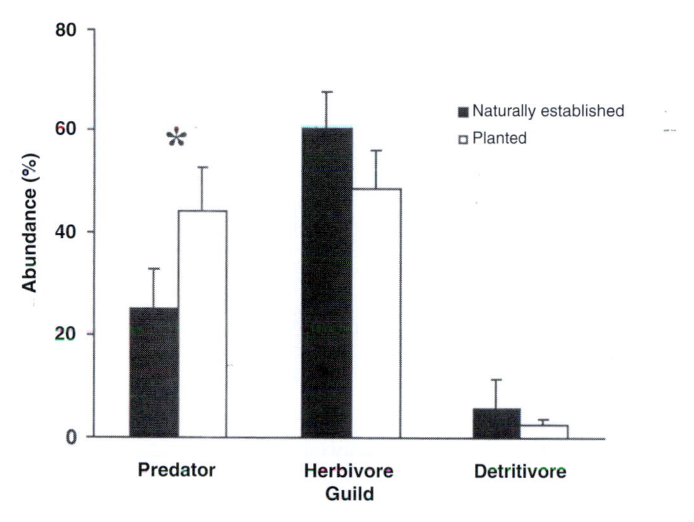

11.7 Percentage of arthropods belonging to detritivore, herbivore and predator (predator per se and parasitoid) guilds in naturally established versus planted shrubs. Values are averages per plant, with 95% confidence intervals. The asterisk indicates a significant difference at the 5% probability level. (Redrawn from Burger *et al.*, 2003, with kind permission of Blackwell Publishing.)

Even though restoration ecology science is still rudimentary, particularly with regards to insect diversity restoration, it is clear that we should restore or rehabilitate wherever we can but yet be very clear on what we are attempting to achieve. But no restoration activity is a substitute for maintaining natural wild areas, which may indeed be a necessary source of plant and animal propagules fundamental to the restoration process.

The maintenance of undegraded and healthy streams appears to be particularly important for ensuring colonists for streams undergoing restoration. This emphasizes that restoration activities operate at multiple spatial scales, with stoneflies (Plecoptera), for example, responding to local riparian effects on food supply and shading, whereas 'long-lived taxa' are more sensitive to a greater part of the river system (Morley and Karr, 2002).

Similarly, countryside-wide scale of restoration can also apply to terrestrial insects and their response may be largely independent of other taxa, such as vertebrates. Outside an African savanna game park, where large indigenous mammals were replaced by domestic livestock, the grasshopper species assemblage was not impoverished, although population densities were lower (Samways and Kreuzinger, 2001). It seems, at least in the case of grasshoppers, that it is the level of trampling and grazing that is important, rather than which type of megaherbivore is doing it. In the case of another game reserve that had been restored from farmland for 62 years, the reserve grasshoppers increased in abundance although there was little change in species richness (Gebeyehu and Samways, 2002) (Figure 11.8). While this is encouraging on the one hand, with

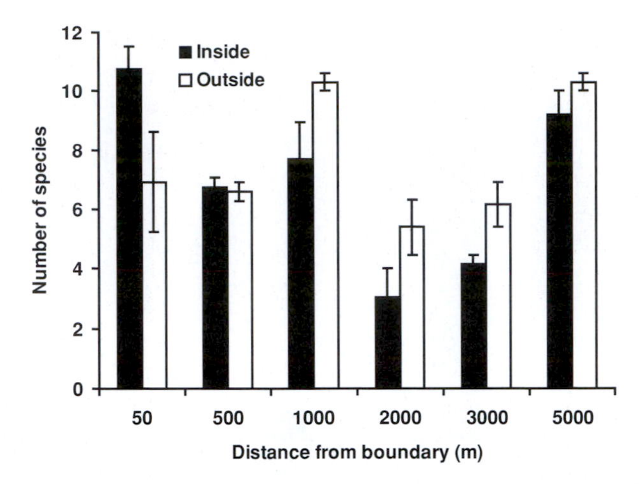

11.8 Mean (± SE) number of grasshopper species inside versus outside an extensively restored area (Mountain Zebra National Park, South Africa). In general, there was no significant difference between mean number of grasshopper species at sites inside and outside the park at various distances from the boundary. (Redrawn from Gebeyehu and Samways, 2002, with kind permission of Kluwer Academic Publishers.)

no difference in species richness, we must remember that extinction is about gradual attrition of populations and that low abundances carry their own extinction risks. Also, parks are still only spots on the greater landscape, and in the case of these grasshoppers, two species were recorded only inside the park and two other species only outside, hence the overall equal species richness across the boundary. This emphasizes that restoration activities must be cognisant of particular species requirements if all species are to be maintained, a point we will return to later in the chapter.

11.5 Insect gardening

Shetler (1998) emphasizes that 'A natural ecosystem is more than the sum of the parts, especially in its historical continuity, and it cannot be concocted like a cake from a mix of ingredients. Sowing, transplanting, or releasing plant and animal species across the land is neither sufficient nor wise a conservation strategy. The urge to play Johnny Appleseed must be resisted . . .' This we must keep uppermost in our minds, and focus on the conservation goal, which, in this homogenized world, surely is the maintenance of ecological integrity. There is no harm in combining aesthetics, particularly butterfly gardening (The Xerces Society and the Smithsonian Institution, 1998; Reid, 2000), with regreening, but where it can be done using indigenous plants (Johnson and Johnson, 1993) and other aspects of the habitat such as shelter and logs (Kirby, 1992) then so much the better.

11.9 Bee 'condos' are artificial nesting sites for bees to enhance local population levels. They may have a single hole size of the diameter preferred by early-spring bees such as the Blue orchard bee *Osmia lignaria*, although the blocks will also be used by some leafcutter bees (a). Tying hollow stems of teasel (*Dipsacus sylvestris*) is an easy way to make nests for wood-nesting bees (b). Bamboo, common reed and sumac stems can also be used in this way. (© 2003 Matthew Shepherd/The Xerces Society.)

For butterflies as well as other species, the aim is to produce conditions for breeding as well as simply for nectar feeders. Nesting and egg-laying sites, for example, can be created for insect pollinators (Shepherd *et al.*, 2003). This includes making bee nesting blocks or bee 'condos', which are wooden blocks with drilled holes as nesting cavities (Figure 11.9). Nevertheless, some nectar-producing plants that are alien and even invaders can have beneficial facets (Vickery, 1998). The infamous invasive *Lantana camara*, for example, provides a reliable nectar supply for two extremely rare hummingbird hawkmoths in the Seychelles (Gerlach, 2000).

Forsyth (1998) points out that attracting functional diversity can also be part of the combining of aesthetics and ecology. Umbel and composite flowers, in particular, attract and sustain populations of predatory and parasitic wasps and flies. As the plant community changes, so does the insect diversity, with both increased taxonomic and trophic diversity following plant succession. This inevitably suggests that insect gardening depends on the goal of the activity which may be to intervene in ecological sucession, so that butterfly gardening becomes 'The art of growing plants that will attract butterflies' (Booth and Allen, 1998) and is essentially a human regenerative activity.

11.6 Species-specific recovery plans

11.6.1 *Success of species-specific recovery plans*

Restoration activities may involve the rearing of particular species for re-introduction or translocation from other sites. Such species recovery plans were originally done with little awareness of the species' genetics, and were done in an ad hoc way. Oates and Warren (1990) overviewed 323 attempts at establishment or population reinforcement of Lepidoptera in Britain, with six of the species not even being indigenous to the country. Nearly half (47%) of the indigenous butterfly releases were genuine attempts at conservation, but 29% were releases of surplus breeding stock, 17% were for amenity purposes and 7% for scientific experimentation. Although reasons for successful establishment are difficult to assess owing to lack of information, there was no correlation between numbers of individuals released and success. Establishment was achieved with as few as three or five mated females. The main reason for failure appeared to be habitat unsuitability. Nevertheless, some introductions have been highly influential on distribution patterns, with the Marsh fritillary *Euphydryas aurinia* in Britain surviving in more introduced colonies than naturally occurring ones.

Holloway *et al.* (2003) have introduced much more rigour into the current 'art' of insect re-introductions. However, this can only be done on ecologically relatively well-known species, such as British butterflies. For these species, recovery targets specify the numbers of populations that should exist by a specific future date to enable survival, but there has been no procedure available to plan strategically to achieve these targets. In response, Holloway *et al.* (2003) developed techniques based upon geographic information systems (GIS) that produce conservation strategy maps (CSM) to assist with achieving recovery targets based on all available and relevant information. Using systematically filtered, relevant habitat, botanical and autoecological data, localities were identified that were suitable for introduction of the Heath fritillary *Mellicta athalia*. The method is sufficiently robust to update information and to be used on many other invertebrates across the world, where sufficient data exist.

Sherley (1998) discusses the translocation of the threatened New Zealand weta, *Deinacrida* sp. (a giant orthopteran) and concluded that the best method of translocation was to put translocated individuals in an enclosure at the release site. This then allows the transferees to breed, with the aim of releasing their progeny.

Pullin (1996) and New (1997) give other examples of specific butterfly recovery plans, with Pullin (1996) recommending that restoration studies first include a detailed study of specific habitat requirements and of life cycles, along with plans on how the habitat will be managed. Management might include, for example, clearing of gorse scrub and careful grazing to provide a mosaic of short, grazed turf and patches of thyme as foodplant for the Large blue butterfly *Maculinea arion*, and fen restoration to enable survival of the Large copper *Lycaena dispar batavus*, particularly making more habitat available for males to hold territories and through removal of barriers to movement (Pullin, 1997). In the case of the threatened European beetle *Osmoderma eremita*, these habitat requirements are very specific, being tree hollows in oaks with openings directed towards the sun and in cavities with large amounts of wood mould (Ranius and Nilsson, 1997).

The synergistic effects of global climate warming and habitat modification (Warren *et al.*, 2001) are likely to make some of the best-intentioned plans difficult to implement with predictive success. Field experimentation at multiple spatial and temporal scales, and in many restoration contexts, appears now to be the best way forward for conserving a multitude of species, as well as underpinning a range of ecological interactions and processes (Zedler, 2000). Interestingly, in the overview by Boersma *et al.* (2001), multispecies plans were found to be more effective in species recovery plans than single species plans, because multispecies plans must adopt a broad view of threats and are more integrative. Nevertheless, certain species require particular management attention. Bushy vegetation, for example, must be removed regularly to encourage the blue lupine food plant of the threatened Karner blue butterfly *Lycaeides melissa samuelis* in the New York area (Smallidge *et al.*, 1996).

Not all insect species recovery plans are for the insects per se, but may be related to plans to boost certain bird populations. Gardiner *et al.* (2003), for example, have emphasized the importance of improving grassland swards in Britain to promote *Chorthippus* spp. grasshopper populations for farmland birds, including the rare Cirl bunting *Emberiza cirlus*.

11.10 At the Zoological Society of London's Invertebrate Conservation Centre, the Field cricket *Gryllus campestris* (a) and Wart-biter bush cricket *Decticus verrucivorus* (b) (next page) are being reared for release in the field. The Field cricket is much easier to rear than the Wart biter, which is an important consideration when captively breeding insects for re-introduction into the field. (Photo by courtesy of Paul Pearce-Kelly.)

11.6.2 Captive breeding

Rearing single species of insects for release in the field is not an easy task. It needs dedicated staff who are sufficiently resourced to ensure high-quality animal care and comprehensive record keeping (Pearce-Kelly *et al.*, 1998). Regular health screening clearly illustrates the essential role that veterinary pathology support has to play in such *ex situ* situations. Close liaison with those managing and monitoring the situation in the field is also essential. One of the reasons that the rearing and release programmes of the British Field cricket *Gryllus campestris* have been successful is that the species can be reared in large numbers in the laboratory: half a dozen field-collected animals brought into captivity can return over 1200 late instar nymphs into the wild in the same season. The same cannot be said for the Wart-biter bush cricket *Decticus verrucivorus*, also a threatened orthopteran in Britain, the rearing of which carries high costs in terms of time and materials (Figure 11.10).

11.10 *(cont.)*

A further challenge with captive breeding of insects, as with other animals, is that to maintain viable captive populations for multiple generations, the captive population must maintain heterozygosity, genetic diversity and remain adaptive to the wild. A difficulty with long-term captive breeding of insects is that genetic adaptation to captive environments can occur rapidly, with the traits responding to selection in captivity varying depending on the biology of the species involved. Lewis and Thomas (2001) suggest that the ideal would be to create large near-natural captive environments and to maintain multiple lines of the same species to increase the chances of success of re-establishment attempts.

11.6.3 *Population viability analysis*

Schultz and Hammond (2003) used population viability analysis to develop recovery criteria for threatened insects, using the North American Fender's blue butterfly *Icaricia icarioides*. They estimate that the butterfly is at high risk of extinction at most of its sites throughout its range, with only one site where the population has a > 90% chance of persisting for 100 years. Nevertheless, at the regional scale, where there is demographic connectivity between nearby populations, the probability that at least one of 12 sites survives through the next century is almost 100%. Of concern, however, is that larger populations were not necessarily more secure. However, to secure the species' future, especially in terms of regional genetic variation as well as local genetic variation, it is important to maintain several independent sites in each area, as a safety net in case of catastrophes such as wildfire, pesticide impacts, etc. A critical factor appears to be that the populations need to maintain a high average growth rate along with having reduced variability in growth rate. To do this, maintenance of habitat quality along with conditions for adequate metapopulation dynamics will be essential, especially as insect populations tend to fluctuate so much. Although such a modelling approach has its pitfalls, principally because continuous, solid data may not be available, Schultz and Hammond (2003) point out that identifying minimum population growth rates is a useful and feasible approach to developing reasonable objectives and measurable recovery criteria for threatened insects.

Baguette and Schtickzelle (2003), working on the arctic-alpine relict Cranberry fritillary *Boloria aquilonaris*, further emphasize that it is essential to take local metapopulation dynamics into account in population viability analysis of species confronted by severe habitat loss and occupying highly fragmented habitat networks. In the case of this butterfly, this means protection of the remaining habitat network. However, even this may be insufficient, because the local population dynamics of this butterfly have been such that there has been local extinction in some patches, and, as with Fender's blue butterfly, local extinction has occurred even in a surprisingly large patch.

11.7 Summary

There is no substitute for wild areas. Nevertheless, it is possible in certain cases to restore disturbed ecosystems to some degree or other.

Restoration depends firstly on deciding on the conservation goal, and in particular whether restoration of the 'original' ecological integrity is the aim and is feasible. It may only be possible (because degradation is so great) to revegetate or rehabilitate to the pre-disturbance condition. Evidence from insects to date suggests time taken for ecological restoration to take place is highly variable depending on taxon or ecosystem under consideration. These differences have probably partly arisen because of differences in measurement of spatial scale and the fact that restoration is an extremely complex process, and must go through a successional process to include the rare specialists. Mature habitats are particularly difficult to restore. It may be necessary to supplement the coarse-filter general restoration process with particular conditions for certain focal species. This may not always be possible where there has been a long history of severe disturbance, with restoration only being attainable to a pre-industrial landscape rather than a pre-historical one.

The process of restoration may involve the planting of certain vegetation in addition to allowing natural regeneration, so as to create a taxonomically and functionally diverse arthropod community. We need to consider the countryside-wide situation, so that wild areas can 'seed' areas being restored. Such a process enhances population levels and genetic variation, rather than simply allowing species to reappear. All the time we also need to consider the spatial heterogeneity of the landscape to cater for the countryside-wide variety of insect diversity.

'Insect gardening' is a particular form of conservation that may be purely aesthetic, such as planting of nectar sources (which may not even be indigenous) for charismatic butterflies. At the other extreme, it blends into the concept of ecological restoration as it involves restoring the complement of indigenous plants to encourage local insects in general.

Specific recovery plans of threatened species have, in some cases, been very successful. Generally this has been because the biology and habitat requirements are well understood. Nevertheless, to ensure greater predictiveness, field experimentation at multiple spatial and temporal scales may be necessary. Indeed, multispecies plans have been found to be highly effective, even in single species plans, because multispecies plans must adopt a broad view of threats and are more integrative.

Rearing of insect species for release in the field is not an easy task, and requires much attention to detailed requirements of species. In particular, maintenance of genetic viability, and adaptiveness to wild conditions to ensure re-establishment success, are essential.

It may be possible, where adequate data are available, to undertake a population viability analysis and determine minimum average population growth rates for a population, and more importantly, a metapopulation, for survival in the long term. This then provides reasonably objective and measurable recovery criteria for threatened insects.

12 Conventions and social issues in insect diversity conservation

> . . . the fact remains that it would only rarely be possible for us to improve the workings of a natural system, whatever wisdom we had. Even if we could, only rarely would we know enough. It is not our job, and it is utterly silly to think that the rest of the world needs us to run its affairs. For the greatest part, the very best thing we can do for a wild animal, a species, or an ecosystem, is to leave it alone.
>
> Lawrence Johnson (1991)

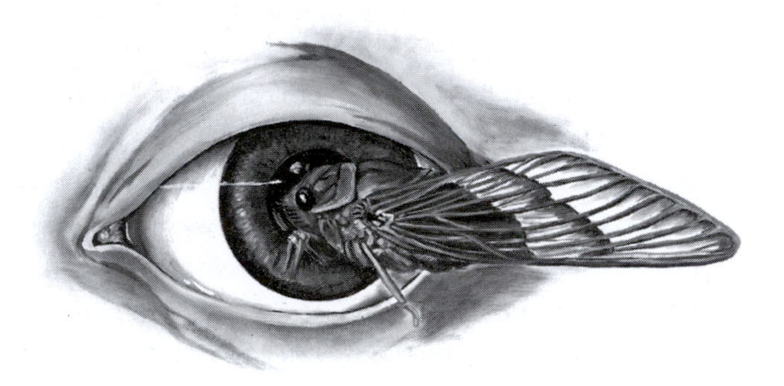

12.1 Introduction

In the last two chapters we overviewed management activities that maintain and enhance insect diversity both at the coarse-filter (habitat or landscape) and fine-filter (species) levels. The fine filter generally overlays the coarse filter to maximize opportunities for future survival of communities in general, as well as for specialist species. We now briefly overview some legal and social issues pertaining to insect diversity conservation.

Various international conventions are in place to promote conservation and sustainable utilization of biodiversity, including insects. These are principally at the level of the coarse filter, although the Convention on International Trade in Endangered Species (CITES) refers to particular species. Implementing these conventions partly involves increasing awareness of the role of invertebrates in general, which have been neglected in comparison with plants and vertebrates. This to some extent complements the IUCN Red List, where particular threatened insect species are given as much credence (space on the list) as any other organism. This is important, because one of the essential ingredients in insect diversity conservation is drawing attention to their plight, which involves employing particular species as icons that 'speak for' the vast multitude of less cryptic species. Education also becomes an important component in this awareness exercise because the young are the conservationists of the future.

12.2 The international arena

There are many dozens of international laws, treaties and protocols concerning conservation to which countries are signatories. Most of these aim to improve sustainability of resources and species. The most significant of these as regards insect diversity conservation is the Convention on Biological Diversity (CBD) which was formulated in 1992 at the first Earth Summit in Rio de Janeiro. The CBD aims to conserve biodiversity worldwide. It also stipulates the sustainable use and fair and equitable sharing of benefits from global utilization of genetic resources to wild organisms. The intention is to benefit current and future generations from the natural resources in particular countries. Most countries are signatories to the CBD and are variously taking measures to maintain their natural ecosystem intact and restore those where possible. The point is that the CBD, like the IUCN Red List, has drawn attention to insects and other invertebrates to a much greater extent than formerly.

Another important international agreement is the Convention on International Trade in Endangered Species (CITES), which aims to regulate trade in whole or parts of organisms where trade is a threatening process. The CITES list has an Appendix I, where the species or parts of it cannot be traded at all; an Appendix II where an export licence is required so as to monitor and control trade, and an Appendix III where permits are required for trade in the species and for conservation monitoring reasons. This convention has been very influential on the international trade in insect livestock, particularly certain papilionid butterflies such as the birdwing butterflies *Ornithoptera alexandrae* (Figure 12.1). It has also created an alert to threats to the endemic South African beetles *Colophon* spp., which occur on isolated mountaintops. While the CITES-listed butterflies may well have collecting and trading as one of their threats, along with habitat destruction, the situation with *Colophon* spp. is rather different. Some species, such as *C. stockoei* and *C. westwoodi*, are not particularly rare species, although localized. They also occur in formally protected areas. In contrast, *C. primosi* is indeed rare and highly threatened, and relatively easily accessible (Figure 12.2). Nevertheless, there are certain aspects of their behaviour that, if generally known, make all these species more vulnerable to harvesting by large-volume collectors. This happened in December 2003, when four beetle dealer-poachers were arrested on charges of illegal possession of over 200 of these beetles as well as other rare insects.

Many threatened insects are on sale in Europe, with 81% of all CITES-listed butterflies traded worldwide being imported for commercial purposes, with more than half coming from ranching or farming programmes, and about 11% having

12.1 A trade-restricted (CITES-listed) insect, the birdwing butterfly *Ornithoptera victoria regis* (a, female; b, male) from the island of Bougainville, Western Pacific. (Photo by courtesty of Henk Geertsema.)

been wild-caught (Melisch, 2003). A fairly large proportion of the species on offer were subject to domestic legislation prohibiting capture and trade without prior permission. This applies to the Corsican swallowtail *Papilio hospiton*, endemic to Sardinia and Corsica, and protected in France, and the butterfly *Atrophaneura luchti*, which is protected in Indonesia. These two species are Red Listed as Endangered and Vulnerable respectively.

Melisch (2003) concluded that there is a lack of statistics available on the trade in insect species. Furthermore, legal restrictions were not generally

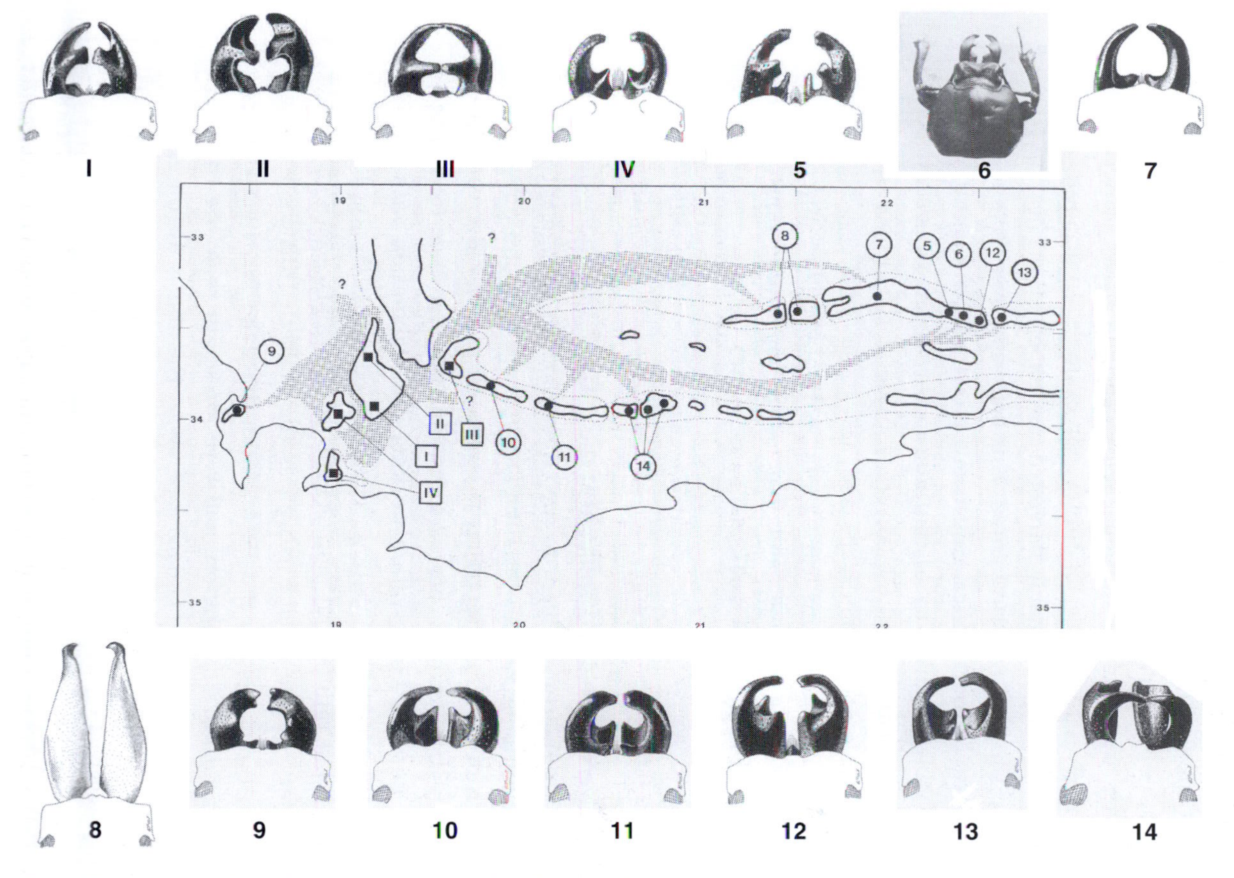

12.2 Hypothetical dispersion and present-day distribution of *Colophon* spp. stag beetles. There are two different lineages represented here by Roman or Arabic numerals. I, *C. cameroni*; II, *C. haughtoni*; III, *C. eastmani*; IV, *C. stokoei*; 5, *C. cassoni*; 6, *C. berrisfordi*; 7, *C. neli*; 8, *C. primosi*; 9, *C. westwoodi*; 10, *C. barnardi*; 11, *C. thunbergi*; 12, *C. whitei*; 13, *C. montisatris*; 14, *C. izaradi*. The whole genus is protected under international law (CITES) and is highly sought after. Evidence is accumulating that there are further *Colophon* species awaiting description. (From Endrödy-Younga, 1988.)

observed at most of the insect Trade Fairs surveyed, and inspections by enforcement authorities and by the insect fair hosts were sporadic and of inadequate quality. This is being addressed both through legal channels as well as by entomological associations who can assist by using self-regulation.

We must always be aware that these collecting and trade regulations are a fine-filter species activity which, although important for a few charismatic species, are not the mainstay of insect diversity conservation which is largely about maintenance of the habitat intact.

There are other aspects of international significance for insect conservation. A practical example is the export of timber, for example, from South Africa. To ensure that timber is sustainably harvested and exported, ratification from the Forest Stewardship Council is required. This body approves certain logging projects. In the case of South Africa, this means growing alien plantation trees in a way that enables biodiversity to continue to thrive. The important point here is that the assessment of the biodiversity value of the timber operation is at the spatial scale of the landscape and not at the smaller scale of the pine patch. In practice, this means leaving a network of corridors and nodes of natural grassland between the pine stands which encourages indigenous 'quality' biodiversity (Pryke and Samways, 2001, 2003) (see Figure 1.2).

12.3 National issues

Insect diversity is being considered at the national level in various countries, arising out of the commitment to the Convention on Biological Diversity. The level of commitment varies substantially from one area and one country to another. Much of insect conservation is embraced in the wider concept of landscape conservation, through partnerships. This is being done for example in Britain, with Butterfly Conservation working in collaboration with English Nature to conserve biodiversity through the UK Biodiversity Action Plan. For example, major input into habitat management is taking place on Exmoor for the Heath fritillary *Mellicta athalia*, as well as the Malvern Hills for the High brown fritillary *Argynnis adippe*, and Devon Culm grassland for the Marsh fritillary *Euphydryas aurinia*. Chiddingford and Wye forests, in turn, are the focus for the Pearl-bordered fritillary *Clossiana euphrosyne*. The aim is to conserve the habitats as much as possible to ensure survival in the long term of a variety of plants and animals.

Meanwhile, it is essential to continue to monitor for species, not only to discover new sites but also to continue to ascertain the status of both common and rare species. While this is progressing well and is feasible for UK butterflies, it is extremely difficult in less developed countries, even for butterflies. Insect diversity in these countries, which is generally so rich, then depends on maintenance of wild areas, and low-impact agriculture. In the arid Karoo of South Africa, this may involve some restoration (Gebeyehu and Samways, 2002) for countryside wide conservation.

The Entry Level Agri-Environment Scheme (ELS) in the UK is an interesting development to encourage farmers to protect and improve wildlife habitats on their farms through activities such as creating grassy field margins, maintaining hedges and reducing frequency of hedge-cutting. The overall aim is to reverse the loss of biodiversity in the wider countryside through incentives to farmers. It is complementary to existing conservation standards such as Countryside Stewardship, and Sites of Special Scientific Interest (SSSIs) which are sites of high natural value. This model is important for British butterflies because farmland is a key habitat for over half of the national species and many species of moths (www.defra.gov.uk/erdp/reviews/grienvs/entrylevel.htm). Interestingly, across Europe in general, arthropods overall appear to have benefited from agri-environment schemes suggesting that such approaches are making a genuine contribution to the conservation of biodiversity (Kleijn and Sutherland, 2003).

12.4 Overcoming the perception challenge

One of the challenges with insect diversity conservation is overcoming the public perception that insects, like many other invertebrates, do not have sufficient charisma to warrant conservation priority. While being recognized as an intrinsic component of compositional and functional biodiversity by experts, insects nevertheless remain relatively low on most conservation agendas. Exceptions do exist, with butterflies and dragonflies receiving particular attention (Hill and Twist, 1998). This relates to human aesthetic perceptions, despite, essentially, a bug and a butterfly having the same nervous systems, and at least theoretically, should enjoy the same attention under the ethic of intrinsic value. Nevertheless, insects on the Red List (or in national Red Data Books), no matter how insignificant, do enjoy particular focus, and feature strongly in recovery plans.

The important point is that rare, threatened and charismatic insect species are icons as any other organism might be. Conservation awareness of insects in general appears to be more enhanced when such icons are given attention than when insects are mentioned as a group, which might include a fly and a cockroach. Awareness trails can be developed for a particular taxon, as was done for dragonflies in South Africa (Suh and Samways, 2001) (Figure 12.3). Such charismatic insects at least to some extent 'stand in' for insects in general in the eyes of the public. They also have important educational value in that children in particular relate very well to these small animals. A vital corollary is that for these icons to win favour with the public, conservationists must refer to them by their common names, rather than their scientific names, so as to give the conservation mission warmth and familiarity.

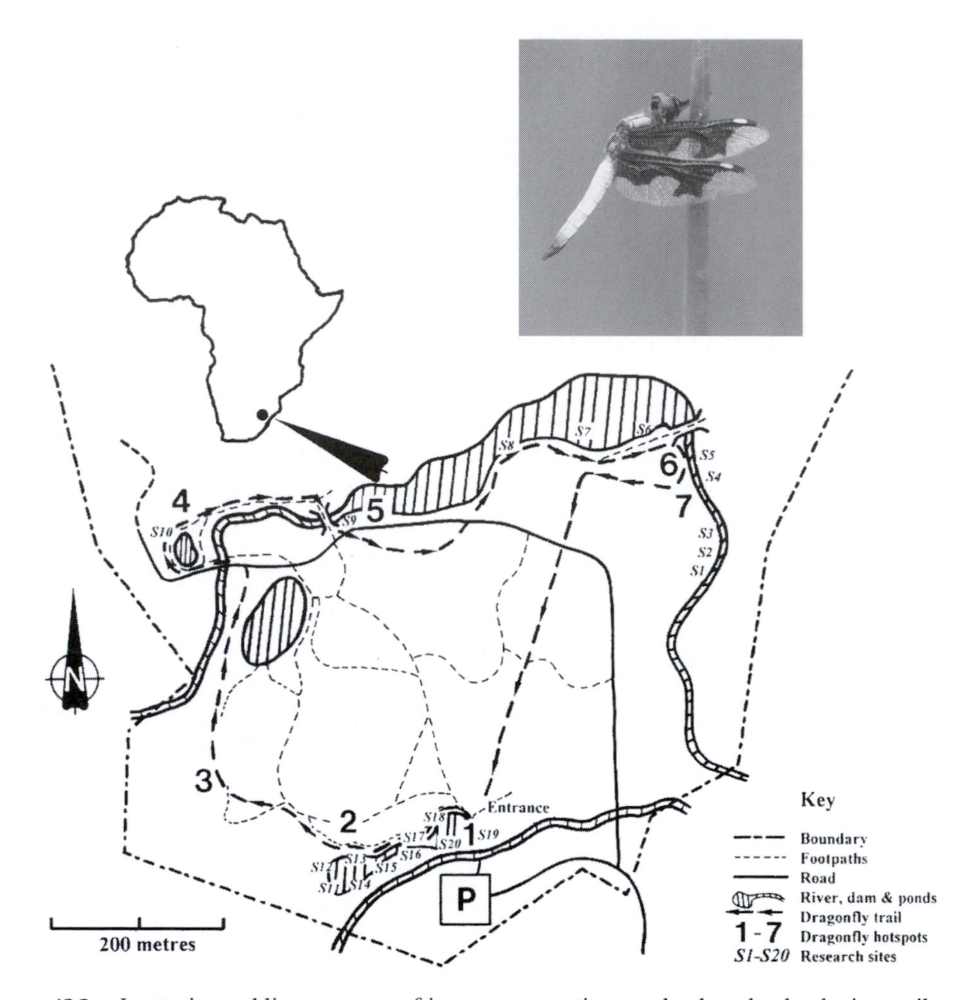

12.3 Increasing public awareness of insect conservation can be done by developing trails where the public can observe live insects in their natural surroundings. A dragonfly trail was developed at the National Botanical Gardens, Pietermaritzburg, South Africa, from 20 a priori research sites. These sites were short-listed to seven viewing or 'hotspot' sites that incorporated the major habitats, yet linked to make a circular walk. It was found that children and the elderly in particular related well to the trail. Nevertheless, 'reliability' is crucial: that is, the species must be present so as not to disappoint. For this reason only 24 'core resident species' were advertised. These were residents, leaving another eight or so possible vagrants as a bonus for serious dragonfly twitchers. The insert is the Portia widow *Palpopleura portia*. (From Suh and Samways (2001). Copyright 2001, with permission from Elsevier.)

In a way, we return to the fine-filter versus coarse-filter approaches. Powerful imaging comes from iconization of insects, and at least when photographed, they are the same size on the page as any vertebrate. This fine-filter approach complements the coarse-filter approach of landscape conservation, where insects are effectively a 'black box', and are included when landscapes are conserved,

12.4 The Karkloof blue (*Orachrysops ariadne*), one of the very rare and threatened lycaenid butterflies in southern Africa which has a reserve especially created for it and its ant host. It is a flagship species, as it is the logo of the local tourist route, the Midlands Meander. (Photo by courtesy of Sheng-Shan Lu.)

so long as the landscapes are the appropriate surrogates for all the insects. If not, then there are two approaches. Either set aside particular reserves for a specific insect species (Figure 12.4), as the recent reserve established for the threatened Coega copper butterfly *Aloeides clarki* at Coega in South Africa (Pringle, 2002). Alternatively, there is no choice but to have some sort of triage (Samways, 1999b), where insects may or may not be included in the decision-making process. This, however, returns us to the particularly difficult question of appropriate surrogacy measures that we discussed in Chapter 8.

12.5 Butterfly houses and increasing conservation awareness

Butterfly houses (Figure 12.5) have become an important medium at the interface between insects and the public. While some houses have displayed over 300 species of Lepidoptera, only about two dozen species are the core attraction. Some houses show up to 70 species at one time, and about 150 over the British summer season. About 500 000 individual butterflies were used in 1986, one-third being bred on site and two-thirds bought from various countries, mostly

12.5 Outside (a) and inside (b) a 'butterfly house' in Singapore. Such 'butterfly houses' are playing a huge role in increasing public awareness and breeding of insects for re-introduction.

tropical. There is no evidence that threatened species are used, in contrast to the deadstock trade (Morris *et al.*, 1991).

Probably the greatest value of butterfly houses is in providing strong positive experiences for many of the visiting public (Weissmann *et al.*, 1995). As swallowtail butterflies are often the largest and showiest of species, they have played a central role. Not only do butterfly houses give the general public a close-up view of both indigenous and foreign species, but they also create or reinforce individual appreciation for the beauty and unique qualities of butterflies. Education about their biology and behaviour is critical to preserving and supplementing individual habitats. As Weissmann *et al.* (1995) emphasize, visitors to butterfly houses are often empowered to champion habitat conservation causes in the tropics and even in their own back gardens. Butterfly houses often provide the impetus as well as expertise and raw materials for planting a butterfly garden. Most butterfly houses enhance the grounds around their enclosure to attract local butterflies. Several also sell gardening books and even seeds and perennial plants, which further encourages positive action.

Butterfly houses are playing an increasing role in research as well. Studies on life history, behaviour, chemical ecology and genetics of many species, including rare and threatened species are being conducted in butterfly house facilities. At Butterfly World in Florida, research includes culture techniques for the Jamaican *Papilio homerus* and the impact of pesticide use on populations of the Schaus swallowtail *Papilio aristodemus ponceanus* (Emmel and Boender, 1991). This can be important when the habitat is under pressure, as when Hurricane Andrew swept through the primary habitat of the Schaus swallowtail in 1992.

As the number of butterfly gardens and butterfly houses increases, public awareness of these animals also increases. This, in turn, engenders greater public awareness in ecotourism, which to name one example, is being used to better protect the Monarch butterfly overwintering sites in Mexico. Ecotourism is also becoming increasingly important in tropical conservation programmes, such as butterfly farms in Costa Rica. Taiwan attracts about 500 000 butterfly tourists per year, mostly from Japan. Butterflies seen by the public may also serve as indicators of species diversity and environmental quality. Visitors to butterfly houses and gardens frequently ask why they see fewer butterflies, especially some of the formerly common garden butterflies in Britain. Answers to these questions in turn encourages awareness not just 'at home' but also elsewhere on the plight of butterflies and even other insects in distant places.

12.6 Deadstock trade

The deadstock trade, which may or may not be associated with butterfly houses, runs into tens of millions of dollars annually. Most of this trade is for the decorative ornament and display trade, using large numbers of mostly bred, common butterflies. The smaller specialist trade is aimed at serious collectors who demand good quality individuals of rare species, US $1500 having been paid for a male hybrid birdwing butterfly *Ornithoptera 'allottei'* in Paris in 1966. Today a specimen of *Ornithoptera meridionalis* can sell for US $1500. The relatively non-charismatic *Colophon* spp. stag beetles, endemic to the Southwestern Cape mountains of South Africa, have even been on offer for US $5000–US $11 000 per specimen (Figure 12.6). Sales of butterflies from Taiwan were US $24 million

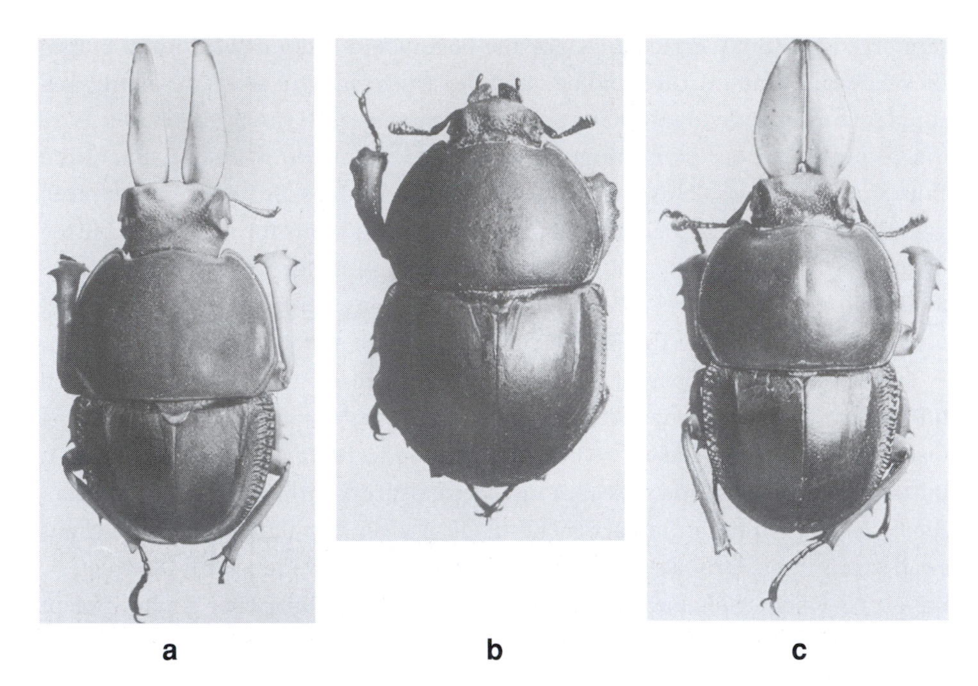

12.6 *Colophon primosi*, a CITES-listed stag beetle from southern Africa. Collectors have been charged with illegal collection and possession of these beetles. Specimens in this genus, all species of which are CITES-protected, have a black market value of up to €600 each. (From Endrödy-Younga, 1988.)

in one year, and about 20 000 people were employed in the butterfly industry in 1975, with worldwide retail sales of butterflies possibly being as high as US $100 million (Parsons, 1995) (Figure 12.7).

12.7 Butterfly farming

Parsons (1995) has pointed out that swallowtail butterflies are particularly showy. Endemic and rare ones are an extremely valuable sustainable resource, but collected specimens are still predominating over farmed ones. Nevertheless, farming projects are under way. With its great financial and educational returns, butterfly farming can play a direct and important supportive role in the protection of swallowtail habitats, and indirectly other insect habitats in tropical forests (Figure 12.8). Several key factors must be in place during the crucial set up phase of farming projects, particularly adequate funding and the constant input of able expertise and tuition that this permits. There must also be an appropriate level of government backing and associated well-formulated and dedicated legislation, which includes not just protection of particular species under CITES but also for protection of the forest habitats. Exposure by the public to swallowtail butterflies, for example, can increase the lobbying power that can instigate new laws.

12.7 Butterfly deadstock on sale in eastern Asia. Many such specimens are captively bred, but occasionally protected, wild-caught specimens are on sale, and purchasers should be cautious as to what they are buying. A certificate or letter of bona fide should be requested should the purchaser have any doubt.

One must not underestimate the power of economics in relation to conservation awareness. Commercial collecting of butterflies in Papua New Guinea can have a potential beneficial impact on conservation as part of an overall economic, as well as ecological, package that includes the extractive reserve concept, so long as the collecting activity does not increase the probability of extinction of a given collected species. It also assumes that the effects of externalities such as commercial logging and other forms of habitat modification are not having an adverse impact. Villagers in Papua New Guinea are often given the chance to make 'quick money' from foreign companies for clear-cutting their forest. If villagers instead could make money from their forest without cutting it, even if it increased the probability of extinction for some collected species, it would still be preferable ecologically to the more certain extinctions of many species due to deforestation (Slone *et al.*, 1997).

12.8 Summary

The Convention on Biological Diversity (CBD) is the most important of the many international laws, treaties and protocols which aim to improve sustainable use of resources and species. The Convention on International Trade in Endangered Species (CITES), in turn, aims to protect species threatened by

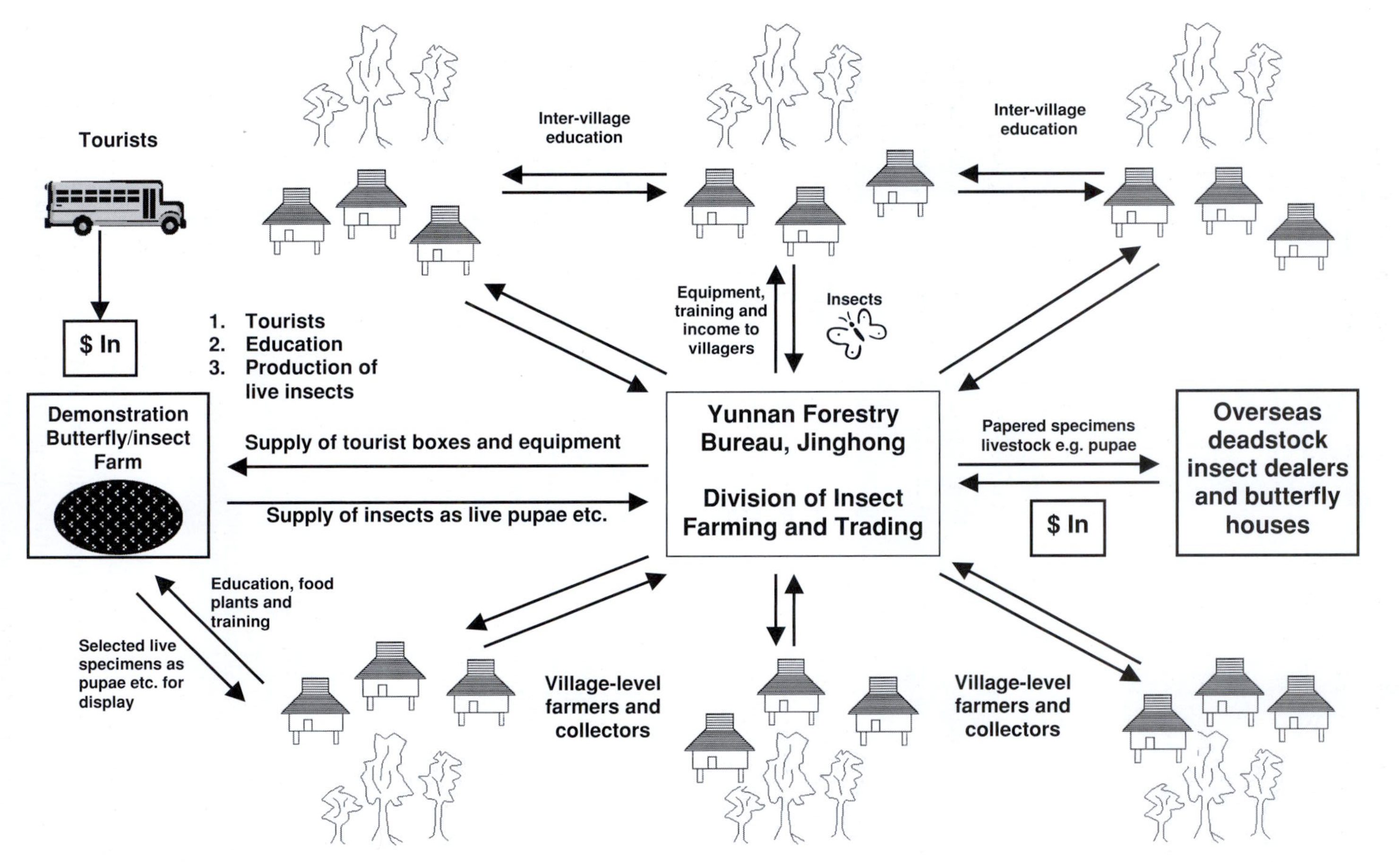

12.8 Diagrammatic representation of the overall integration of butterfly farming with tropical forest conservation as demonstrated by the Xishuangbanna project in southern China. (From Parsons, 1995.)

international trade. However, rarely is collecting and trade per se the most threatening process for insects, but rather habitat destruction. Besides, legal trade restrictions are generally not observed. Timber-importing countries may stipulate that forest products are produced in a way that maintains natural biodiversity, as in networks of remnant grassland corridors being left between planted stands of pine.

Many countries have national and regional regulations. Some countries such as Britain have a Biodiversity Action Plan which involves both conservation of specific species and their habitats, as well as landscapes, to ensure survival in the long term of a variety of plants and animals. Activities in Britain also now involve a commitment to improving wildlife habitats on farms.

While butterflies and dragonflies are often sufficiently charismatic to curry conservation attention, this may not be the case with less cryptic species. Nevertheless, the Red List does recognize intrinsic value, and gives equal weighting to all species on the list, no matter how small and brown the species. Such listed rarities become icons in their own right, although the more charismatic species, especially if rare, like the birdwing butterflies, are instant icons and 'stand in' for other insects both in terms of habitat protection and in the eyes of the public. This emphasizes, as we have observed in earlier chapters, the important mutualism between the fine-filter species approach and the coarse-filter habitat/landscape one. We need both approaches for effective conservation.

Butterfly houses have become an important medium at the interface between insects and the public. There is no evidence that threatened species are used, and much of the material is captively bred. These butterfly houses create in the public a sense of appreciation for the insects, and makes them feel empowered to champion habitat conservation locally and abroad. Additionally, butterfly houses play a role in the rearing for release of certain rare species. They also engender a sense of ecotourism and an increased awareness of the plight of insects in general.

The deadstock trade runs into tens of millions of dollars annually, and often rare and threatened species are traded. Indeed, some, such as the rare South African *Colophon* stag beetles, which are not at all glamorous, sell purely on their perceived rarity status.

Butterfly farming projects in indigenous areas can play an important role in the protection of swallowtail butterflies through taking pressure off wild-caught specimens. However, it is essential that such farming practices are set up with adequate financial backing to be feasible and effective in taking pressure off wild forests. Nevertheless, wild catching of butterflies, although risking extinction of certain species, is preferable to wholesale logging of forests and consequent local or total extinction of many species.

References

Abbadie, L., Lepage, M. and Le Roux, X. (1992) Soil fauna at the forest-savanna boundary: role of termite mounds in nutrient cycling. In Furley, P. A., Proctor, J. and Ratter, J. A. (eds.), *Nature and Dynamics of Forest-Savanna Boundaries*, pp. 473–84. London: Chapman and Hall.

Abensperg-Traun, M. and Smith, G. T. (1999) How small is too small for small animals? Four terrestrial arthropods species in different-sized remnant woodlands in agricultural Western Australia. *Biodiversity and Conservation* **8**, 709–26.

Abensperg-Traun, M., Smith, G. T., Arnold, G. W. and Steven, D. E. (1996) The effects of habitat fragmentation and livestock-grazing on animal communities in remnants of gimlet *Eucalyptus salubris* woodland in Western Australian wheat belt. I. Arthropods. *Journal of Applied Ecology* **33**, 1281–301.

Acheta domestica, M. E. S. (1849) *Episodes of Insect Life*. London: Reeve, Benham and Reeve.

Adis, J. (1990) Thirty million arthropod species – too many or too few? *Journal of Tropical Ecology* **6**, 115–18.

Adsersen, H. (1995) Research on islands: classic, recent, and prospective approaches. In Vitousek, P. M., Loope, L. L. and Adsersen, H. (eds.), *Islands: Biological Diversity and Ecosystem Function*, pp. 7–21. Berlin: Springer-Verlag.

Akçakaya, H. R. and Ferson, S. (2001) *RAMAS® Red List: Threatened Species Classifications under Uncertainty*. Version 2.0. Applied Biomathematics, New York.

Akçakaya, H. R., Ferson, S., Burgman, M. A., Keith, D. A., Mace, G. M. and Todd, C. A. (2000) Making consistent IUCN classifications under uncertainty. *Conservation Biology* **14**, 1001–13.

Alaruikka, D., Kotze, D. J., Matveinen, K. and Niemelä, J. (2002) Carabid beetle and spider assemblages along a forested urban-rural gradient in southern Finland. *Journal of Insect Conservation* **6**, 195–206.

Albariño, R. J. and Balseiro, E. G. (2002) Leaf litter breakdown in Patagonian streams: native versus exotic trees and the effect of invertebrate size. *Aquatic Conservation* **12**, 181–92.

Allen-Wardell, G., Bernhardt, P., Bitner, R. *et al.* (1998) The potential consequences of pollinator declines on the conservation of biodiversity and stability of food crops yields. *Conservation Biology* **12**, 8–17.

Andelman, S. J. and Fagan, W. F. (2000) Umbrellas and flagships: efficient conservation surrogates or expensive mistakes? *Proceedings of the National Academy of Sciences, USA* **97**, 5954–9.

Andersen, A. N. (1997a) Using ants as bioindicators: multiscale issues in ant community ecology. *Conservation Ecology* **1**(1), 8. (Online URL: http.//www.consecol.org/vol1/iss1/art8.)

(1997b) Ants as indicators of ecosystem restoration following mining: a functional group approach. In Hale, P. and Lamb, D. (eds.), *Conservation outside Nature Reserves*, pp. 319–25. University of Queensland, Brisbane: Centre for Conservation Biology.

(1999) My bioindicator or yours? Making the selection. *Journal of Insect Conservation* **3**, 61–4.

Andersen, A. N., Hoffmann, B. D., Müller, W. J. and Griffiths, A. D. (2002) Using ants as bioindicators in land management: simplifying assessment of ant-community responses. *Journal of Applied Ecology* **39**, 8–17.

Andersen, A. N., Hoffmann, B. D. and Some, J. (2003) Ants as indicators of minesite restoration: community recovery at one of eight rehabilitation sites in central Queensland. *Ecological Management and Restoration* **4** (Suppl.), 512–19.

Andersen, A. N. and Sparling, G. P. (1997) Ants as indicators of restoration success: relationship with soil microbial biomass in Australian seasonal tropics. *Restoration Ecology* **5**, 109–14.

Andersen, J. (2000) What is the origin of the carabid beetle fauna of dry, anthropogenic habitats in Western Europe? *Journal of Biogeography* **27**, 795–806.

Anderson, J. M. (2000) Food web functioning and ecosystem processes: problems and perceptions of scaling. In Coleman, D. C. and Hendrix, P. F. (eds.), *Invertebrates as Webmasters in Ecosystems*, pp. 3–24. Wallingford: CAB International.

Andrieu-Ponel, V. and Ponel, P. (1999) Human impact on Mediterranean wetland Coleoptera: an historical perspective at Tourves (Var, France). *Biodiversity and Conservation* **8**, 391–407.

Angelibert, S. and Giani, N. (2003) Dispersal characteristics of three odonate species in a patchy habitat. *Ecography* **26**, 13–20.

Anonymous (1986/87) The Assisi event. *The New Road* No. 1. Gland, Switzerland: WWF-International.

Aoki, T. (1997) Northward expansion of *Ictinogomphus pertinax* (Selys) in eastern Shikoku and western Kinki Districts, Japan (Anisoptera: Gomphidae). *Odonatologica* **26**, 121–33.

Appelt, M. and Poethke, H. J. (1997) Metapopulation dynamics in a regional population of the blue-winged grasshopper (*Oedipoda caerulescens*, Linnaeus, 1758). *Journal of Insect Conservation* **1**, 205–14.

Araújo, M. B. (2002) Biodiversity hotspots and zones of ecological transition. *Conservation Biology* **16**, 1662–3.

Araújo, M. B. and Williams, P. H. (2001) The bias of complementary hotspots toward marginal populations. *Conservation Biology* **15**, 1710–20.

Armstrong, A. J. and van Hensbergen, H. J. (1997) Evaluation of afforestable montane grasslands for wildlife conservation in the north-eastern Cape, South Africa. *Biological Conservation* **81**, 179–90.

Ashmole, N. P., Nelson, J. M., Shaw, M. R. and Garside, A. (1983) Insects and spiders on snowfields in the Cairngorms, Scotland. *Journal of Natural History* **17**, 599–613.

Ashworth, A. C. (1996) The response of arctic Carabidae (Coleoptera) to climate change based on the fossil record of the Quaternary Period. *Annals Zoologici Fennici* **33**, 125–31.

Ashworth, A. C. and Hoganson, J. W. (1993) The magnitude and rapidity of the climate change marking the end of the Pleistocene in the mid-latitudes of South America. *Palaeogeography, Palaeoclimatology, Palaeoecology* **101**, 263–70.

Askew, R. R. (1971) *Parasitic Insects*. Heinemann, London, UK.

Assmann, T. and Janssen, J. (1999) The effects of habitat changes on the endangered ground beetle *Carabus nitens* (Coleoptera: Carabidae). *Journal of Insect Conservation* **3**, 107–16.

Aubert, B. and Quilici, S. (1983) Novel equilibre biologique observé à la Réunion sur les populations de psyllides après l'introduction et l'établissement d'hymenopteres chalcidiens. *Fruits* **38**, 771–80.

Auerbach, M. J., Connor, E. F. and Mopper, S. (1995) Minor miners and major miners: population dynamics of leaf-mining insects. In Cappuccino, N. and Price, P. W. (eds.), *Population Dynamics: New Approaches and Synthesis*, pp. 83–110. San Diego, CA: Academic Press.

Austin, A. D. (1999) The importance of "species" in biodiversity studies: lessons from a mega-diverse group – the parasitic Hymenoptera. In Ponder, W. and Lunney, D. (eds.), *The Other 99%: the Conservation and Biodiversity of Invertebrates*, pp. 159–65. Mosman, Australia: Royal Zoological Society of New South Wales.

Ayres, M. P. and Lombardero, M. J. (2000) Assessing the consequence of global change for forest disturbance from herbivores and pathogens. *Science of the Total Environment* **262**, 263–86.

Bäckman, J.-P. C. and Tiainen, J. (2002) Habitat quality of field margins in a Finnish farmland area for bumblebees (Hymenoptera: *Bombus* and *Psithyrus*). *Agriculture, Ecosystems and Environment* **89**, 53–68.

Baguette, M. and Schtickzelle, N. (2003) Local population dynamics are important to the conservation of metapopulations in highly fragmented landscapes. *Journal of Applied Ecology* **40**, 404–12.

Bailey, S. A., Haines-Young, R. H. and Watkins, C. (2002) Species presence in fragmented landscapes: modelling the species requirements at the national level. *Biological Conservation* **108**, 307–16.

Balleto, E. and Casale, A. (1991) Mediterranean insect conservation. In Collins, M. N. and Thomas, J. A. (eds.), *The Conservation of Insects and their Habitats*, pp. 121–42. London: Academic Press.

Balmer, O. and Erhardt, A. (2000) Consequences of succession on extensively grazed grasslands for Central European butterfly communities: rethinking conservation practices. *Conservation Biology* **14**, 746–57.

Barbalat, S. (1996) Influence de l'exploitation forestière sur trois familes de coléoptères lies au bois dans les Gorges de l'Areuse. *Revue Suisse de Zoologie* **103**, 553–64.

Barbero, E., Palestrini, C. and Rolando, A. (1999) Dung beetle conservation: effects of habitat and resource selection (Coleoptera: Scarabaeoidea). *Journal of Insect Conservation* **3**, 75–84.

Barr, T. C. and Holsinger, J. R. (1985) Speciation in cave faunas. *Annual Review of Ecology and Systematics* **16**, 313–17.

Barratt, B. I. P., Ferguson, C. M. and Evans, A. A. (1996) Implications of nontarget effects of introduced biological control agents particularly with respect to beneficial species. *Proceedings of the 20th International Congress of Entomology*, p. 634. Firenze, Italy.

(2001) Nontarget effects of introduced biological control agents and some implications for New Zealand. In Lockwood, J. A., Holwarth, F. G. and Purcell, M. F. (eds.), *Balancing Nature: Assessing the Impact of Importing Non-native Biological Control Agents (an International Perspective)*, pp. 41–54. Lanham, MA: Entomological Society of America.

Bartlett, R., Pickering, J., Gauld, I. and Windsor, D. (1999) Estimating global biodiversity: tropical beetles and wasps send different signals. *Ecological Entomology* **24**, 118–21.

Basset, Y. (1996) Local communities of arboreal herbivores in Papua New Guinea: predictors of insect variables. *Ecology* **77**, 1906–19.

Basset, Y., Novotny, V., Miller, S. E. and Pyle, R. (2000) Quantifying biodiversity: experience with parataxonomists and digital photography in Papua New Guinea and Guyana. *BioScience* **50**, 899–908.

Beardsley, J. W. Jr (1991) Introduction of arthropod pests into the Hawaiian islands. *Micronesia (Supplement)* **3**, 1–4.

Bei-Bienko, G. Y. (1970) Orthopteroid insects of the national park areas near Kursk and their significance as indices of the local landscape. *Zhurnal Obschei Biologii* **31**, 30–46.

Beier, P. and Noss, R. (1998) Do habitat corridors provide connectivity? *Conservation Biology* **12**, 1241–52.

Bell, H. A., Edwards, J. P., Gatehouse, J. A. and Gatehouse, A. M. R. (2002) The interactions between a genetically modified crop expressing snowdrop lectin (GNA) and beneficial biological control agents. *Antenna* **26**, 97–8.

Belovsky, G. E. (2000) Grasshoppers as integral elements of grasslands. In Lockwood, J. A., Latchininsky, A. V. and Sergeev, M. G. (eds.), *Grasshoppers and Grassland Health*, pp. 7–29. Dordrecht, The Netherlands: Kluwer.

Belshaw, R. and Bolton, B. (1993) The effect of forest disturbance on the leaf litter ant fauna in Ghana. *Biodiversity and Conservation* **2**, 656–66.

Beneš, V., Kepka, P. and Konvička, M. (2003) Limestone quarries as refuges for European xerophilous butterflies. *Conservation Biology* **17**, 1058–69.

Bennett, A. F. (1999) *Linkages in the Landscape: The Role of Corridors and Connectivity in Wildlife Conservation*. Gland, Switzerland and Cambridge: IUCN.

Benton, T. G., Bryant, D. M., Cole, L. and Crick, H. Q. P. (2002) Linking agricultural practice to insect and bird populations: a historical study over three decades. *Journal of Applied Ecology* **39**, 673–87.

Berggren, Å., Birath, B. and Kindvall, O. (2002) Effect of corridors and habitat edges on dispersal behavior, movement rates, and movement angles in Roesel's bush-cricket (*Metrioptera roeseli*). *Conservation Biology* **16**, 1562–9.

Bergman, K.-O. and Landin, J. (2002) Population structure and movements of a threatened butterfly (*Lopinga achine*) in a fragmented landscape in Sweden. *Biological Conservation* **108**, 361–9.

Bernays, E. A. (1998) Evolution of feeding behaviour in insect herbivores. *BioScience* **48**, 35–44.

Berry, R. E. and Taylor, L. R. (1968) High altitude migration of aphids in maritime and continental climates. *Journal of Animal Ecology* **37**, 713–22.

Berryman, A. A. (1996) What causes population cycles of forest Lepidoptera? *Trends in Ecology and Evolution* **11**, 28–32.

Betzholtz, P.-E. (2002) Population structure and movement patterns within an isolated and endangered population of the moth *Dysauxes ancilla* L. (Lepidoptera: Ctenuchidae): implications for conservation. *Journal of Insect Conservation* **6**, 57–66.

Bhattacharya, M., Primack, R. B. and Gerwien, J. (2003) Are roads and railroads barriers to bumblebee movement in a temperate suburban conservation area? *Biological Conservation* **109**, 37–45.

Biedermann, R. (2000) Metapopulation dynamics of the froghopper *Neopilaenus albipennis* (F., 1798) (Homoptera, Cercopidae) – what is the minimum viable population size? *Journal of Insect Conservation* **4**, 99–107.

Bieringer, G. and Zulka, K. P. (2003) Shading out species richness: edge effect of a pine plantation on the Orthoptera (Tettigoniidae and Acrididae) assemblage of an adjacent dry grassland. *Biodiversity and Conservation* **12**, 1481–95.

Bignell, D. E., Eggleton, P., Nunes, L. and Thomas, K. L. (1997) Termites as mediators of carbon fluxes in tropical forests: budgets for carbon dioxide and methane emissions. In Watt, A. D., Stork, N. E. and Hunter, M. D. (eds.), *Forests and Insects*, pp. 109–34. London: Chapman and Hall.

Blackmore, S. (1999) *The Meme Machine*. Oxford: Oxford University Press.

Blair, R. B. (1999) Birds and butterflies along an urban gradient: surrogate taxa for assessing biodiversity? *Ecological Applications* **9**, 164–70.

Blair, R. B. and Launer, A. E. (1997) Butterfly diversity and human land use: species assemblages along an urban gradient. *Biological Conservation* **80**, 113–25.

Blake, S., Foster, G. N., Fisher, G. E. J. and Ligertwoow, G. L. (1996) Effects of management practices on the carabid faunas of newly-established wildflower meadows in southern Scotland. *Annals Zoologici Fennici* **33**, 139–47.

Boecklen, W. J. (1997) Nestedness, biogeographic theory, and the design of nature reserves. *Oecologia* **112**, 123–42.

Boersma, P. D., Kareiva, P., Fagan, W. F., Clark, J. A. and Hoeksta, J. M. (2001) How good are endangered species recovery plans? *BioScience* **51**, 643–9.

Boettner, G. H., Elkington, J. S. and Boettner, C. J. (2000) Effects of a biological control introduction on three nontarget native species of saturniid moths. *Conservation Biology* **14**, 1798–806.

Bolger, D. T., Suarez, A. V., Crooks, K. R., Morrison, S. A. and Case, T. J. (2000) Arthropods in urban habitat fragments in southern California: area, age, and edge effects. *Ecological Applications* **10**, 1230–48.

Bonaszak, J. (1992) Strategy for conservation of wild bees in an agricultural landscape. *Agriculture, Ecosystems and Environment* **40**, 179–92.

Bond, W. J. (1994) Do mutualisms matter? Assessing the impact of pollinator and disperser disruption on plant extinction. *Philosophical Transactions of the Royal Society of London, B* **344**, 83–90.

Bond, W. J. and Slingsby, P. (1984) Collapse of an ant-plant mutualism: the Argentine ant (*Iridomyrmex humilis)* and myrmecochorous Proteaceae. *Ecology* **65**, 1031–7.

Booij, C. J. H. and Noorlander, J. (1992) Farming systems and insect predators. *Agriculture, Ecosystems and Environment* **40**, 125–35.

Booth, M. and Allen, M. M. (1998) Butterfly garden design. In *Butterfly Gardening*, created by The Xerces Society in association with The Smithsonian Institution, pp. 69–93. San Francisco, CA: Sierra Club.

Borges, P. A. V. (1999) Plant and arthropod species composition of sown and semi-natural pasture communities of three Azorean Islands (Santa Maria, Terceira and Pico) *Arquipélago. Life and Marine Sciences* **17A**, 1–21.

Bourtzis, K. and O'Neill, S. L. (1998) *Wolbachia* infections and arthropod reproduction. *BioScience* **48**, 287–93.

Bradley, J. D. and Mere, R. M. (1966) Natural history of the garden of Buckingham Palace. Further records and observations 1964–1965. Lepidoptera. *Proceeding of the South London Entomological and Natural History Society* **1966**,15–17.

Bradshaw, A. D., Goode, D. A. and Thorp, E. (eds.) (1986) *Ecology and Design in Landscape*. Oxford: Blackwell Scientific Publications.

Braker, E. (1991) Natural history of a neotropical gap-inhabiting grasshopper. *Biotropica* **23**, 41–50.

Brändle, M., Amarell, U., Auge, H., Klotz, S. and Brandl, R. (2001) Plant and insect diversity along a pollution gradient: understanding species richness across trophic levels. *Biodiversity and Conservation* **10**, 1497–511.

Brändle, M., Durka, W. and Altmoos, M. (2000) Diversity of surface dwelling beetle assemblages in open-cast lignite mines in Central Germany. *Biodiversity and Conservation* **9**, 1297–311.

Briggs, C. J. and Godfray, H. C. J. (1996) Dynamics of insect-pathogen interactions in seasonal environments. *Theoretical Population Biology* **50**, 149–77.

Briggs, J. C. (1991) Global species diversity. *Journal of Natural History* **25**, 1403–6.

Brommer, J. E. and Fred, M. S. (1999) Movement of the Apollo butterfly *Parnassius apollo* related to host plant and nectar plant patches. *Ecological Entomology* **24**, 125–31.

Brooks, D. R., Bohan, D. A., Champion, G. T. *et al.* (2003) Invertebrate responses to the management of genetically modified herbicide-tolerant and conventional spring crops. I. Soil–surface-active invertebrates. *Proceedings of the Royal Society of London, B* **358** 1847–62.

Brower, L. P. (1977) Monarch migration. *Natural History* **86**, 40–53.

Brower, L. P., Castillegja, G., Peralta, A., Lopez-Garcia, J., Bojorquez-Tapia, L., Diz, S., Melgarejo, D. and Missrie, M. (2002) Quantitative changes in forest quality in a principal overwintering area of the Monarch butterfly in Mexico, 1971–1999. *Conservation Biology* **16**, 346–59.

Brown, J. C. and Albrecht, C. (2001) The effect of tropical deforestation on stingless bees of the genus *Melipona* (Insecta: Hymenoptera: Apidae: Meliponini) in central Rondonia, Brazil. *Journal of Biogeography* **28**, 623–34.

Brown, K. S., Jr (1997) Diversity, disturbance, and sustainable use of an Indochinese montane rainforest. *Biological Conservation* **80**, 9–15.

Brown, K. S., Jr and Freitas, A. V. L. (2002) Butterfly communities of urban forest fragments in Campinas, São Paulo, Brazil: structure, instability, environmental correlates, and conservation. *Journal of Insect Conservation* **6**, 217–31.

Brown, K. S., Jr and Hutchings, H. R. W. (1997) Disturbance, fragmentation and the dynamics of diversity in Amazonian forest butterflies. In Laurance, W. F. and Bierregaard, R. O., Jr (eds.), *Tropical Forest Remnants: Ecology, Management and Conservation of Fragmented Communities*, pp. 91–110. Chicago, IL: University of Chicago Press.

Brown, V. K. (1995) Insect herbivores and gaseous air pollutants – current knowledge and predictions. In Harrington, R. and Stork, N. E. (eds.), *Insects in a Changing Environment*, pp. 219–49. London: Academic Press.

Brown, V. K., Hendrix, S. D. and Dingle, M. (1987) Plants and insects in early old-field succession: comparison of an English site and an American site. *Biological Journal of the Linnean Society* **31**, 59–74.

Brühl, C. A., Eltz, T. and Linsenmair, E. K. (2003) Effect of tropical rainforest fragmentation on the leaf litter ant community in Sabah, Malaysia. *Biodiversity and Conservation* **12**, 1371–89.

Bruner, A. G., Gullison, R. E., Rice, R. E. and da Fonseca, G. A. B. (2001) Effectiveness of parks in protecting biodiversity. *Science* **291**, 125–8.

Burchard, I. (1998) Anthropogenic impact on the climate since man began to hunt. *Palaeogeography, Palaeoclimatology, Palaeoecology* **139**, 1–14.

Burel, F. and Baudry, J. (1995) Farming landscapes and insects. In Glen, D. M., Greaves, M. P. and Anderson, H. M. (eds.), *Ecology and Integrated Farming Systems*, pp. 203–20. Chichester: John Wiley & Sons.

Burger, J. C., Redak, R. A., Allen, E. B., Rotenberry, J. T. and Allen, M. F. (2003) Restoring arthropod communities in coastal sage scrub. *Conservation Biology* **17**, 460–7.

Buse, A. and Good, J. E. (1996) Synchronization of larvae emergence in winter moth (*Operophtera brumata* L.) and budburst in pedunculate oak (*Quercus robur* L.) under simulated climate change. *Ecological Entomology* **21**, 335–43.

Bush, A. O., Fernández, J. C., Esch, G. W. and Seed, J. R. (2001) *Parasitism: The Diversity and Ecology of Animal Parasites*. Cambridge: Cambridge University Press.

Butterfield, J. (1997) Carabid community succession during the forestry cycle in conifer plantations. *Ecography* **20**, 614–25.

Cagnolo, L., Molina, S. I. and Valladares, G. R. (2002) Diversity and guild structure of insect assemblages under grazing and exclusion regimes in a montane grassland from Central Argentina. *Biodiversity and Conservation* **11**, 407–20.

Caley, M. J. and Schluter, D. (1997) The relationship between local and regional diversity. *Ecology* **78**, 70–80.

Callicot, J. B. (1990) Whither conservation ethics? *Conservation Biology* **4**, 15–20.

Capra, F. (1996) *The Web of Life*. Flamingo, London.

Caro, T. M. and Laurenson, M. K. (1994) Ecological and genetic factors in conservation: a cautionary tale. *Science* **263**, 485–6.

Caro, T. M. and O'Doherty, G. (1999) On the use of surrogate species in conservation biology. *Conservation Biology* **13**, 805–14.

Carpenter, J. E., Bloem, K. A. and Bloem, S. (2001) Applications of F1 sterility for research and management of *Cactoblastis cactorum* (Lepidoptera: Pyralidae). *Florida Entomologist* **84**, 531–6.

Carpintero, S., Reyes-López, J. and de Reyna, L. A. (2003) Impact of human dwellings on the distribution of the exotic Argentine ant: a case study in the Doñana National Park, Spain. *Biological Conservation* **115**, 279–89.

Carreck, N. L. and Williams, I. H. (2002) Food for insect pollinators on farmland: insect visits to flowers of annual seed mixtures. *Journal of Insect Conservation* **6**, 13–23.

Carvell, C. (2002) Habitat use and conservation of bumblebees (*Bombus* spp.) under diffent grassland management regimes. *Biological Conservation* **103**, 33–49.

Castaño-Meneses, G. and Palacios-Vargas, J. G. (2003) Effects of fire and agricultural practices on neotropical ant communities. *Biodiversity and Conservation* **12**, 1913–19.

Cartagena, M. C. and Galante, E. (2002) Loss of Iberian island tenebrionid beetles and conservation management recommendations. *Journal of Insect Conservation* **6**, 73–81.

Chambers, B. Q. and Samways, M. J. (1998) Grasshopper responses to a 40-year experimental burning and mowing regime, with recommendations for invertebrate conservation management. *Biodiversity and Conservation* **7**, 985–1012.

Chardon, J. P., Adriaensen, F. and Matthysen, E. (2003) Incorporating landscape elements into a connectivity measure: a case study for the speckled wood butterfly (*Parange aegeria* L.). *Landscape Ecology* **18**, 561–73.

Chase, J. M. and Leibold, M. A. (2002) Spatial scale dictates the productivity – biodiversity relationship. *Nature* **416**, 427–30.

Cherrill, A. J. and Brown, V. K. (1990) The life cycle and distribution of the wart-biter *Decticus verrucivorus* (L.) (Orthoptera: Tettigoniidae) in a chalk grassland in southern England. *Biological Conservation* **53**, 125–43.

Chown, S. L. and Block, W. (1997) Comparative nutritional ecology of grass-feeding in a sub-Antarctic beetle: the impact of introduced species on *Hydromedion sparsutum* from South-Georgia. *Oecologia* **111**, 216–24.

Chown, S. L., McGeoch, M. A. and Marshall, D. J. (2002) Diversity and conservation of invertebrates on the sub-Antarctic Prince Edward Islands. *African Entomology* **10**, 67–82.

Chust, G., Pretus, J. L., Ducrot, D., Bedôs, A. and Deharveng, L. (2003) Identification of landscape units from an insect perspective. *Ecography* **26**, 257–68.

Cilgi, T. and Jepson, P. (1995) Pesticide spray drift into field boundaries and hedgerows: toxicity to non-target Lepidoptera. *Journal of Environment and Pollution* **87**, 1–9.

Cilliers, S. S., van Wyk, E. and Bredenkamp, G. J. (1999) Urban nature conservation: vegetation of natural areas in the Potchefstroom area, North West Province, South Africa. *Koedoe* **42** /**1**,1–30.

Cincotta, R. P. and Engelman, R. (2000). *Nature's Place: Human Population and the Future of Biological Diversity*. Washington, DC: Population Action International.

Clausnitzer, V. (2003) Dragonfly communities in coastal habitats of Kenya: indication of biotope quality and the need of conservation measures. *Biodiversity and Conservation* **12**, 333–56.

Clark, T. E. and Samways, M. J. (1997) Sampling arthropod diversity for urban ecological landscaping in a species-rich southern hemisphere botanic garden. *Journal of Insect Conservation* **1**, 221–34.

Clarke, R. T., Thomas, J. H., Elmes, G. W. and Hachberg, M. E. (1997) The effects of spatial patterns in habitat quality on community dynamics within a site. *Proceedings of the Royal Society of London, B* **264**, 347–54.

Clout, M. N. (1999) Biodiversity conservation and the management of invasive animals in New Zealand. In Sandlund, O. T., Schei, P. J. and Viken, Å. (eds.), *Invasive Species and Biodiversity Management*, pp. 349–61. Dordrecht, The Netherlands: Kluwer.

Clout, M. N. and Lowe, S. J. (2000) Invasive species and environmental changes in New Zealand. In Mooney, H. A. and Hobbs, R. J. (eds.), *Invasive Species in a Changing World*, pp. 369–83. Washington, DC: Island Press.

Cochrane, M. A. and Schulze, M. D. (1999) Fire as a recurrent event in tropical forests of the eastern Amazon: effects on forest structure, biomass, and species composition. *Biotropica* **31**, 2–16.

Coleman, D. C. and Hendrix, P. F. (eds.) (2000) *Invertebrates as Webmasters in Ecosystems*. Wallingford: CAB International.

Collinge, S. K. (2000) Effects of grassland fragmentation on insect species loss, colonization, and movement patterns. *Ecology* **81**, 2211–26.

Collinge, S. K., Prudic, K. L. and Oliver, J. C. (2003) Effects of local habitat characteristics and landscape context on grassland butterfly diversity. *Conservation Biology* **17**, 178–87.

Collins, N. M. and Morris, M. G. (1985) *Threatened Swallowtail Butterflies of the World*. Gland, Switzerland and Cambridge: IUCN.

Collins, N. M. and Smith, H. M. (1995) Threats and priorities concerning swallowtail butterflies. In Scriber, J. M., Tsubaki, Y. and Lederhouse, R. C. (eds.), *Swallowtail Butterflies: their Ecology and Evolutionary Biology*, pp. 345–57. Gainesville, FL: Scientific Publications.

Colville, J., Picker, M. D. and Cowling, R. M. (2002) Species turnover of monkey beetles (Scarabaeidae: Hopliini) along environmental and disturbance gradients in the Namaqualand region of the succulent Karoo, South Africa. *Biodiversity and Conservation* **11**, 243–64.

Colwell, R. K. and Coddington, J. A. (1994) Estimating terrestrial biodiversity through extrapolation. *Philosophical Transactions of the Royal Society of London, B* **345**, 101–18.

Combes, C. (1996) Parasites, biodiversity and ecosystem stability. *Biodiversity and Conservation* **5**, 953–62.

Connell, J. M. (1978) Diversity in tropical rain forests and coral reefs. *Science* **1999**, 1302–10.

Connor, E. F., Adams-Manson, R. H., Carr, T. G. and Beck, M. W. (1994) The effects of host plant phenology on the demography and population dynamics of the leaf-mining moth *Cameraria hamadryadella* (Lepidoptera: Gracillariidae). *Ecological Entomology* **19**, 111–120.

Connor, E. F., Hafernik, J., Levy, J., Moore, V. L. and Rickman, J. K. (2002) Insect conservation in an urban biodiversity hotspot: the San Francisco Bay Area. *Journal of Insect Conservation* **6**, 247–59.

Conrad, K. F., Willson, K. H., Harvey, I. F., Thomas, C. J. and Sherratt, T. N. (1999) Dispersal characteristics of seven odonate species in an agricultural landscape. *Ecography* **22**, 524–31.

Conradi, M., Brunzel, S. and Plachter, H. (1999) Dispersal and establishment of dung beetles in the genus *Aphodius* (Scarabaeidae, Coleoptera). *Verhandlungen de Gesellschaft für Ökologie* **29**, 349–55.

Conradt, L., Bodsworth, E. J., Roper, T. J. and Thomas, C. (2000) Nonrandom dispersal in the butterfly *Maniola jurtina*: implications for metapopulation model. *Proceedings of the Royal Society of London, B* **267**, 1505–10.

Cook, J. L. (2003) Conservation of biodiversity in an area impacted by the red imported fire ant, *Solenopsis invicta* (Hymenoptera: Formicidae). *Biodiversity and Conservation* **12**, 187–95.

Coope, G. R. (1995) Insect faunas in ice age environments: why so little extinction? In Lawton, J. H. and May, R. M. (eds.), *Extinction Rates*, pp. 55–74. Oxford: Oxford University Press.

Coope, G. R. and Brophy, J. A. (1972) Late Glacial environmental changes indicated by a coleopteran succession from North Wales. *Boreas* **1**, 97–142.

Corbet, S. A. (2000) Conserving compartments in pollination webs. *Conservation Biology* **14**, 1229–32.

Corke, D. (1999) Are honeydew/sap-feeding butterflies (Lepidoptera: Rhopalocera) affected by particulate air pollution? *Journal of Insect Conservation* **3**, 5–14.

Cory, J. (2002) The influence of host behaviour on transmission of recombinant baculovirus insecticides. *Antenna* **26**, 100–1.

Cottrell, C. B. (1985). The absence of co-evolutionary associations with Capensis Floral Element plants in the larval/plant relationships of Southwestern Cape butterflies. In Vrba, E. S. (ed.). *Species and Speciation*, pp. 115–24. Pretoria, South Africa: Transvaal Museum.

Couty, A., Ripoll, C., Azzouz, H., Kaiser, L., Pham-Delégue, M. H., Poppy, G. and Jouanin, L. (2002) Direct and indirect efforts of snowdrop lectin (GNA) on aphid parasitoids: impact on adult parasitoids. *Antenna* **26**, 95–7.

Couvet, D. (2002) Deleterious effects of restricted gene flow in fragmented populations. *Conservation Biology* **16**, 369–76.

Cowley, M. J. R., Thomas, C. D., Thomas, J. A. and Warren, M. S. (1999) Flight areas of British butterflies: assessing species status and decline. *Proceedings of the Royal Society of London, B* **266**, 1587–92.

Crist, T. O. and Wiens, J. A. (1996) The distribution of ant colonies in a semiarid landscape: implications for community and ecosystem processes. *Oikos* **76**, 301–11.

Crooks, J. A. and Soulé, M. E. (1999) Lag times in population explosions of invasive species: causes and implications. In Sandlund, O. T., Schei, P. J. and Viken, Å. (eds.), *Invasive Species and Biodiversity Management*, pp. 103–25. Dordrecht, The Netherlands: Kluwer.

Crome, F. H. J., Moore, L. A. and Richards, G. R. (1992) A study of logging damage in upland rainforest in north Queensland. *Forest Ecology and Management* **49**, 1–29.

Croxton, P. J., Carvell, C., Mountford, J. O. and Sparks, T. H. (2002) A comparison of green lanes and field margins as bumblebee habitat in an arable landscape. *Biological Conservation* **107**, 365–74.

Culver, D. C., Master, L. L., Christman, M. C. and Hobbs, H. H. III (2000) Obligate cave fauna of the 48 contiguous United States. *Conservation Biology* **14**, 386–401.

Curry, J. P. (1994) *Grassland Invertebrates: Ecology, Influence on Soil Fertility and Effects on Plant Growth*. London: Chapman and Hall.

Czechura, G. V. (1994) Is the public really interested in invertebrates? What the Queensland Museum Reference Centre enquiries from 1986–1993 tell us. *Memoirs of the Queensland Museum* **36**, 41–6.

Dale, V. H., Joyce, L. A., McNulty, S. *et al.* (2001) Climate change and forest disturbances. *BioScience* **51**, 723–34.

Dansgaard, W., White, J. W. C. and Johnsen, S. J. (1989) The abrupt termination of the younger Dryas event. *Nature* **339**, 532–3.

Davis, A. J., Jenkinson, L. S., Lawton, J. H., Shorrocks, B. and Wood, S. (1995) Global warming, population dynamics and community structure in a model insect assemblage. In Harrington, R. and Stork, N. E. (eds.), *Insects in a Changing Environment*, pp. 431–9. London: Academic Press.

Davis, A. J. and Sutton, S. L. (1998) The effects of rainforest canopy loss on arboreal dung beetles in Borneo: implications for the measurement of biodiversity in derived tropical ecosystems. *Diversity and Distributions* **4**, 167–73.

Davis, B. N. K. (1978) Urbanisation and the diversity of insects. In Mound, L. A. and Waloff, N. (eds.), *Diversity of Insect Faunas*, pp. 126–38. Oxford: Blackwell Scientific Publications.

(1989) Habitat creation for butterflies on a landfill site. *The Entomologist* **108**, 109–22.

De Kock, A. E. and Giliomee, J. H. (1989) A survey of the Argentine ant, *Iridomyrmex humilis* (Mayr), (Hymenoptera: Formicidae) in South African Fynbos. *Journal of Entomological Society of Southern Africa* **52**, 157–64.

De Maynadier P. and Hunter, M. L. (1994) Keystone support. *BioScience* **44**, 2.

De Moor, F. C. (1994) Aspects of the life history of *Simulium chutteri* and *S. bovis* (Diptera; Simuliidae) in relation to changing environmental conditions in South African rivers. *Verhandlungen Internationale Vereiningung für Theoretische und Angewandte Limnologie* **25**, 1817–21.

de Snoo, G. R. (1999) Unsprayed field margins: effects on environment, biodiversity and agricultural practice. *Landscape and Urban Planning* **46**, 151–60.

De Vries, H. H., den Boer, P. J. and van Dijk, Th. S. (1996) Ground beetles species in heathland fragments in relation to survival, dispersal, and habitat preference. *Oecologia* **107**, 332–42.

De Vries P. J., Murray, D. and Lande, R. (1997) Species diversity in vertical, horizontal and temporal dimensions of a fruit-feeding butterfly community in an Ecuadorian rainforest. *Biological Journal of Linnean Society* **62**, 343–64.

Death, R. G. (1996) The effect of patch disturbance on stream invertebrate community structure: the influence of disturbance history. *Oecologia* **108**, 567–76.

Dempster, J. P. (1989) Insect introductions: natural dispersal and population persistence in insects. *The Entomologist* **108**, 5–13.

Dennis, P. (1997) Impact of forest and woodland structure on insect abundance and diversity. In Watt, A. D., Stork, N. E. and Hunter, M. D. (eds.), *Forests and Insects*, pp. 321–40. London: Chapman and Hall.

Dennis, P., Young, M. R. and Gordon, I. J. (1998) Distribution and abundance of small insect and arachnids in relation to structural heterogeneity of grazed, indigenous grasslands. *Ecological Entomology* **23**, 253–64.

Dennis, R. L. H. (1977) *The British Butterflies: their Origins and Distribution*. Faringdon, UK: Classey.

(1993) Butterflies and Climate Change. Manchester: Manchester University Press.

(2001) Progressive bias in species status is symptomatic of fine-grained mapping units subject to repeated sampling. *Biodiversity and Conservation* **10**, 483–94.

Dennis, R. L. H. and Eales, H. T. (1997) Patch occupancy in *Coenonympha tullia* (Müller, 1764) (Lepidoptera: Satyrinae): habitat quality matters as much as patch size. *Journal of Insect Conservation* **1**, 167–76.

Dennis, R. L. H and Hardy, P. B. (1999) Targeting squares for survey: predicting species richness and incidence of species for a butterfly atlas. *Global Ecology and Biogeography Letters* **8**, 443–54.

Dennis, R. L. H. and Shreeve, T. G. (1991) Climate change and British butterfly fauna: opportunities and constraints. *Biological Conservation* **55**, 1–16.

Dennis, R. L. H., Shreeve, T. G., Olivier, A. and Coutsis, J. G. (2000) Contemporary geography dominates butterfly diversity gradients within the Aegean archipelago (Lepidoptera: Papilionoidea, Hesperioidea). *Journal of Biogeography* **27**, 1365–83.

Dennis, R. L. H., Shreeve, T. G. and Sparks, T. H. (1998) The effects of island area, isolation and source population size on the presence of the grayling butterfly *Hipparchia semele* (L.) (Lepidoptera: Satyridae) on British and Irish offshore islands. *Biodiversity and Conservation* **7**, 765–76.

Dennis R. L. H., Sparks, T. H. and Hardy, P. B. (1999) Bias in butterfly distribution maps: the effects of sampling effort. *Journal of Insect Conservation* **3**, 33–42.

Denno, R. F., Roderick, G. K., Peterson, M. A., Huberty, A. F., Döbel, H. G., Eubanks, M. D., Losey, J. E. and Langellotta, G. A. (1996) Habitat persistence underlies intraspecific variation in the dispersal strategies of planthoppers. *Ecological Monographs* **66**, 389–408.

Denys, C. and Schmidt, H. (1998) Insect communities on experimental mugwort (*Artemisia vulgaris* L.) plots along an urban gradient. *Oecologica* **113**, 269–77.

Derraik, J. G. B., Closs, G. P., Dickinson, K. J. M., Sirvid, P., Barratt, B. I. P. and Patrick, B. H. (2002) Arthropod morphospecies versus taxonomic species: a case study with Araneae, Coleoptera and Lepidoptera. *Conservation Biology* **16**, 1015–23.

Deutschländer, M. S. and Bredenkamp, G. J. (1999) Importance of vegetation analysis in the conservation management of the endangered butterfly *Aloeides dentatis dentatis* (Swierstra) (Lepidoptera, Lycaenidae). *Koedoe* **42/2**, 1–12.

Diaz, S. and Cabido, M. (2001) Vive la difference: plant functional diversity matters to ecosystem processes. *Trends in Ecology and Evolution* **16**, 646–55.

Didham, R. K., Ghazoul, J., Stork, N. E. and Davis, A. J. (1996) Insects in fragmented forests: a functional approach. *Trends in Ecology and Evolution* **11**, 255–60.

Dinnin, M. (1997) Holocene beetle assemblages from the Lower Trent floodplain at Bole Ings, Nottinghamshire, UK. In Ashworth, A. C., Buckland, P. C. and Sadler, J. P. (eds.), *Studies in Quaternary Entomology – an Inordinate Fondness for Insects*, pp. 83–104. Chichester: John Wiley & Sons.

Dixon, A. F. G. and Kindlmann, P. (1990) Role of plant abundance in determining the abundance of herbivorous insects. *Oecologia* **83**, 281–83.

Dixon, D. (1990) *Man after Man*. London: Blandford.

Docherty, M., Salt, D. T. and Holopainen, J. K. (1997) The impacts of climate change and pollution on forest pests. In Watt, A. D., Stork, N. E. and Hunter, M. D. (eds.), *Forests and Insects*, pp. 229–47. London: Chapman and Hall.

Dolek, M. and Geyer, A. (1997) Influence of management on butterflies of rare grassland ecosystems in Germany. *Journal of Insect Conservation* **1**, 125–30.

Dolling, W. P. (1991) *The Hemiptera*. New York, NY: Oxford University Press.

Donaldson, J., Nänni, I., Zachariades, C. and Kemper, J. (2002) Effects of habitat fragmentation on pollinator diversity and plant reproductive success in Renosterveld shrublands of South Africa. *Conservation Biology* **16**, 1267–76.

Donnelly, D. and Giliomee, J. H. (1985) Community structure of epigaeic ants in a pine plantation and newly-burnt fynbos. *Journal of the Entomological Society of Southern Africa* **48**, 259–65.

Dover, J. W. (1991) The conservation of insects on arable farmland. In Collins, N. M. and Thomas, J. A. (eds.), *The Conservation of Insects and their Habitats*, pp. 293–318. London: Academic Press.

(1997) Conservation headlands: effects on butterfly distribution and behaviour. *Agriculture, Ecosystems and Environment* **63**, 31–49.

(2001) Butterflies and hedges. *Butterfly Conservation News* **78**, 2–3.

Dover, J. W. and Fry, G. L. A. (2001) Experimental simulation of some visual and physical components of a hedge and the effects on butterfly behaviour in an agricultural landscape. *Entomologia Experimentalis et Applicata* **10**, 221–33.

Dover, J. W., Sparks, T. H. and Greatorex-Davis, J. N. (1997) The importance of shelter for butterflies in open landscapes. *Journal of Insect Conservation* **1**, 89–97.

Drake, V. A. and Gatehouse, A. G. (1995) *Insect Migration: Tracking Resources through Space and Time*. Cambridge: Cambridge University Press.

Drayton, B. and Primack, R. B. (1996) Plant species cost in an isolated conservation area in Metropolitan Boston from 1894 to 1993. *Conservation Biology* **10**, 30–9.

Drummond, H. (1889) *Tropical Africa*. London: Hodder and Stoughton.

Dudley, R. (1998) Atmospheric oxygen, giant paleozoic insects and the evolution of aerial locomotor performance. *Journal of Experimental Biology* **201**, 1043–50.

Duelli, P., Studer, M., Marchand, I. and Jakob, S. (1990) Population movements of arthropods between natural and cultivated areas. *Biological Conservation* **54**, 193–207.

Duffey, E. (1968) Ecological studies on the large copper butterfly *Lycaena dispar* Haw. *batavus* Obth. at Woodwalton Fen National Nature Reserve, Huntingdonshire. *Journal of Applied Ecology* **5**, 69–96.

Duffy, J. E. (2003) Biodiversity loss, trophic skew and ecosystem functioning. *Ecology Letters* **6**, 680–7.

Dufrêne, M. and Legendre, P. (1997) Species assemblages are indicator species: the need for a flexible asymmetrical approach. *Ecological Monographs* **67**, 345–66.

Dutton, A., Klein, H., Romeis, J. and Bigler, F. (2002) Assessing bt-toxin content of piercing-sucking herbivores on transgenic maize and consequences for *Chrysoperla carnea*. *Antenna* **26**, 105–6.

Dwyer, G., Dushoff, J., Elkinton, J. S. and Levin, S. A. (2000) Pathogen-driven outbreaks in forest defoliators revisited: building models form experimental data. *American Naturalist* **156**, 105–20.

Dyer, M. I. (2000) Herbivores, biochemical messengers and plants: aspects of intertrophic transduction. In Coleman, D. C. and Hendrix, P. F. (eds.), *Invertebrates as Webmasters in Ecosystems*, pp. 3–24. Wallingford: CAB International.

Edge, D. A. (2002) Some ecological factors influencing the breeding success of the Brenton blue butterfly, *Orachrysops niobe* (Trimen) (Lepidoptera: Lycaeidae). *Koedoe* **45/2**, 19–34.

Eggleton, P., Bignell, D. E., Sands, W. A., Mawdsley, N. A., Lawton, J. H., Wood, T. G. and Bignell, N. C. (1996) The diversity, abundance and biomass of termites under differing levels of disturbance in the Mbalmayo Forest Reserve, Southern Cameroon. *Philosophical Transactions of the Royal Society, B* **351**, 51–68.

Eggleton, P., Bignell, D. E., Sands, W. A., Waite, B., Wood, T. G. and Lawton, J. H. (1995) The species richness of termites (Isoptera) under different levels of forest disturbance in the Mbalmayo Forest Reserve, southern Cameroon. *Journal of Tropical Ecology* **11**, 85–98.

Eggleton, P., Williams, P. H. and Gaston, K. J. (1994) Explaining global termite diversity: productivity orhistory? *Biodiversity and Conservation* **3**, 318–30.

Ehrlich, P. R. and Raven, P. H. (1964) Butterflies and plants: a study in coevolution. *Evolution* **18**, 586–608.

Elias, S. A. (1994) *Quaternary Insects and their Environments*. Washington, DC: Smithsonian Institution.

Ellsbury, M. M., Powell, J. E., Forcella, F., Woodson, W. D., Clay, S. A. and Riedell, W. E. (1998) Diversity and dominant species of ground beetle assemblages (Coleoptera: Carabidae) in crop rotation and chemical input systems for the northern Great Plains. *Annals of the Entomological Society of America* **91**, 619–25.

Elligsen, H., Beinlich, B. and Plachter, H. (1997) Effects of large-scale cattle grazing on populations of *Coenonympha glycerion* and *Lesiommata megera* (Lepidoptera: Satyridae). *Journal of Insect Conservation* **1**, 13–23.

(1998) Large-scale grazing systems and species protection in the Eastern Carpathians of Ukraine. *La Cañada* **9**, 10–12.

Emmel, T. C. and Boender, R. (1991) Wings in paradise: Florida's Butterfly World. *Wings* **15(3)**, 7–12.

Endrödy-Younga, S. (1988) Evidence of the low-altitude origin of the Cape mountain biome derived from systematic revision of the genus *Colophon* Gray (Coleoptera, Lucanidae). *Annals of the South African Museum* **96**, 359–424.

Englund, R. A. (1999) The impacts of introduced poeciliid fish and Odonata on endemic *Megalagrion* (Odonata) damselflies on Oahu Island, Hawaii. *Journal of Insect Conservation* **3**, 225–43.

Englund, R. A. and Polhemus, D. A. (2001) Evaluating the effects of introduced rainbow trout (*Oncorhynchus mykiss)* on native stream insects on Kauai Island, Hawaii. *Journal of Insect Conservation* **5**, 265–81.

Erichsen, C., Samways, M. J. and Hattingh, V. (1991) Reaction of the ladybird *Chilocorus nigritus* (F.) (Col. Coccinellidae) to a doomed food resource. *Journal of Applied Entomology* **112**, 493–8.

Erwin, T. L. (1982) Tropical forests: their richness in Coleoptera and other arthropod species. *Coleopterists' Bulletin* **36**, 74–5.

(1988) The tropical forest canopy – the heart of biotic diversity. In Wilson, E. O. (ed.), *Biodiversity*, pp. 123–9. Washington, DC: National Academy Press.

(1991) An evolutionary basis for conservation strategies. *Science* **253**, 750–2.

Fabricius, C., Burger, M. and Hockey, P. A. R. (2003) Comparing biodiversity between protected areas and adjacent rangeland in xeric succulent thicket, South Africa: arthropods and reptiles. *Journal of Applied Ecology* **40**, 392–403.

Fagan, W. F. and Kareiva, P. M. (1997) Using compiled species lists to make biodiversity composition among regions: a test case using Oregon butterflies. *Biological Conservation* **80**, 249–59.

Fagan, W. F., Meir, E., Prendergast, J., Folarin, A. and Kareiva, P. (2001) Characterizing population vulnerability for 758 species. *Ecology Letters* **4**, 132–8.

Fahrig, L. (2003) Effects of habitat fragmentation on biodiversity. *Annual Review of Ecology and Systematics* **34**, 487–515.

Faith, D. P. (2002) Quantifying biodiversity: a phylogenetic perspective. *Conservation Biology* **16**, 248–54.

Faith, D. P. and Walker, P. A. (1996) How do indicator groups provide information about the relative biodiversity of different sets of areas?: on hotspots, complementarity and pattern-based approaches. *Biodiversity Letters* **3**, 18–25.

Faragher, R. A. (1980) Life cycle of *Hemicordulia tau* Selys (Odonata: Corduliidae) in Lake Eucumbene S.W, with notes on predation on it by two trout species. *Journal of the Australian Entomological Society* **19**, 269–70.

Feber, R. E., Firbank, L. G., Johnson, P. J. and Macdonald, D. W. (1997) The effects of organic farming on pest and non-pest butterfly abundance. *Agriculture, Ecosystems and Environment* **64**, 133–9.

Feber, R. E. and Smith, M. (1995) Butterfly conservation on arable farmland. In Pullin, A. S. (ed.), *Ecology and Conservation of Butterflies*, pp. 84–97. London: Chapman and Hall.

Feber, R. E., Smith, M. and Macdonald, D. W. (1996) The effects on butterfly abundance of the management of uncropped edges of arable fields. *Journal of Applied Ecology* **33**, 1191–205.

Fermon, H., Schultze, C. H., Waltert, M. and Mühlenberg, M. (2001) The butterfly fauna of the Noyau Central, Lama Forest (Republic of Benin), with notes on its ecological composition and geographic distribution. *African Entomology* **9**, 177–85.

Fermon, H., Waltert, M., Larsen, T. B., Dall' Asta, U. and Mühlenber, M. (2000). Effects of forest management on diversity and abundance of fruit-feeding nymphalid butterflies in South-eastern Côte d'Ivoire. *Journal of Insect Conservation* **4**, 173–89.

Ferreira, R. L. and Silva, M. S. (2001) Biodiversity under rocks: the role of microhabitats in structuring invertebrate communities in Brazilian outcrops. *Biodiversity and Conservation* **10**, 1171–83.

Ferrier, S. (2002) Mapping spatial pattern in biodiversity for regional conservation planning: where to from here? *Systematic Biology* **51**, 331–63.

Ferrier, S., Watson, G., Pearce, J. and Drielsma, M. (2002a) Extended statistical approaches to modelling spatial pattern in biodiversity in northeast New South Wales. I. Species-level modelling. *Biodiversity and Conservation* **11**, 2275–307.

Ferrier, S., Drielsma, M., Manion, G. and Watson, G. (2002b) Extended statistical approaches to modelling spatial pattern in biodiversity in northeast New South Wales. II. Community-level modelling. *Biodiversity and Conservation* **11**, 2309–38.

Fincke, O. M., Yanoviak, S. P. and Hanschu, R. D. (1997) Predation by odonates depresses mosquito abundance in water-filled tree holes in Panama. *Oecologia* **112**, 244–53.

Findlay, C. S. and Bourdages, J. (2000) Response time of wetland biodiversity to road construction on adjacent lands. *Conservation Biology* **14**, 86–94.

Firbank, L. G. (2003) Introduction: the Farm Scale Evaluation of spring-sown genetically modified crops. *Proceedings of the Royal Society of London, B* **358**, 1777–8.

Firbank, L. G., Telfer, M. G., Eversham, B. C. and Arnold, H. R. (1994) The use of species decline statistics to help target conservation policy for set-aside arable land. *Journal of Environmental Management* **42**, 415–22.

Fischer, K., Beinlich, B. and Plachter, H. (1999) Population structure, mobility and habitat preferences of the violet copper *Lycaena helle* (Lepidoptera: Lycaenidae) in western Germany: implications for conservation. *Journal of Insect Conservation* **3**, 43–52.

Fischer, S. F., Poschold, P. and Beinlich, B. (1996) Experimental studies on the dispersal of plants and animals on sheep in calcareous grasslands. *Journal of Applied Ecology* **33**, 1206–22.

Flannery, T. F. (1994) *Future Eaters*. New York, NY: Braziller.

Fleishman, E. and MacNally, R. (2002) Topographic determination of faunal nestedness in Great Basin butterfly assemblages: application to conservation planning. *Conservation Biology* **16**, 422–9.

Folgarait, P. J. (1998) Ant biodiversity and its relationship to ecosystem functioning: a review. *Biodiversity and Conservation* **7**, 1221–44.

Follett, P. A. and Duan, J. J. (eds.) (2000) *Non-target Effects of Biological Control*. Dordrecht, The Netherlands: Kluwer.

Forman, R. T. T. (1995) *Land Mosaics*. Cambridge: Cambridge University Press.

Forman, R. T. T. and Deblinger, R. D. (2000) The ecological road effect zone of a Massachusetts (U.S.A.) suburban highway. *Conservation Biology* **14**, 36–46.

Forsyth, A. (1998) Predator patches: moving beyond butterflies. In *Butterfly Gardening*, created by The Xerces Society in association with The Smithsonian Institution, pp. 95–109. San Francisco, CA: Sierra Club.

Fox, W. (1993) Why care about the world around us? *Resurgence* **161**, 10–12.

Frankham, R., Ballov, J. D. and Briscoe, D. A. (2002) *Introduction to Conservation Genetics*. Cambridge: Cambridge University Press.

Frankie, G. W. and Ehler, L. E. (1978) Ecology of insects in urban environments. *Annual Review of Entomology* **23**, 367–87.

Franklin, J. F. (1993) Preserving biodiversity: species, ecosystems or landscapes? *Ecological Applications* **3**, 202–5.

Fredericksen, N. J. and Fredericksen, T. S. (2002) Terrestrial wildlife responses to logging and fire in a Bolivian tropical humid forest. *Biodiversity and Conservation* **1**, 25–42.

Freedman, B. (1989) *Environmental Ecology: The Impact of Pollution and other Stresses on Ecosystem Structure and Function*. New York, NY: Academic Press.

French, K. (1999) Spatial variability in species composition in birds and insects. *Journal of Insect Conservation* **3**, 183–9.

Fries, C., Carlsson, M., Lämas, T. and Sallnäs, O. (1998) A review of conceptual landscape planning models for multiobjective forestry in Sweden. *Canadian Journal of Forest Research* **28**, 159–67.

Fry, G. (1995) Landscape ecology of insect movement in arable ecosystems. In Glen, D. M., Greaves, M. P. and Anderson, H. M. (eds.), *Ecology and Integrated Farming Systems*, pp. 177–202. Chichester: John Wiley & Sons.

Gadeberg, R. M. E. and Boomsma, J. J. (1997) Genetic population structure of the large blue butterfly *Maculinea alcon* in Denmark. *Journal of Insect Conservation* **1**, 99–111.

Gagné, W. C. and Howarth, F. G. (1985) Conservation status of endemic Hawaiian Lepidoptera. *Proceedings of the 3rd Congress of European Lepidoptera, Cambridge, 1982*, pp. 74–84.

Gandar, M. V. (1982) The dynamics and trophic ecology of grasshoppers (Acridoidea) in a South African savanna. *Oecologia* **54**, 370–8.

García, A. (1992) Conserving the species-rich meadows of Europe. *Agriculture, Ecosystems and Environment* **40**, 219–32.

Gärdenfors, U. (2001) Classifying threatened species at national versus global levels. *Trends in Ecology and Evolution* **16**, 511–16.

Gärdenfors, U., Hilton-Taylor, C., Mace, G. M. and Rodriguez, J. P. (2001) The application of IUCN Red Listed Criteria at regional levels. *Conservation Biology* **15**, 1206–12.

Gardiner, T., Pye, M., Field, R. and Hill, J. (2003) The influence of sward height and vegetation composition in determining the habitat preferences of three *Chorthippus* species (Orthoptera: Acrididae) in Chlemsford, Essex, UK. *Journal of Orthoptera Research* **11**, 207–13.

Garner, A. (2003) Spirituality and sustainability. *Conservation Biology* **17**, 946.

Gascon, C., Williamson, G. B. and da Fonseca, G. A. B. (2000) Receding forest edges and vanishing reserves. *Science* **288**, 1356–8.

Gaston, K. J. (1991) The magnitude of global insect species richness. *Conservation Biology* **5**, 283–96.

(1992) Regional numbers of insect and plant species. *Functional Ecology* **6**, 243–7.

(1994) *Rarity*. London: Chapman and Hall.

Gaston, K. J., Jones, A. G., Hänel, C. and Chown, S. L. (2003) Rates of species introduction to a remote oceanic island. *Proceedings of the Royal Society of London, B* **270**, 1091–8.

Gaston, K. J., Rodrigues, A. S. L., van Rensburg, B. J., Koleff, P. and Chown, S. L. (2001) Complementary representation and zones of ecological transition. *Ecology Letters* **4**, 4–9.

Gathorne-Hardy, F. J., Jones, D. T. and Syaukani (2002) A regional perspective on the effects of human disturbance on the termites of Sundaland. *Biodiversity and Conservation* **11**, 1991–2006.

Gebeyehu, S. and Samways, M. J. (2002) Grasshopper response to a restored national park (Mountain Zebra National Park, South Africa). *Biodiversity and Conservation* **11**, 283–304.

(2003) Responses of grasshopper assemblages to long-term grazing management in a semi-arid African savanna. *Agriculture, Ecosystems and Environment* **95**, 613–22.

Geertsema, H. (2000) Range expansion, distribution records and abundance of some western Cape insects. *South African Journal of Science* **96**, 396–8.

Gehring, C. A., Cobb, N. S. and Whitham, T. G. (1997) Three-way interactions among ectomycorrhizal mutualists, scale insects and resistant and susceptible pinyon pines. *American Naturalist* **149**, 824–41.

Geist, H. J. and Lambin, E. F. (2002) Proximate causes and underlying driving forces of tropical deforestation. *BioScience* **52**, 143–50.

Gerlach, J. (2000) The rediscovery of the Seychelles hummingbird hawkmoth *Macroglossum alluaudi* Joannis, 1893 (Lepidoptera: Sphingidae). *Phelsuma* **8**, 79–80.

Gibb, H. and Hochuli, D. F. (1999) Nesting analysis of arthropod assemblages in habitat fragments in the Sydney region. In Ponder, W. and Lunney, D. (eds.), *The Other 99%: The Conservation and Biodiversity of Invetebrates*, pp. 77–81. Mosman, Australia: The Royal Zoological Society of New South Wales.

Gilbert, O. L. (1989) *The Ecology of Urban Habitats*. London: Chapman and Hall.

Gillespie, R. G. (1999) Naiveté and novel perturbations: conservation of native spiders on oceanic island system. *Journal of Insect Conservation* **3**, 263–72.

Girling, M. A. (1982) Fossil insect faunas from forest sites. In Bell, M. and Limbrey, S. (eds.), *Archeological Aspects of Woodland Ecology. Symposia of the Association for Environmental Archaeology* No. 2, BAR International Series 146. Oxford: John & Erica Hedges Ltd.

Godfray, H. C. J. and Lawton, J. H. (2001) Scale and species numbers. *Trends in Ecology and Evolution* **16**, 400–4.

Goldsmith, O. (1866) *A History of the Earth and Animated Nature*. Glasgow: Blackie.

Goméz, C., Casellas, D., Oliveras, J. and Bas, J. M. (2003) Structure of ground-foraging ant assemblages in relation to land-use change in the northwestern Mediterrarean region. *Biodiversity and Conservation* **12**, 2135–46.

Goulson, D., Stout, J. C. and Kells, A. R. (2002) Do exotic bumblebees and honeybees compete with native flower-visiting insects in Tasmania? *Journal of Insect Conservation* **6**, 179–89.

Grant, C. C. and Van der Walt, J. L. (2000) Towards an adaptive management approach for the conservation of rare antelope in the Kruger National Park – Outcome of a workshop held in May 2000. *Koedoe* **43/2**, 103–11.

Greatorex-Davis, J. N., Sparks, T. H. and Hall, M. L. (1994) The response of Heteroptera and Coleoptera species to shade and aspect in rides of coniferised lowland. *Biological Conservation* **12**, 1099–111.

Greenslade, P. (1993) Australian native steppe-type landscapes: neglected areas for invertebrate conservation in Australia. In Gaston, K. J., New, T. R. and Samways, M. J. (eds.), *Perspectives on Insect Conservation*, pp. 51–73. Andover: Intercept.

Griffith, M. B., Barrows, E. M. and Perry, S. A. (1998) Lateral dispersal of adult aquatic insects (Plecoptera, Trichoptera) following emergence from headwater streams in forested Appalachian catchments. *Annals of the Entomological Society of America* **91**, 195–201.

Groffman, P. M., Bain, D. J., Band, L. E., Belt, K. T., Brush, G. S., Grove, J. M., Pouyat, R. V., Yesilonis, I. C. and Zipperer, W. C. (2003) Down by the riverside: urban riparian ecology. *Frontiers in Ecology and Environment* **1**, 315–21.

Grove, S. J. (2002) Saproxylic insect ecology and the sustainable management of forests. *Annual Review of Ecology and Systematics* **33**, 1–23.

Grove, S. J. and Stork, N. E. (1999) The conservation of saproxylic insects in tropical forests: a research agenda. *Journal of Insect Conservation* **3**, 67–74.

Haddad, N. (1999) Corridor and distance effects on interpatch movements: a landscape experiment with butterflies. *Ecological Applications* **9**, 612–22.

(2000) Corridor length and patch colonization by a butterfly, *Junonia coenia*. *Conservation Biology* **14**, 738–45.

Haddad, N. M., Rossenberg, D. K. and Noon, B. R. (2000) On experimentation and the study of corridors: response to Beier and Noss. *Conservation Biology* **14**, 1543–5.

Hails, R. S. (2000) Genetically modified plants – the debate continues. *Trends in Ecology and Evolution* **15**, 14–18.

(2002) Assessing the risks associated with new agricultural practices. *Nature* **418**, 2–5.

Hambler, C. and Speight, M. R. (1996) Extinction rates in British nonmarine invertebrates. *Conservation Biology* **10**, 892–6, and **11**, 304.

Hamer, K. C. and Hill, J. K. (2000) Scale-dependent effects of habitat disturbance on species richness in tropical forests. *Conservation Biology* **14**, 1435–40.

Hamer, K. C., Hill, J. K., Benedick, S., Mustaffa, N., Sherratt, T. N., Maryati, M. and Chey, V. K. (2003) Ecology of butterflies in natural and selectively logged forests of northern Borneo: the importance of habitat heterogeneity. *Journal of Applied Ecology* **40**, 150–62.

Hammond, P. M. (1992) Species inventory. In Groombridge, B. (ed.), *Global Biodiversity: Status of the Earth's Living Resources*, pp. 17–39. London: Chapman and Hall.

Hannah, L., Carr, J. L. and Lankerani, A. (1995) Human disturbance and natural habitat: a biome level analysis of a global data set. *Biodiversity and Conservation* **4**, 128–55.

Hannah, L., Migdley, G., Lovejoy, T., Bond, W. J., Bush, M., Lovett, J. C., Scott, D. and Woodward, F. I. (2002) Conservation of biodiversity in a changing climate. *Conservation Biology* **16**, 264–8.

Hanski, I. (1998) Metapopulation dynamics. *Nature* **396**, 41–9.

(1999) *Metapopulation Ecology*. Oxford: Oxford University Press.

Hanski, I. and Gilpin, M. E. (1997) *Metapopulation Biology*. San Diego, CA: Academic Press.

Hanski, I., Pakkala, T., Kuusaari, M. and Lei, E. (1995) Metapopulation persistence of an endangered butterfly in a fragmented landscape. *Oikos* **72**, 21–8.

Hardy, P. B. and Dennis, R. L. H. (1999) The impact of urban development on butterflies within a city region. *Biodiversity and Conservation* **8**, 1261–79.

Harrington, R., Bale, J. S. and Tatchell, G. M. (1995) Aphids in a changing climate. In Harrington, R. and Stork, N. E. (eds.), *Insects in a Changing Environment*, pp. 125–55. London: Academic Press.

Harrington, R., Woiwod, I. and Sparks, T. (1999) Climate change and trophic interactions. *Trends in Ecology and Evolution* **14**, 146–50.

Harrison, I. J. and Stiassny, M. L. J. (1999) The quiet crisis: a preliminary listing of the freshwater fishes of the world that are extinct or "missing in action". In MacPhee, R. D. E. (ed.), *Extinctions in Near Time*, pp. 271–331. New York, NY: Kluwer/Plenum.

Harrison, S. and Hastings, A. (1996) Genetic and evolutionary consequences of metapopulation structure. *Trends in Ecology and Evolution* **11**, 180–3.

Harrison, T. (1964) Borneo caves with special reference to Niah Great Cave. *Studies in Speleology* **1**, 26–32.

Hartley, S. E., Gardner, S. M. and Mitchell, R. J. (2003) Indirect effects of grazing and nutrient addition on the hemipteran community of heather moorlands. *Journal of Applied Ecology* **40**, 793–803.

Haskell, D. G. (2000) Effects of forest roads on macroinvertebrate soil fauna of the southern Appalachian mountains. *Conservation Biology* **14**, 57–63.

Haslett, J. R. (1994) Community structure and the fractal dimensions of mountain habitats. *Journal of Theoretical Biology* **167**, 407–11.

(2001) Biodiversity and conservation of Diptera in heterogenous land mosaics: A fly's eye view. *Journal of Insect Conservation* **5**, 71–5.

Hassall, M., Hawthorne, A., Maudsley, M., White, P. and Cardwell, C. (1992) Effects of headland management on invertebrate communities in cereal fields. *Agriculture, Ecosystems and Environment* **40**, 155–78.

Hattingh, V. and Samways, M. J. (1995) Visual and olfactory location of biotopes, prey patches, and individual prey by the ladybeetle *Chilocorus nigritus*. *Entomologia Experimentalis et Applicata* **75**, 87–98.

Haughton, A. J., Champion, G. T., Hawes, C. *et al.* (2003) Invertebrate responses to the management of genetically modified herbicide-tolerant and conventional spring crops. II. Within-field epigeal and aerial arthropods. *Proceedings of the Royal Society of London, B* **358**, 1863–77.

Hawes, C., Haughton, A. J., Osborne, J. L. *et al.* (2003) Responses of plants and invertebrate trophic groups to contrasting herbicide regimes in the Farm Scale Evaluations of genetically modified herbicide-tolerant crops. *Proceedings of the Royal Society of London, B* **358**, 1899–913.

Hawkins, B. A. (1993) Refuges, host population dynamics and the genesis of parasitoid diversity. In LaSalle, J. and Gauld, I. D. (eds.), *Hymenoptera and Biodiversity*, pp. 235–56. Wallingford: CAB International.

Hawthorne, D. J. and Via, S. (2001) Genetic linkage of ecological specialization and reproductive isolation in pea aphids. *Nature* **412**, 904–7.

Haysom, K. A. and Coulson, J. C. (1998) The Lepidoptera fauna associated with *Calluna vulgaris*: effects of plant architecture on abundance and diversity. *Ecological Entomology* **23**, 377–85.

Heard, S. B. and Mooers, A. Ø. (2000) Phylogenetically patterned speciation rates and extinction risks change the loss of evolutionary history during extinction. *Proceedings of the Royal Society of London, B* **267**, 613–20.

Hee, J. J., Holway, D. A., Suarez, A. V. and Case, T. J. (2000) Role of propagule size in the success of incipient colonies of the invasive Argentine ant. *Conservation Biology* **14**, 559–63.

Hendersen, N. (1992) Wilderness and nature conservation ideal: Britain, Canada, and the United States contrasted. *Ambio* **21**, 394–9.

Hengeveld, R. (1999) Modelling the impact of biological invasions. In Sandlund, O. T., Schei, P. J. and Viken, Å. (eds.), *Invasive Species and Biodiversity Management*, pp. 127–38. Dordrecht, The Netherlands: Kluwer.

Henning, S. F. and Henning, G. A. (1989) *South African Red Data Book – Butterflies*. Pretoria, South Africa: Foundation for Research Development.

Hess, G. R. and Fischer, R. A. (2001) Communicating clearly about conservation corridors. *Landscape and Urban Planning* **55**, 195–208.

Higgs, E. S. (1997) What is good ecological restoration? *Conservation Biology* **11**, 338–48.

Hight, S. D., Carpenter, J. E., Bloem, K. A., Bloem, S., Pemberton, R. W. and Stiling, P. (2002) Expanding geographical range of *Cactoblastis cactorum* (Lepidoptera: Pyralidae) in North America. *Florida Entomologist* **85**, 527–9.

Hill, B., Beinlich, B. and Plachter, H. (1999) Habitat preference of *Lestes barbarus* (Fabricius, 1798) (Odonata, Lestidae) on a low-intensity cattle pasture in the Sava floodplain (Croatia). *Verhandlungen der Gesselschaft für Ökologie* **29**, 539–45.

Hill, C. J. (1995) Conservation corridors and rainforest insects. In Watt, A. D., Stork, N. E. and Hunter, M. D. (eds.), *Forests and Insects*, pp. 381–93. London: Chapman and Hall.

Hill, J. K. (1999) Butterfly spatial distribution and habitat requirements in atropical forest: impacts of selective logging. *Journal of Applied Ecology* **36**, 564–72.

Hill, J. K., Hamer, K. C., Lace, L. A. and Banham, W. M. T. (1995) Effects of selective logging on tropical forest butterflies on Buru, Indonesia. *Journal of Applied Ecology.* **32**, 754–60.

Hill, J. K., Thomas, C. D., Fox, R., Telfer, M. G., Willis, S. G., Asher, J. and Huntley, B. (2002) Responses of butterflies to twentieth century climate warming: implications for future ranges. *Proceedings of the Royal Society of London, B* **269**, 2163–71.

Hill, J. K., Thomas, C. D. and Lewis, O. T. (1996) Effects of habitat patch size and isolation on dispersal by *Hesperia comma* butterflies: implications for metapopulation structure. *Journal of Animal Ecology* **65**, 725–35.

Hill, M., Holm, K., Vel, T., Shah, N. J. and Matyot, P. (2003) Impact of the introduced crazy ant *Anoplolepis gracilipes* on Bird Island, Seychelles. *Biodiversity and Conservation* **12**, 1969–84.

Hill, P. and Twist, C. (1998) *Butterflies and Dragonflies: A Site Guide, 2nd edn*. Chelmsford, UK: Arlequin.

Hillier, J. and Birch, N. (2002) The use of strategic mathematical models in risk assessment. *Antenna* **26**, 101–2.

Hilton-Taylor, C. (compiler) (2000) *2000 IUCN Red Lists of Threatened Species*. Gland, Switzerland and Cambridge: IUCN.

Hjermann, D. Ø. and Ims, R. A. (1996) Landscape ecology of the wart-biter *Decticus verrucivorus* in a patchy landscape. *Journal of Animal Ecology* **65**, 768–80.

Hodkinson, I. D. and Casson, D. (1991) A lesser predilection for bugs: Hemiptera (Insecta) diversity in tropical rain forests. *Biological Journal of the Linnean Society* **43**, 101–9.

Hoffmann, A. A. and Hercus, M. J. (2000) Environmental stress as an evolutionary force. *BioScience* **50**, 217–26.

Hoffmann, A. A. and Parsons, P. A. (1997) *Extreme Environmental Change and Evolution*. Cambridge: Cambridge University Press.

Holl, K. D. (1995) Nectar resources and their influence on butterfly communities on reclaimed coal surface mines. *Restoration Ecology* **3**, 76–85.

(1996) The effect of coal surface mine reclamation on diurnal Lepidoptera conservation. *Journal of Applied Ecology* **33**, 225–36.

Holloway, G. J., Griffiths, G. H. and Richardson, P. (2003) Conservation strategy maps: a tool to facilitate biodiversity action planning illustrated using the heath fritillary butterfly. *Journal of Applied Ecology* **40**, 413–21.

Holloway, J. D., Kirk-Spriggs, A. H. and Khen, C. V. (1992) The response of some rain forest insect groups to logging and conversion to plantation. *Philosophical Transactions of the Royal Society* **335**, 425–36.

Holmes, J. C. (1996) Parasites as threats to biodiversity in shrinking ecosystems. *Biodiversity and Conservation* **5**, 975–83.

Homes, V., Hering, D. and Reich, M. (1999) The distribution and macrofauna of ponds in stretches of an alpine floodplain differently impacted by hydrological engineering. *Regulated Rivers: Research and Management* **15**, 405–17.

Holsinger, K. E. (2000) Demography and extinction in small populations. In Young, A. G. and Clarke, G. M. (eds.), *Genetics, Demography and Viability of Fragmented Populations*, pp. 55–74. Cambridge: Cambridge University Press.

Holway, D. A. (1995) Distribution of the Argentine ant (*Linepithema humile*) in northern California. *Conservation Biology* **9**, 1634–7.

Holway, D. A., Lach, L., Suarez, A. V., Tsutsui, N. D. and Case, T. J. (2002) The causes and consequences of ant invasions. *Annual Review of Ecology and Systematics* **33**, 181–233.

Hope, A. and Johnson, B. (2002) Transgenetic insect-resistant crops: a route to more sustainable agriculture systems? *Antenna* **26**, 98–9.

Hopkin, S. P. (1995) Deficiency and excess of essential and non-essential metals in terrestrial insects. In Harrington, R. and Stork, N. E. (eds.), *Insects in a Changing Environment*, pp. 251–70. London: Academic Press.

(1998) Collembola: the most abundant insects on earth. *Antenna* **22**, 117–21.

Hopkinson, P., Travis, J. M. J, Evans, J., Gregory, R. D., Telfer, M. and Williams, P. H. (2001) Flexibility and the use of indicator taxa in the selection of reserves. *Biodiversity and Conservation* **10**, 271–85.

Horgan, F. G. (2002) Shady field boundaries and the colonisation of dung by coprophagous beetles in Central American pastures. *Agriculture, Ecosystems and Environment* **91**, 25–36.

Horn, D. J. (1991) Potential impact of *Coccinella septempuctata* on endangered Lycaenidae (Lepidoptera) in Northern Ohio. In Polár, L., Chambers, R. J., Dixon, A. F. G. and Hodek, I. (eds.), *Behaviour and Impact of Aphidophaga*, pp. 159–62. The Hague, The Netherlands: Academic Publications.

Horner-Devine, M. C., Daily, G. C., Ehrlich, P. R. and Boggs, C. L. (2003) Countryside biogeography of tropical butterflies. *Conservation Biology* **17**, 168–77.

Horwitz, P., Recher, H. and Majer, J. (1999) Putting invertebrates on the agenda: political and bureaucratic challenges. In Ponder, W. and Lunney, D. (eds.), *The Other 99%: The Conservation of Biodiversity of Invertebrates*, pp. 398–406. Mosman, Australia: The Royal Zoological Society of New South Wales.

Houck, M. A., Clark, J. B., Peterson, K. R. and Kidwell, M. G. (1991) Possible horizontal transfer of *Drosophila* genes by the mite *Proctolaelaps regalis*. *Science* **253**, 1125–9.

Howard, P. C., Viskanic, P., Davenport, T. R. B., Kigenyi, F. W., Baltzer, M., Dickinson, C. J., Lwanga, J. S., Matthews, R. A. and Balmford, A. (1998) Complementarity and the use of indicator groups for reserve selection in Uganda. *Nature* **394**, 472–5.

Howarth, F. G. (1981) The conservation of cave invertebrates. In Mylroie, J. E. (ed.), *Proceedings of the First International Cave Management Symposium, Murray, Kentucky*, 15–18 July, pp. 57–63.

(1987) The evaluation of non-relictual tropical troglobites. *International Journal of Speleology* **16**, 1–16.

(1991) Environmental impact of classical biological control. *Annual Reviews of Entomology* **36**, 485–509.

(2001) Environmental issuses concerning the importation of non-indigenous biological control agents. In Lockwood, J. A., Howarth, F. G. and Purcell, M. F. (eds.), *Balancing Nature: Assessing the Impact of Importing Non-native Biological Control Agents (an International Perspective)*, pp. 70–99. Lanham, MA: Entomological Society of America.

Hughes, L. (2000) Biological consequences of global warming: is the signal already apparent? *Trends in Ecology and Evolution* **15**, 56–61.

Hughes, J. B., Daily, G. C. and Ehrlich, P. R. (2000) Conservation of insect diversity: a habitat approach. *Conservation Biology* **14**, 1788–97.

Human, K. G. and Gordon, D. M. (1997) Effects of Argentine ants on invertebrate biodiversity in northern California. *Conservation Biology* **11**, 1242–8.

Human, K. G., Weiss, A., Sandler, B. and Gordon, D. M. (1998) Effect of abiotic factors on the distribution and activity of the invasive Argentine ant (Hymenoptera: Formicidae). *Environmental Entomology* **27**, 822–33.

Humphreys, G. S. (1994) Bioturbation, biofabrics and the biomantle: an example from the Sydney Basin. In Ring-Voase, A. J. and Humphreys, A. S. (eds.), Soil *Micromorphology: Studies in Management and Genesis*, pp. 421–436;. Amsterdam: Elsevier.

Humphries, C. J. (2000) Form, space and time: which came first? *Journal of Biogeography* **27**, 11–15.

Hunter, M. L. Jr (1996) *Fundamentals of Conservation Biology*. Cambridge, MA: Blackwell Science.

(2000a) *Fundamentals of Conservation Biology, 2nd edn*. Oxford: Blackwell Science.

(2000b) Refining normative concepts in conservation. *Conservation Biology* **14**, 573–4.

Huston, M. A. (1994) *Biological Diversity: the Coexistence of Species on Changing Landscapes*. Cambridge: Cambridge University Press.

Hutton, S. A. and Giller, P. S. (2003) The effects of the intensification of agriculture on northern temperate dung beetle communities. *Journal of Applied Ecology* **40**, 994–1007.

Ingham, D. S. and Samways, M. J. (1996) Application of fragmentation and variegation models to epigaeic invertebrates in South Africa. *Conservation Biology* **10**, 1353–8.

IIBC (1996) *Annual Report (1995) of the International Institute of Biological Control*. Wallingford: CAB International.

Ims, R. A. (1995) Movement patterns related to spatial structures. In Hansson, L., Fahrig, L. and Merriam, G. (eds.), *Mosaic Landscapes and Ecological Processes*, pp. 85–109. London: Chapman and Hall.

IPCC (2001) *Climate Change 2001: the Scientific Basis*. Cambridge: Cambridge University Press.

Irmler, U., Heller, K., Meyer, H. and Reinke, H.-D. (2002) Zonation of ground beetles (Coleoptera: Carabidae) and spiders (Araneida) in salt marshes at the North and the Baltic Sea and the impact of the predicted sea level increase. *Biodiversity and Conservation* **11**, 1129–47.

Irmler, U. and Hoernes, U. (2003) Assignment and evaluation of ground beetle (Coleoptera: Carabidae) assemblages to sites on different scales in a grassland landscape. *Biodiversity and Conservation* **12**, 1405–19.

IUCN (1993) *Parks for Life: Report of the IVth World Congress on National Parks and Protected Areas*. Gland, Switzerland: IUCN.

(2001) *IUCN Red List of Categories and Criteria: Version 3.1*. Gland, Switzerland and Cambridge: IUCN.

(2003a) http://redlist.org

(2003b) *Guidelines for Application of IUCN Red List Criteria at Regional Levels: Version 3.0*. IUCN Species Survival Commission, Gland, Switzerland and Cambridge: IUCN.

IUCN/UNEP/WWF (1991) *Caring for the Earth. A Strategy for Sustainable Living*. Gland, Switzerland: IUCN/UNEP/WWF.

Jacobson, S. K. (1990) Graduate education in conservation biology. *Conservation Biology* **4**, 431–40.

Janzen, D. H. (1978) The ecology and evolutionary biology of seed chemistry as relates to seed predation. In Harborne, J. B. (ed.), *Biochemical Aspects of Plant and Animal Coevolution*, pp. 163–206. London: Academic Press.

(1998) How to grow a wildland: the gardenification of nature. *Insect Science and Application* **17**, 269–76.

(1999) Gardenification of tropical conserved wildlands: multitasking, multicropping, and multiusers. *Proceedings of the National Academy of Sciences, USA* **96**, 5987–94.

Jeanbourquin, P. and Turlings, T. C. J. (2002) The relative attractiveness of bt maize plants to two parasitoids of lepidopteran pests. *Antenna* **26**, 106.

Jesse, L. C. H. and Obrycki, J. J. (2000) Field deposit of Bt transgenic corn pollen: lethal effects on the monarch butterfly. *Oecologia* **125**, 241–8.

Jin, X.-B. and Yen, A. L. (1998) Conservation and the cricket culture in China. *Journal of Insect Conservation* **2**, 211–16.

Johannesen, J., Samietz, J., Wallaschek, M., Seitz, A. and Veith, M. (1999) Patch connectivity and genetic variation in two congeneric grasshopper species with different habitat preferences. *Journal of Insect Conservation* **3**, 201–9.

Johns, A. G. (1997) *Timber Production and Biodiversity Conservation in Tropical Rain Forests*. Cambridge: Cambridge University Press.

Johnson, D. and Johnson, S. (1993) *Gardening with Indigenous Trees and Shrubs*. Johannesburg: Southern Book Publisher.

Johnson, L. E. (1991) *A Morally Deep World*. Cambridge: Cambridge University Press.

Johnson, S. D. and Steiner, K. E. (2000) Generalization versus specialization in plant pollinator systems. *Trends in Ecology and Evolution* **15**, 140–3.

Jolivet, P. (1998) *Interrelationship between Insects and Plants*. Boca Raton, FL: CRC Press.

Jones, A. G., Chown, S. L. and Gaston, K. J. (2002) Terrestrial invertebrates of Gough Island: an assemblage under threat? *African Entomology* **10**, 83–91.

Jones, C. G., Lawton, J. H. and Shachak, M. (1997) Positive and negative effects of organisms as physical ecosystem engineers. *Ecology* **78**, 1946–57.

Jones, D. T., Susilo, F. X., Bignell, D. E., Hardiwinoto, S., Gillison, A. N. and Eggleton, P. (2003) Termite assemblage collapse along a land-use intensification gradient in lowland central Sumatra, Indonesia. *Journal of Applied Ecology* **40**, 380–91.

Jones, J. A., Swanson, F. J., Wample, B. C. and Snyder, K. U. (2000) Effects of roads on hydrology, geomorphology, and disturbance patches in stream networks. *Conservation Biology* **14**, 76–85.

Jonsell, M., Nordlander, G. and Jonsson, M. (1999) Colonization patterns of insects breeding in wood decaying fungi. *Journal of Insect Conservation* **3**, 145–61.

Joy, J. (1997) *Bracken Management for Fritillary Butterflies in the West Midlands and Gloucestershire Region. Occasional Paper No. 8*. Colchester, UK: Butterfly Conservation.

Joy, J. and Pullin, A. S. (1999) Field studies on flooding and survival of overwintering large heath butterfly *Coenonympha tullia* larvae on Fennis and Whixall Mossesin Shropshire and Wrexham, UK. *Ecological Entomology* **24**, 426–31.

Kaila, L., Martikainen, P. and Punttila, P. (1997) Dead trees left in clear-cuts benefit saproxylic Coleoptera adapted to natural disturbances in boreal forest. *Biodiversity and Conservation* **6**, 1–18.

Karg, J. (1991) Monitoring of insect diversity and abundance in large areas. *Laufener Seminarbeiträge* **7**, 61–7.

Keals, N. and Majer, J. D. (1991) The conservation status of ant communities along the Wubin-Perenjori corridor. In Saunders, D. A. and Hobbs, R. J. (eds.), *Nature Conservation 2, The Role of Corridors*, pp. 387–93. Chipping Norton, Australia: Surrey Beatty.

Keane, R. M. and Crawley, M. J. (2002) Exotic plant invasions and the enemy release hypothesis. *Trends in Ecology and Evolution* **17**, 164–70.

Kearns, C., Inouye, D. and Waser, N. M. (1998) Endangered mutualisms: the conservation of plant pollinator interactions. *Annual Review of Ecology and Systematics* **29**, 83–112.

Keller, L. F. and Waller, D. M. (2002) Inbreeding effects in wild populations. *Trends in Ecology and Evolution* **17**, 230–41.

Kellert, S. R. (1986) Social and perceptual factors in the preservation of animal species. In Norton, B. G. (ed.). *The Preservation of Species*, pp. 50–73. Princeton, NJ: Princeton University Press.

(1993) Values and perceptions of invertebrates. *Conservation Biology* **4**, 845–55.

Kells, A. R. and Goulson, D. (2003) Preferred nesting sites of bumblebee queens (Hymenoptera: Apidae) in agroecosystems in the UK. *Biological Conservation* **109**, 165–74.

Kelly, J. A. and Samways, M. J. (2003) Diversity and conservation of forest-floor arthropods on a small Seychelles island. *Biodiversity and Conservation* **12**, 1793–813.

Kerr, J. T., Sugar, A. and Packer, L. (2000) Indicator taxa, rapid biodiversity assessment, and nestedness in an endangered ecosystem. *Conservation Biology* **14**, 1726–34.

Kindvall, O. (1995) The impact of extreme weather on habitat preference and survival in a metapopulation of the bush cricket *Metrioptera bicolor* in Sweden. *Biological Conservation* **73**, 51–8.

Kindvall, O. and Ahlén, I. (1992) Geometrical factors and metapopulation dynamics of the bush cricket, *Metrioptera bicolor* Philippi (Orthoptera: Tettigoniidae). *Conservation Biology* **6**, 520–9.

Kinvig, R. and Samways, M. J. (2000) Conserving dragonflies (Odonata) along steams running through commercial forestry. *Odonatologica* **29**, 195–208.

Kirby, K. J., Webster, S. D. and Antzak, A. (1991) Effects of forest management on island structure and the quality of fallen dead wood: some British and Polish examples. *Forest Ecology and Management* **43**, 167–74.

Kirby, P. (1992) (reprinted 2001) *Habitat Management for Invertebrates: A Practical Handbook*. Sandy, Bedfordshire, UK: Royal Society for the Protection of Birds.

Kitahara, M. and Fujii, K. (1994) Biodiversity and community structure of temperate butterfly species within a gradient of human disturbance: an analysis based on the concept of generalist vs. specialist strategies. *Researches on Population Ecology* **36**, 187–99.

Kitahara, M. and Sei, K. (2001) A comparison of the diversity and structure of butterfly communities in semi-natural and human-modified grassland habitats at the foot of Mt. Fuji, central Japan. *Biodiversity and Conservation* **10**, 331–51.

Kitching, R. L., Li, D. and Stork, N. E. (2001) Assessing biodiversity 'sampling packages': how similar are arthropod assemblages in different tropical forests? *Biodiversity and Conservation* **10**, 793–813.

Klass, K.-D., Zompro, O., Kristensen, N. P. and Adis, J. (2002) Mantophasmatodea: a new insect order with extant members in the Afrotropics. *Science* **296**, 1456–9.

Kleijn, D. and Sutherland, W. J. (2003) How effective are European agri-environment schemes in conserving and promoting biodiversity? *Journal of Applied Ecology* **40**, 947–69.

Klein, A. M., Steffan-Dewenter, I., Buchori, D. and Tscharntke, T. (2002) Effects of land-use intensity in tropical agroforestry systems on coffee flower-visiting and trap-nesting bees and wasps. *Conservation Biology* **16**, 1003–14.

Klein, B. C. (1989) Effects of forest fragmentation on dung and carrion beetle communities in Central Amazonia. *Ecology* **70**, 1715–25.

Kluge, R. L. and Caldwell, P. M. (1992) Microsporidian diseases and biological weed control agents: to release or not to release? *Biocontrol News and Information* **13**, 43N–47N.

Kochér, S. D. and Williams, E. H. (2000) The diversity and abundance of North American butterflies vary with habitat disturbance and geography. *Journal of Biogeography* **22**, 785–94.

Koivula, M., Kukkanen, J. and Niemelä, J. (2002) Boreal carabid-beetle (Coleoptera, Carabidae) assemblages along clear-cut originated succession gradient. *Biodiversity and Conservation* **11**, 1269–88.

Kotze, D. J., Niemelä, J., O'Hara, R. B. and Turin, H. (2003) Testing abundance-range size relationships in European carabid beetles (Coleoptera, Carabidae). *Ecography* **26**, 553–66.

Kotze, D. J., Niemelä, J. and Nieminen, M. (2000) Colonization success of carabid beetles on Baltic islands. *Journal of Biogeography* **27**, 807–19.

Kotze, D. J. and O'Hara, R. B. (2003) Species decline – but why? Explanations of carabid beetle (Coleoptera, Carabidae) declines in Europe. *Oecologia* **135**, 138–48.

Kotze, D. J. and Samways, M. J. (1999a) Support for the multi-taxa approach in biodiversity assessment, as was shown by epigaeic invertebrates in an Afromontane forest archipelago. *Journal of Insect Conservation* **3**, 125–43.

(1999b) Invertebrate conservation at the interface between the grassland matrix and natural Afromontane forest fragments. *Biodiversity and Conservation* **8**, 1339–63.

(2001) No general edge effects for invertebrates at Afromontane forest/grassland ecotones. *Biodiversity and Conservation* **10**, 1027–37.

Koricheva, J., Mulder, C. P. H., Schmid, B., Joshi, J. and Huss-Danell, K. (2000) Numerical responses of different trophic groups of invertebrates to manipulations of plant diversity in grasslands. *Oecologia* **125**, 271–82.

Kozlov, M. V. (1996) Patterns of forest insect distribution within a large city: microlepidoptera in St Petersburg, Russia. *Journal of Biogeography* **23**, 95–103.

Kozlov, M. V., Jalava, J., Lvovsky, A. L. and Mikkola, K. (1996) Population densities and diversity of Noctuidae (Lepidoptera) along an air pollution gradient on the Kola Peninsula, Russia. *Entomologica Fennica* **7**, 9–15.

Kremen, C., Coldwell, R. K., Erwin, T. L., Murphy, D. D., Noss, R. F. and Sanjayan, M. A. (1993) Terrestrial arthropod assemblages, their use in conservation planning. *Conservation Biology* **1**, 796–808.

Kremen, C. and Ricketts, T. (2000) Global perspectives on pollination disruptions. *Conservation Biology* **14**, 1226–8.

Kromp, B. and Steinberger, K. H. (1992) Grassy margins and arthropod diversity: a case study on ground beetles and spiders in eastern Austria (Coleoptera: Carabidae; Arachnida, Opiliones). *Agriculture, Ecosystems and Environment* **40**, 71–93.

Kruess, A. (2003) Effects of landscape structure and habitat type on a plant-herbivore-parasitoid community. *Ecography* **26**, 283–90.

Kruess, A. and Tscharntke, T. (2002) Grazing intensity and the diversity of grasshoppers, butterflies and trap-nesting bees and wasps. *Conservation Biology* **16**, 1570–80.

Krupnick, G. A. and Kress, W. J. (2003) Hotspots and ecoregions: a test of conservation priorities using taxonomic data. *Biodiversity and Conservation* **12**, 2237–53.

Kuchlein, J. H. and Ellis, W. N. (1997) Climate-induced changes in the microlepidoptera fauna of the Netherlands and the implications for nature conservation. *Journal of Insect Conservation* **1**, 73–80.

Kuussaari, M., Nieminen, M. and Hanski, I. (1996) An experimental study of migration in the Glanville fritillary butterfly *Melitaea cinxia*. *Journal of Animal Ecology* **65**, 791–801.

Kwak, M. M., Velterop, O. and Boerigter, E. J. M. (1996) Insect diversity and the pollination of rare plant species. In Matheson, A., Buchmann, S. L., O' Toole, C., Westrich, P. and Williams, I. H. (eds.), *The Conservation of Bees*, pp. 115–24. London: Academic Press.

Labandeira, C. C., Johnson, K. R. and Wilf, P. (2002) Impact of the terminal Cretaceous event on plant-insect associations. *Proceedings of the National Academy of Sciences* **99**, 2061–6.

Labandeira, C. C. and Sepkoski, J. J. Jr (1993) Insect diversity and the fossil record. *Science* **261**, 310–15.

Lagerlöf, J., Stark, J. and Svensson, B. (1992) Margins of agricultural fields as habitats for pollinating insects. *Agriculture, Ecosystems and Environment* **40**, 117–24.

Lambeck, R. J. (1997) Focal species: a multi-species umbrella for nature conservation. *Conservation Biology* **11**, 849–56.

Landau, D., Prowell, D. and Carlton, C. E. (1999) Intensive versus long-term sampling to assess lepidopteran diversity in a southern mixed mesophytic forest. *Annals of the Entomological Society of America* **92**, 435–41.

Lapchin, L. (2002) Host-parasitoid association and diffuse coevolution: when to be a generalist? *American Naturalist* **160**, 245–54.

LaSalle, J. (1993) Parasitic Hymenoptera, biological control and biodiversity. In LaSalle, J. and Gauld, I. D. (eds.), *Hymenoptera and Biodiversity*, pp. 197–215. Wallingford: CAB International.

Latchininsky, A. V. (1998) Moroccan locust *Dociostaurus maroccanus* (Thunberg, 1815): a faunistic rarity or an important economic pest? *Journal of Insect Conservation* **2**, 167–78.

Laurance, W. F. (2000) Cut and run: the dramatic rise of transnational logging in the tropics. *Trends in Ecology and Evolution* **15**, 433–4.

Laurance, W. F. and Yensen, E. (1991) Predicting the impacts of edge effects in fragmented habitats. *Biological Conservation* **55**, 77–92.

Lawler, J. L., White, D., Sifneos, J. C. and Master, L. L. (2003) Rare species and the use of indicator groups for conservation planning. *Conservation Biology* **17**, 875–82.

Lawrence, J. M. and Samways, M. J. (2001) Hilltopping in *Hyalites encedon* (Lepidoptera: Acraeinae) butterflies. *Metamorphosis* **12**, 41–8.

(2002) Influence of hilltop vegetation type on an African butterfly assemblage and its conservation. *Biodiversity and Conservation* **11**, 1163–71.

Lawton, J. H. (1993) Range, population abundance and conservation. *Trends in Ecology and Evolution* **8**, 409–13.

Lawton, J. H., Bignell, D. E., Bolton, B., Bloemers, G. F., Eggleton, P., Hammond, P. M., Hodda, M., Holt, R. D., Larsen, T. B., Mawdsley, N. A., Stork, N. E., Srivastava, D. S. and Watt, A. D. (1998) Biodiversity inventories, indicator taxa and effects of habitat modification in tropical forest. *Nature* **391**, 72–6.

Legg, J. T., Rousch, R., De Salle, R., Vogler, A. P. and May, B. (1996) Genetic criteria for establishing evolutionarily significant units in Cryan's buckmoth. *Conservation Biology* **10**, 85–98.

Lejeune, R. R., Fell, W. M. and Burbridge, D. P. (1955) The effects of flooding on development and survival of the Larch Sawfly *Pristiphora erichsonii* (Tenthredinidae). *Ecology* **36**, 63–70.

LeMaitre, S. (2002) Food and density limitations of the Seychelles Magpie Robin *Copsychus sechellarum*, on Cousine Island. *Ostrich* **73**, 119–26.

Léon-Cortes, J. L., Cowley, M. J. R. and Thomas, C. D. (2000) The distribution and decline of a widespread butterfly *Lycaena phlaeas* in a pastoral landscape. *Ecological Entomology* **25**, 285–94.

Leopold, A. (1949) *A Sand County Almanac and Sketches Here and There.* Oxford: Oxford University Press.

(1953) *Round River*. New York, NY: Oxford University Press.

Lewis, O. T. and Bryant, S. R. (2002) Butterflies on the move. *Trends in Ecology and Evolution* **17**, 351–2.

Lewis, O. T. and Thomas, C. D. (2001) Adaptations to captivity in the butterfly *Pieris brassicae* (L.) and implications for ex situ conservation. *Journal of Insect Conservation* **5**, 55–63.

Lien, V. V. and Yuan, D. (2003) The differences of butterfly (Lepidoptera, Papilionidae) communities in habitats with various degrees of disturbance and altitudes in tropical forests of Vietnam. *Biodiversity and Conservation* **12**, 1099–111.

Lines, J. (1995) The effects of climatic and land-use changes on insect vectors of human disease. In Harrington, R. and Stork, N. E. (eds.), *Insects in a Changing Environment*, pp. 157–75. London: Academic Press.

Lobo, J. M. and Martín-Piera, F. (2002) Searching for a predictive model for species richness of Iberian dung beetle based on spatial and environmental variables. *Conservation Biology* **16**, 158–73.

Lockwood, J. A. (1987) The moral standing of insects and the ethics of extinction. *Florida Entomologist* **70**, 70–89.

Lockwood, J. A. (2001) The ethics of 'Classical' biological control and the value of *Place*. In Lockwood, J. A., Howarth, F. G. and Purcell, M. F. (eds.), *Balancing Nature: Assessing the Impact of Importing Non-native Biological Control Agents (an International Perspective)*, pp. 100–19. Lanham, MA: Entomological Society of America.

Lockwood, J. A. and DeBrey, L. D. (1990) A solution for the sudden and unexplained extinction of the Rocky Mountain grasshopper (Orthoptera: Acrididae). *Environmental Entomology* **19**, 1194–205.

Lockwood, J. A., Howarth, F. G. and Purcell, M. F. (2001) Introduction. In Lockwood, J. A., Howarth, F. G. and Purcell, M. F. (eds.), *Balancing Nature: Assessing the Impact of Importing Non-native Biological Control Agents (an International Perspective)*, pp. 1–2. Lanham, MA: Entomological Society of America.

Lockwood, J. A., Latchininsky, A. V. and Sergeev, M. G. (eds.) (2000) *Grasshoppers and Grassland Health*. Dordrecht, The Netherlands: Kluwer.

Lockwood, J. A. and Sergeev, M. G. (2000) Comparative biogeography of grasshoppers (Orthoptera: Acrididae) in North America and Siberia: applications to the conservation of biodiversity. *Journal of Insect Conservation* **4**, 161–72.

Lombard, A. T. (1995) The problem with multi-species conservation: do hotspots, ideal reserves and existing reserves coincide? *South African Journal of Zoology* **30**, 145–63.

Lomolino, M. V. (2000) Ecology's most general yet protean pattern: the species-area relationship. *Journal of Biogeography* **27**, 17–26.

Lomolino, M. V. and Creighton, J. C. (1996) Habitat selection, breeding success and conservation of the endangered American burying beetle, *Nicrophorus americanus*. *Biological Conservation* **77**, 235–41.

Longley, M. and Sotherton, N. W. (1997) Factors determining the effects of pesticides upon butterflies inhabiting arable farmland. *Agriculture, Ecosystems and Environment* **61**, 1–12.

Losey, J. E., Rayor, L. S. and Carter, M. E. (1999) Transgenic pollen harms monarch larvae. *Nature* **399**, 214.

Louda, S. M., Arnett, A. E., Rand, T. A. and Russell, F. L. (2003) Invasiveness of some biological control insects and adequacy of their ecological risk assessment and regulation. *Conservation Biology* **17**, 73–82.

Louda, S. M. and O'Brien, C. W. O. (2001) Unexpected ecological effects of distributing the exotic weevil, *Larius planus* (F.), for the biological control of Canada thistle. *Conservation Biology* **16**, 717–27.

Lövei, G. L. and Cartellieri, M. (2000) Ground beetles (Coleoptera, Carabidae) in forest fragments of the Manawatu, New Zealand: collapsed assemblages? *Journal of Insect Conservation* **4**, 239–44.

Lövei, G., Pedersen, B. P., Felkl, G., Brodsgaard, H. and Hansen, L. M. (2002) Developing a test system for evaluating environmental risks of transgenic plants: the polyphagous predator module. *Antenna* **26**, 104–5.

Lu, S.-S. and Samways, M. J. (2002) Conservation management recommendations for the Karkloof blue butterfly, *Orachrysops ariadne* (Lepidoptera: Lycaenidae). *African Entomology* **10**, 149–59.

Lyons, K. G. and Schwartz, M. W. (2001) Rare species loss alters ecosystem function - invasion resistance. *Ecology Letters* **4**, 358–65.

MacArthur, R. H. and Wilson, E. O. (1967) *Theory of Island Biogeography*. Princeton, NJ: Princeton University Press.

Maddison, A. (1995) *Monitoring the World Economy, 1820–1992*. Paris: Organization for Economic Cooperation and Development.

Maddock, A. and Du Plessis, A. (1999) Can species data only be appropriately used to conserve biodiversity? *Biodiversity and Conservation* **8**, 603–15.

Mader, H. J. (1984) Animal habitat isolation by roads and agricultural fields. *Biological Conservation* **29**, 81–96.

Maeto, K., Sato, S. and Miyata, H. (2002) Species diversity of longicorn beetles in humid warm-temperature forests: the impact of forest management practices on old-growth forest species in southwest Japan. *Biodiversity and Conservation* **11**, 1919–37.

Magagula, C. N. and Samways, M. J. (2001) Maintenance of ladybeetle diversity across a heterogenous African agricultural/savanna land mosaic. *Biodiversity and Conservation* **10**, 209–22.

Maggini, R., Guisan, A. and Cherix, D. (2002) A stratified approach for modelling the distribution of a threatened ant species in the Swiss National Park. *Biodiversity and Conservation* **11**, 2117–41.

Magura, T., Ködöböcz, V. and Tóthmérész, B. (2001) Effects of habitat fragmentation on carabids in forest patches. *Journal of Biogeography* **28**, 129–38.

Maina, G. G. and Howe, H. F. (2000) Inherent rarity in community restoration. *Conservation Biology* **14**, 1335–40.

Majer, J. D. (1996) Ant recolonization of rehabilitated bauxite mines at Trombetas, Pará, Brazil. *Journal of Tropical Ecology* **12**, 257–73.

Majer, J. D., Brennan, K. E. C. and Bisevac, L. (2002) Terrestrial invertebrates. In Perrow, M. R. and Davy, A. J. (eds.), *Handbook of Ecological Restoration. Vol. 1. Principles of Restoration*, pp. 279–99. Cambridge: Cambridge University Press.

Majer, J. D., Recher, H. F. and Ganesh, S. (2000) Diversity patterns of eucalypt canopy arthropods in eastern and western Australia. *Ecological Entomology* **25**, 295–306.

Majerus, M. E. N. (1994) *Ladybirds*. London: HarperCollins.

Malhi, Y. and Grace, J. (2000) Tropical forests and atmospheric carbon dioxide. *Trends in Ecology and Evolution* **15**, 332–7.

Mallet, J. (1997) A species definition for the modern synthesis. *Trends in Ecology and Evolution* **10**, 294.

Mallet, J. and Gilbert, L. E. (1995) Why are there so many mimicry rings? Correlations between habitat, behaviour and mimicry and *Heliconius* butterflies. *Biological Journal of the Linnean Society* **55**, 159–80.

Marden, J. H. and Kramer, M. G. (1994) Surface-skimming stoneflies: a possible intermediate stage in insect flight evolution. *Science* **266**, 427–30.

Marino, P. C. and Landis, D. A. (1996) Effect of landscape structure on parasitoid diversity and parasitism in afroecosystems. *Ecological Applications* **6**, 276–84.

Marsault, D., Pierre, J. and Pham-Delégue, M. H. (2002) Biodiversity and foraging behaviour of pollinating insects on herbicide-tolerant transgenic oilseed-rape genotypes under field conditions. *Antenna* **26**, 103–4.

Marshall, E. J. P., Baudry, J., Moonen, C., Fevre, E. and Thomas, C. F. G. (1996) Factors affecting floral diversity in European field margin networks. In Simpson, I. A. and Dennis, P. (eds.), *The Spatial Dynamics of Biodiversity*, pp. 97–104. Sterling, UK: The UK Region of the International Association for Landscape Ecology.

Martikainen, P. and Kouki, J. (2003) Sampling the rarest: threatened beetles in boreal forest biodiversity inventories. *Biodiversity and Conservation* **12**, 1815–31.

Martín-Piera, F. (2001) Area networks for conserving Iberian insects: a case study of dung beetles (Col., Scarabaeoidea). *Journal of Insect Conservation* **5**, 233–52.

Masters, G. J., Brown, V. K., Clarke, I. P., Whittaker, J. B. and Hollier, J. A. (1998) Direct and indirect effects of climate change on insect herbivores: (Auchenorrhynca, Homoptera). *Ecological Entomology* **23**, 45–52.

Matter, S. F. (1996) Interpatch movement of the red milkweed beetle, *Tetraopes tetraopthalmus:* individual responses to patch size and isolation. *Oecologia* **105**, 447–453.

Mattoni, R., Longcore, T. and Novotny, V. (2000) Arthropod monitoring for fine-scale habitat analysis: a case study of the El Segundo sand dunes. *Environmental Management* **25**, 445–52.

Mawdsley, N. A., Compton, S. G. and Whittaker, R. J. (1998) Population persistence, pollination mutualism, and figs in fragmented tropical landscapes. *Conservation Biology* **12**, 1416–20.

Mawdsley, N. A. and Stork, N. E. (1995) Species extinctions in insects: ecological and biogeographical considerations. In Harrington, R. and Stork, N. E. (eds.), *Insects in a Changing Environment*, pp. 321–69. London: Academic Press.

May, R. M. (1989) How many species. In Friday, L. and Laskey, R. (eds.), *The Fragile Environment*, pp. 61–81. Cambridge: Cambridge University Press.

(1990) How many species? *Philosophical Transactions of the Royal Society, B* **330**, 293–304.

Mayhew, P. J. (2002) Shifts in hexapod diversification and what Haldane could have said. *Proceedings of the Royal Society of London, B* **269**, 969–76.

Mayle, F. E., Burbridge, R. and Lilleen, T. J. (2000) Millennial scale dynamics of southern Amazonian rain forests. *Science* **290**, 2291–4.

McCafferty, W. P. (1981) *Aquatic Entomology: The Fishermens' and Ecologists' Illustrated Guide to Insects and their Relatives*. Boston, MASS: Jones and Bertlett.

McDonnell, M. J. and Rickett, S. T. A. (1990) Ecosystem structure and function along urban-rural gradients: an unexploited opportunity for ecology. *Ecology* **71**, 1232–7.

McGeoch, M. A. (1998) The selection, testing and application of terrestrial insects as bioindicators. *Biological Reviews* **73**, 181–201.

McGeoch, M. A. and Chown, S. L. (1997) Impact of urbanization on gall-inhabiting Lepidoptera assemblage: the importance of reserves in urban areas. *Biodiversity and Conservation* **6**, 979–93.

(1998) Scaling up the value of bioindicators. *Trends in Ecology and Evolution* **13**, 46–7.

McGeoch, M. A. and Gaston, K. J. (2000) Edge effects on the prevalence and mortality factors of *Phytomyza ilicus* (Diptera, Agromyzidae) in a suburban woodland. *Ecology Letters* **3**, 23–9.

McGeoch, M. A., van Rensburg, B. J. and Botes, A. (2002) The verification and application of bioindicators: a case study of dung beetles in a savanna ecosystem. *Journal of Applied Ecology* **39**, 661–72.

McIntyre, N. E., Rango, J., Fagan, W. F. and Faeth, S. H. (2001) Ground arthropod community structure in a heterogenous urban environment. *Landscape and Urban Planning* **52**, 257–74.

McKee, J. K., Scivlli, P. W., Fooce, C. D. and Waite, T. A. (2003) Forecasting global biodiversity threats associated with human population growth. *Biological Conservation* **115**, 161–4.

McKenna, D. D., McKenna, K. M., Malcolm, S. B. and Berenbaum, M. R. (2001) Mortality of Lepidoptera along roadways in central Illinois. *Journal of the Lepidopterists' Society* **55**, 63–8.

McKibben, B. (1990) *The End of Nature*. London: Viking.

McKinney, M. L. (1999) High rates of extinction and threat in poorly studied taxa. *Conservation Biology* **13**, 1273–81.

McKone, M. J., McLauchank, K. K., Lebrun, E. G. and McCall, A. C. (2001) An edge effect caused by adult corn-rootworm beetles on sunflowers in tallgrass prairie remnants. *Conservation Biology* **15**, 1315–24.

McNeely, J. A. (1999) The great reshuffling: how alien species help feed the global economy. In Sandlund, O. T., Schei, P. J. and Viken, Å. (eds.), *Invasive Species and Biodiversity Management*, pp. 11–31. Dordrecht, The Netherlands: Kluwer.

Melisch, R. (2003) Butterflies and beetles in Germany. *TRAFFIC Bulletin* **18(3)**, 1–3.

Memmott, J. (2000) Food webs as a tool for studying nontarget effects in biological control. In Follett, P. A. and Duan, J. J. (eds.), *Nontarget Effects of Biological Control*, pp. 147–63. Boston, MASS: Kluwer.

Memmott, J. and Godfray, H. C. J. (1993) Parasitoid webs. In LaSalle, J. and Gauld, I. D. (eds.), *Hymenoptera and Biodiversity*, pp. 217–34. Wallingford: CAB International.

Merritt, R., Moor, N. W. and Eversham, B. C. (1996) *Atlas of the Dragonflies of Britain and Ireland*. London: HMSO.

Messing, R. H. and Purcell, M. F. (2001) Regulatory constraints to the practice of biological control in Hawaii. In Lockwood, J. A., Howarth, F. G. and Purcell, M. F. (eds.), *Balancing Nature: Assessing the Impact of Importing Non-native Biological Control Agents (an International Perspective)*, pp. 3–14. Lanham, MA: Entomological Society of America.

Michaels, K. F. and McQuillan, P. B. (1995) Impact of commercial forest management on geophilous carabid beetles (Coleoptera, Carabidae) in tall, wet *Eucalyptus oblique* forest in southern Tasmania. *Australian Journal of Ecology* **20**, 316–23.

Miller, J. C. (1990) Field assessment of the effects of a microbial pest control agent on nontarget Lepidoptera. *American Entomologist* **36**, 135–9.

(1993) Insect natural history, multi-species interactions and biodiversity in ecosystems. *Biodiversity and Conservation* **2**, 233–41.

Mills, L. S., Soulé, M. E. and Doak, D. F. (1993) The keystone-species concept in ecology and conservation. *BioScience* **43**, 219–24.

Milton, S. J. (1995) Effects of rain, sheep and tephritid flies on seed production of two arid Karoo shrubs in South Africa. *Journal of Applied Ecology* **32**, 137–44.

Milton, S. J. and Dean, W. R. J. (1996) Rates of wood and dung disintegration in arid South African rangelands. *African Journal of Range and Forage Science* **13**, 89–93.

Mitchell, C. E., Turner, M. G. and Pearson, S. M. (2002) Effects of historical land use and forest patch size on myrmecochores and ant communities. *Ecological Applications* **12**, 1364–77.

Moller, H. (1996) Lessons for invasion theory from social insects. *Biological Conservation* **78**, 125–42.

Mönkkönen, M. and Mutanen, M. (2003) Occurrence of moths in boreal forest corridors. *Conservation Biology* **17**, 468–75.

Moore, N. W. (1987) *The Bird of Time: The Science and Politics of Nature Conservation*. Cambridge: Cambridge University Press.

Moore, N. W. (Compiler) (1997) *Status Survey and Conservation Action Plan: Dragonflies*. IUCN/SSC Odonata Specialist Group. Gland, Switzerland and Cambridge: IUCN.

Moran, N. A. and Telang, A. (1998) Bacteriocyte-associated symbionts of insects. *BioScience* **48**, 295–304.

Moritz, C. (1994) Defining 'evolutionarily significant units' for conservation. *Trends in Ecology and Evolution* **9**, 373–5.

(2002) Strategies to protect biological diversity and the evolutionary processes that sustain it. *Systematic Biology* **51**, 238–54.

Morley, S. A. and Karr, J. R. (2002) Assessing and restoring the health of urban streams in the Puget Sound Basin. *Conservation Biology* **16**, 1498–509.

Morris, M. G. (1981) Responses of grassland invertebrates to management by cutting. IV. Positive responses of Auchenorrhyncha. *Journal of Applied Ecology* **18**, 763–71.

(1991) The management of reserves and protected areas. In Spellerberg, I. F., Goldsmith, F. B. and Morris, M. G. (eds.), *The Scientific Management of Temperate Communities for Conservation*, pp. 323–47. Oxford: Blackwell Scientific Publications.

Morris, M. G., Collins, N. M., Vane-Wright, R. I. and Waage, J. (1991) The utilization and value of non-domesticated insects. In Collins, N. M. and Thomas, J. A. (eds.), *The Conservation of Insects and their Habitats*, pp. 319–47. London: Academic Press.

Mullen, K., Fahy, O. and Gormally, M. (2003) Ground flora and associated arthropod communities of forest road edges in Connemara, Ireland. *Biodiversity and Conservation* **12**, 87–101.

Munguira, M. L. and Thomas, J. A. (1992) Use of road verges by butterfly and burnet populations, and the effect of roads on adult dispersal and mortality. *Journal of Applied Ecology* **29**, 316–29.

Myers, N. (1996a) The biodiversity crisis and the future of evolution. *The Environmentalist* **16**, 27–47.

(1996b) Two key challenges for biodiversity: discontinuities and synergisms. *Biodiversity and Conservation* **5**, 1025–34.

Myers, N., Mittermeier, R. A., Mittermeier, C. G., da Fonseca, G. A. B. and Kent, J. (2000) Biodiversity hotspots for conservation priorities. *Nature* **403**, 853–8.

Naess, A. (1989) *Ecology, Community and Lifestyle*. Cambridge: Cambridge University Press.

Nafus, D. M. (1993) Extinction, biological control, and insect conservation on islands. In Gaston, K. J., New, T. R. and Samways, M. J. (eds.), *Perspectives on Insect Conservation*, pp. 139–54. Andover, UK: Intercept.

Naskrecki, P. (2000) *Katydids of Costa Rica Vol. 1. Systematics and Bioacoustics of the Conehead Katydids*. CD ROM. Philadelphia, PA: The Orthopterists' Society and the Academy of Natural Sciences of Philadelphia.

National Research Council (1999) *Our Common Journey: A Transition Toward Sustainability*. Washington, DC: National Academy Press.

Naveh, Z. and Lieberman, A. S. (1990) *Landscape Ecology: Theory and Application*. Berlin: Springer-Verlag.

Naylor, R. L. (2000) The economics of alien species invasions. In Mooney, H. A. and Hobbs, R. J. (eds.), *Invasive Species in a Changing World*, pp. 241–59. Washington, DC: Island Press.

Nekola, J. C. (2002) Effects of fire management on the richness and abundance of central North American grassland land snail faunas. *Animal Biodiversity and Conservation* **25.2**, 53–66.

Nettleton, G. (1996) Gypsy moth. *Antenna* **20**, 98–9.

Néve, G., Barascud, B., Hughes, R., Aubert, J., Descimon, H., Lebrun, M. and Baguette, M. (1996) Dispersal, colonization power and metapopulation structure in the vulnerable butterfly *Proclossiana eunomia* (Lepidoptera: Nymphalidae). *Journal of Applied Ecology* **33**, 14–22.

New, T. R. (1984) *Insect Conservation: An Australian Perspective*. Dordrecht, The Netherlands: Junk.

(1991) *Insects as Predators*. Kensington, Australia: The New South Wales University Press.

New, T. R. (ed.) (1993) *Conservation Biology of Lycaenidae (Butterflies)*. Gland, Switzerland and Cambridge: IUCN.

New, T. R. (1994) *Exotic Insects in Australia*. Adelaide, Australia: Gleneagles Publishing.

(1996) Taxonomic focus and quality control in insect surveys for biodiversity conservation. *Australian Journal of Entomology* **35**, 97–106.

(1997) *Butterfly Conservation, 2nd edn*. Melbourne: Oxford University Press.

(1998) *Invertebrate Surveys for Conservation*. New York, NY: Oxford University Press.

(1999) Descriptive taxonomy as a facilitating discipline in invertebrate conservation. In Ponder, W. and Lunney, D. (eds.), *The Other 99%: The Conservation and Biodiversity of Invertebrates*, pp. 154–8. Mosman, Australia: Royal Zoological Society of New South Wales.

(2000a) How to conserve the 'meek inheritors'. *Journal of Insect Conservation* **4**, 151–2.

(2000b) How useful are ant assemblages for monitoring habitat disturbances on grassland in southeastern Australia. *Journal of Insect Conservation* **4**, 153–9.

(2000c) *Conservation Biology*. Melbourne: Oxford University Press.

New, T. R., Pyle, R. M., Thomas, J. A., Thomas, C. D. and Hammond, P. C. (1995) Butterfly conservation management. *Annual Reviews of Entomology* **40**, 57–83.

New, T. R. and Sands, D. P. A. (2002) Conservation concerns for butterflies in urban areas of Australia. *Journal of Insect Conservation* **6**, 207–15.

New, T. R. and Thornton, I. WB. (1988) A pre-vegetation population of crickets subsisting on allochthonous aeolian debris on Anak Krakatau. *Philosophical Transactions of the Royal Society of London, B* **322**, 481–5.

Ney-Nifle, M. and Mangel, M. (2000) Habitat loss and changes in the species-area relationship. *Conservation Biology* **14**, 893–8.

Niemelä, J. (1997) Invertebrates and boreal forest management. *Conservation Biology* **11**, 601–10.

Niemelä, J., Kotze, J., Ashworth, A., Brandmayr, P., Desender, K., New, T., Penev, L., Samways, M. J. and Spence, J. (2000) The search for common anthropogenic impacts on biodiversity: a global network. *Journal of Insect Conservation* **4**, 3–9.

Niemelä, J., Kotze, J. H., Venn, S., Penev, L., Stoyanov, I., Spence, J., Hartley, D. and Montes de Oca, E. (2002) Carabid beetle assemblages (Coleoptera, Carabidae) across urban-rural gradients: an international comparison. *Landscape Ecology* **17**, 387–401.

Nieminen, M. (1996) Migration of moth species in a network of small islands. *Oecologia* **108**, 643–51.

Nieminen, M. and Hanski, I. (1998) Metapopulations of moths on islands: a test of two contrasting models. *Journal of Animal Ecology* **67**, 149–60.

Nieminen, M., Nuorteva, P. and Tulisalo, E. (2001) The effect of metals on the mortality of *Parnassius apollo* larvae (Lepidoptera: Papilionidae). *Journal of Insect Conservation* **5**, 1–7.

Nilsson, S. G. and Baranowski, R. (1997) Habitat predictability and the occurrence of wood beetles in old-growth beech forests. *Ecography* **20**, 491–8.

Nixon, K. C. and Wheeler, Q. D. (1992) Measures of phylogenetic diversity. In Novacek, M. J. and Wheeler, Q. D. (eds.), *Evolution and Phylogeny*, pp. 216–34. New York, NY: Columbia University Press.

Nordén, B. and Appelqvist, T. (2001) Conceptual problems of ecological continuity and its bioindicator. *Biodiversity and Conservation* **10**, 779–91.

Norton, B. G. (2000) Biodiversity and environmental values: in search of a universal earth ethic. *Biodiversity and Conservation* **9**, 1029–44.

Noss, R. F. (1990) Indicators for monitoring biodiversity: a hierarchial approach. *Conservation Biology* **4**, 355–64.

Noss, R. F. and Beier, P. (2000) Arguing over little things: response to Haddad *et al. Conservation Biology* **14**, 1546–8.

Novacek, M. J. and Wheeler, Q. D. (1992) Extinct taxa: accounting for 99.999. . . . . .% of the earth's biota. In Novacek, M. J. and Wheeler, Q. D. (eds.), *Extinction and Phylogeny*, pp. 1–16. New York, NY: Columbia University Press.

Novotny, V. and Basset, Y. (2000) Rare species in communities in tropical insect herbivores: pondering the mystery of singletons. *Oikos* **89**, 564–72.

Oates, M. R. and Warren, M. S. (1990) *A Review of Butterfly Introductions in Britain and Ireland. Report for the Joint Committee for the Conservation of British Insects*. Godalming, Surrey, UK: World Wide Fund for Nature.

Obrycki, J. J., Losey, J. E., Taylor, O. R. and Jesse, L. C. T. (2001) Transgenic insecticidal corn: beyond insecticidal toxicity to ecological complexity. *BioScience* **51**, 353–61.

Ødegaard, F., Diserud, O. H., Engen, S. and Aagaard, K. (2000) The magnitude of local host specificity for phytophagous insects and its implications for estimates of global species richness. *Conservation Biology* **14**, 1182–6.

Økland, B. (1996) Unlogged forests: important sites for preserving the diversity of mycetophilids (Diptera: Sciaroidea). *Biological Conservation* **76**, 297–310.

Økland, B., Bakke, A., Haguar, S. and Kvamme, T. (1996) What factors influence the diversity of saproxylic beetles? A multiscaled study from a spruce forest in southern Norway. *Biodiversity and Conservation* **5**, 75–100.

Oliver, I. and Beattie, A. J. (1993) A possible method for the rapid assessment of biodiversity. *Conservation Biology* **7**, 562–8.

(1996) Invertebrate morphospecies as surrogates for species: a case study. *Conservation Biology* **10**, 99–109.

Oliver, I., Beattie, A. J. and York, A. (1998) Spatial fidelity of plant, vertebrate, and invertebrate assemblages in multiple-use forest in eastern Australia. *Conservation Biology* **12**, 822–35.

Oliver, I., Dangerfield, J. M. and York, A. (1999) When and how to conduct a biodiversity assessment of terrestrial invertebrates. In Ponder, W. and Lunney, D. (eds.), *The Other 99%: The Conservation and Biodiversity of Invertebrates*, pp. 8–18. Mosman, Australia: Royal Zoological Society of New South Wales.

Oliver, I., Pik, A., Britton, D., Dangerfield, J. M., Colwell, R. K. and Beattie, A. J. (2000) Virtual biodiversity assessment systems: The application of bioinformatics technologies to the accelerated accumulation of biodiversity information. *BioScience* **50**, 441–50.

Ollerton, J. (1996) Reconciling ecological processes with phylogenetic patterns: the apparent paradox of plant-pollinator systems. *Journal of Ecology* **84**, 767–9.

Olsen, D. M. and Dinerstein, E. (1998) The Global 200? A representation approach to conserving the Earth's distinctive ecoregions. *Conservation Biology* **12**, 505–15.

Orr, D. W. (2002) Four challenges of sustainability. *Conservation Biology* **16**, 1457–60.

Ortman, E. E., Barry, B. D. and Bushman, L. L. *et al.* (2001) Transgenic insecticidal corn: the agronomic and economic rationale for its use. *BioScience* **51**, 900–3.

Osbourne, P. J. (1997) Insects, man and climate in the British Holocene. In Ashworth, A. C., Buckland, P. C. and Sadler, J. P. (eds.), *Studies in Quaternary Entomology – An Inordinate Fondness for Insects*, pp. 193–8, Chichester, UK: John Wiley & Sons.

Ott, J. (1995) Do dragonflies have a chance to survive in an industrialized country like Germany? In Corbet, P. S., Dunkle, S. W. and Ubukata, H. (eds.), *Proceedings of the International Symposium on the Conservation of Dragonflies and their Habitats*, pp. 28–44. Kushiro, Japan: Japanese Society for Preservation of Birds.

Owen, J. (1991) *The Ecology of a Garden: the First Fifteen Years*. Cambridge: Cambridge University Press.

Pagel, M. and Payne, R. J. H. (1996) How migration affects estimation of the extinction threshold. *Oikos* **76**, 323–9.

Paine, R. T. (1969) A note on trophic complexity and community stability. *American Naturalist* **103**, 91–3.

Palmer, M. and Finlay, V. (2003) *Faith in Conservation: New Approaches to Religions and the Environment*. Washington, DC: The World Bank.

Panzer, R. (2002) Compatibility of predescribed burning with the conservation of insects in small isolated prairie reserves. *Conservation Biology* **16**, 1296–307.

(2003) Importance of in situ survival, recolonization, and habitat gaps in the postfire recovery of fire-sensitive prairie insect species. *Natural Areas Journal* **23**, 14–21.

Panzer, R. and Schwartz, M. W. (1998) Effectiveness of a vegetation-based approach to insect conservation. *Conservation Biology* **12**, 693–02.

Panzer, R., Stillwargh, D., Gnaediger, R. and Derkovitz, G. (1995) Prevalence of remnant dependence among prairie and savanna-inhabiting insects of the Chicago region. *Natural Areas Journal* **15**, 101–16.

Paquin, P. and Coderre, D. (1997) Changes in soil macroarthropod communities in relation to forest maturation through three successional stages in the Canadian boreal forest. *Oecologia* **112**, 104–11.

Parendes, L. A. and Jones, J. A. (2000) Role of light availability and dispersal in exotic plant invasion along roads and streams in the H. J. Andrews Experimental Forest, Oregon. *Conservation Biology* **14**, 64–75.

Parmesan, C. (1996) Climate and species' range. *Nature* **382**, 765–6.

Parmesan, C., Ryrholm, N., Stefanescy, C. *et al.* (1999) Poleward shifts in geographical ranges of butterfly species associated with regional warming. *Nature* **399**, 579–83.

Parsons, M. J. (1995) Butterfly farming and trading in the Indo-Australian region, and its benefits in the conservation of swallowtails and their tropical forest habitats. In Scriber, J. M., Tsubaki, Y. and Lederhouse, R. C. (eds.), *Swallowtail Butterflies: their Ecology and Evolutionary Biology*, pp. 393–400. Gainesville, FL: Scientific Publishers.

Pascarella, J. B., Waddington, K. D. and Neal, P. R. (2001) Non-apoid flower-visiting fauna of Everglades National Park, Florida. *Biodiversity and Conservation* **10**, 551–66.

Pearce-Kelly, P., Jones, R., Clarke, D., Walker, C., Atkin, P. and Cunningham, A. A. (1998) The captive rearing of threatened Orthoptera: a comparison of the conservation potential and practical considerations of two species' breeding programmes at the Zoological Society of London. *Journal of Insect Conservation* **2**, 201–10.

Peck, S. B. (1994a) Aerial dispersal of insects between islands in the Galápagos Archipelago, Ecuador. *Annals of the Entomological Society of America* **87**, 218–24.

(1994b) Sea-surface (pleuston) transport of insects between islands in the Galápagos Archipelago, Ecuador. *Annals of the Entomological Society of America* **87**, 576–82.

Pellmyr, O. (1992) Evolution of insect pollination and angiosperm diversification. *Trends in Ecology and Evolution* **7**, 46–9.

Perfecto, I. and Vandermeer, J. (2000) Quality of agroecological matrix in a tropical montane landscape: ants in coffee plantations in southern Mexico. *Conservation Biology* **16**, 174–82.

Perrow, M. R. and Davy, A. J. (eds.) (2002) *Handbook of Ecological Restoration, Vols 1 and 2*. Cambridge: Cambridge University Press.

Petanidou, T., Vokou, D. and Margaris, N. S. (1991) *Panaxia quadripunctaria* in the highly touristic valley of butterflies (Rhodes, Greece): Conservation problems and remedies. *Ambio* **20**, 124–8.

Petersen, F. T., Meier, R. and Larsen, M. N. (2003) Testing species richness estimation methods using museum label data on the Danish Asilidae. *Biodiversity and Conservation* **12**, 687–701.

Peterson, A. T., Ortega-Huerta, M. A., Bartley, J., Sánchez-Cordero, V., Soberón, J., Buddemeier, R. H. and Stockwell, D. R. B. (2002) Future projections for Mexican faunas under global climate change scenarios. *Nature* **416**, 626–9.

Peterson, M. A. (1996) Long-distance gene flow in the sedentary butterfly, *Euphilotes enoptes* (Lepidoptera: Lycaenidae). *Evolution* **50**, 1990–9.

Peterson, M. A. and Denno, R. F. (1997) The influence of intraspecific variation in dispersal strategies on the genetic structure of planthopper populations. *Evolution* **51**, 1189–206.

Petraitis, P. S., Latham, R. E. and Niesenbaum, R. P. (1989) The maintenance of species diversity by disturbance. *Quarterly Review of Biology* **64**, 393–418.

Picker, M. D., Colville, J. and Van, Noort, S. (2002) Mantophasmatodea now in South Africa. *Science* **297**, 1475.

Pickett, S. T. A., Cadenasso, M. L., Grove, J. M., Nilon, C. H., Pouyat, R. V., Zipperer, W. C. and Costanza, R. (2001) Urban ecological systems: linking terrestrial ecological, physiological, and socio-economic components of metropolitan areas. *Annual Review of Ecology and Systematics* **32**, 127–57.

Pimentel, D. (1975) Introduction. In Pimentel, D. (ed.), *Insects, Science & Society*, pp. 1–10. New York, NY: Academic Press.

(1991) Diversification of biological control strategies in agriculture. *Crop Protection* **10**, 243–53.

(1995) Ecological theory, pest problems, and biologically based solutions. In Glen, D. M., Greaves, M. P. and Anderson, H. M. (eds.), *Ecology and Integrated Farming Systems*, pp. 69–82. Chichester, UK: John Wiley & Sons.

Pimentel, D., Lach, L., Zuniga, R. and Morrison, D. (2000) Environmental and economic costs of non-indigenous species in the United States. *BioScience* **50**, 53–65.

Pimentel, D., Stachow, U., Takacs, D. A., Brubaker, H. W., Dumas, A. R., Meaney, J. J., O-Neil, J. A. S., Onsi, D. E. and Corzilius, D. B. (1992) Conserving biological diversity in agricultural/forestry systems. *BioScience* **42**, 354–62.

Pimm, S. L. (1996) Lessons from a kill. *Biodiversity and Conservation* **5**, 1059–67.

(1998) Extinction. In Sutherland, W. J. (ed.), *Conservation Science and Action*, pp. 20–38. Oxford: Blackwell Science.

Pollard, E. and Yates, T. J. (1993) *Monitoring Butterflies for Ecology and Conservation*. London: Chapman and Hall.

Ponder, W. F., Carter, G. A., Flemons, P. and Chapman, R. R. (2001) Evaluation of museum collection data for use in biodiversity assessment. *Conservation Biology* **15**, 648–57.

Ponel, P. (1995) Rissian, Eemian and Würmian Coleoptera assemblages from La Grande Pile (Vosges, France). *Palaeogeography, Palaeoclimatology and Palaeoecology* **114**, 1–41.

Ponel, P., Beaulieu, J.-L. de, Tobolski, K. (1992) Holocene palaeoenvironments at the timberline in the Taillefer Massif, French Alps: a study of pollen, plant macrofossils and fossil insects. *The Holocene* **2**, 117–30.

Ponel, P., Orgeas, J., Samways, M. J., Andrieu-Ponel, V., De, Beaulieu L., Reille, M., Roche, P. and Tatoni, T. (2003) 110 000 years of Quaternary beetle diversity change. *Biodiversity and Conservation* **12**, 2077–89.

Poppy, G. M. (2002) GMOs and insects – where to from here? *Antenna* **26**, 88–92.

Porter, J. (1995) The effects of climate change on the agricultural environment for crop insect pests with particular reference to the European corn borer and grain maize. In Harrington, R. and Stork, N. E. (eds.), *Insects in a Changing Environment*, pp. 93–123. London: Academic Press.

Posadas, P., Esquivel, D. R. M. and Crisci, J. V. (2001) Using phylogenetic diversity measures to get priorities in conservation: an example from southern South America. *Conservation Biology* **15**, 1325–34.

Powell, J. A. (1992) Interrelationships of yuccas and yucca moths. *Trends in Ecology and Evolution* **7**, 10–15.

Power, M. E., Stout, R. J., Cushing, C. E. *et al.* (1988) Biotic and abiotic controls in river and stream communities. *Journal of the North American Benthological Society* **7**, 456–79.

Power, M. E., Tilman, D., Estes, J. A. *et al.* (1996) Challenges in the quest for keystones. *BioScience* **46**, 609–20.

Prance, G. T. (1994) A comparison of the efficiency of higher taxa and species numbers in the assessment of biodiversity in the Neotropics. *Philosophical Transactions of the Royal Society of London, B* **354**, 89–99.

Prance, G. T. (2000) The failure of biogeography to convey the conservation message. *Journal of Biogeography* **27**, 51–3.

Prendergast, J. R., Quinn, R. M., Lawton, J. H., Eversham, B. C. and Gibbons, D. W. (1993) Rare species, the coincidence of diversity hotspots and conservation strategies. *Nature* **365**, 335–7.

Pressey, R. L., Humphries, C. J., Margules, C. R., Vane-Wright, R. I. and Williams, P. H. (1993) Beyond opportunism: key principles for systematic reserve selection. *Trends in Ecology and Evolution* **8**, 124–8.

Pressey, R. L. and Logan, V. S. (1997) Inside looking out: Findings of research on reserve selection relevant to 'off-reserve' nature conservation. In Hale, P. and Lamb, D. (eds.), *Conservation Outside Nature Reserves*, pp. 407–18. University of Queensland, Australia: Centre for Conservation Biology.

Price, P. W., Craig, T. P. and Roininen, H. (1995) Working toward theory on galling sawfly population dynamics. In Cappuccino, N. and Price, P. W. (eds.), *Population Dynamics: New Approaches and Synthesis*, pp. 321–38. San Diego, CA: Academic Press.

Priddel, D., Carlile, N., Humphrey, M., Fellenberg, S. and Hiscox, D. (2003) Rediscovery of the 'extinct' Lord Howe Island stick-insect (*Dryocoelus australis* Montrouzier) (Phasmatodea) and recommendations for its conservation. *Biodiversity and Conservation* **12**, 1391–403.

Primack, R., Kobri, H. and Mori, S. (2000) Dragonfly pond restoration promotes conservation awareness in Japan. *Conservation Biology* **14**, 1553–4.

Pringle, E. L. (2002) New butterfly reserve at Coega, Eastern Cape Province, South Africa. *Metamorphosis* **13**, 90–1.

Prinsloo, G. L. and Samways, M. J. (2001) Host specificity among introduced chalcidoid biological control agents in South Africa. In Lockwood, J. A., Howarth, F. G. and Purcell, M. F. (eds.), *Balancing Nature: Assessing the Impact of Importing Non-native Biological Control Agents (An International Perspective)*, pp. 31–40. Lanham, MD: Entomological Society of America.

Pryke, S. R. and Samways, M. J. (2001) Width of grassland linkages for the conservation of butterflies in South African afforested areas. *Biological Conservation* **101**, 85–96.

(2003) Quality of remnant indigenous grassland linkages for adult butterflies (Lepidoptera) in an afforested African landscape. *Biodiversity and Conservation* **12**, 1985–2004.

Pulliam, H. R. (1988) Sources, sinks, and population regulation. *American Naturalist* **132**, 652–61.

Pullin, A. S. (1996) Restoration of butterfly populations in Britain. *Restoration Ecology* **4**, 471–80.

(1997) Habitat requirements of *Lycaena dispar batavus* and implications for re-establishment in England. *Journal of Insect Conservation* **1**, 177–85.

(1999) Changing water levels and insect submergence – a neglected threat? *Journal of Insect Conservation* **3**, 169–70.

Purse, B. V., Hopkins, G. W., Day, K. J. and Thompson, D. J. (2003) Dispersal characteristics and management of a rare damselsfly. *Journal of Applied Ecology* **40**, 716–28.

Quammen, D. (1996) *The Song of the Dodo*. Hutchinson, London.

Rainio, J. and Niemelä, J. (2003) Ground beetles (Coleoptera: Carabidae) as bioindicators. *Biodiversity and Conservation* **12**, 487–506.

Ramsay, G. (2002) Understanding the role of insects in gene flow. *Antenna* **26**, 99–100.

Ranius, T. (2002) *Osmoderma eremita* as an indicator of species richness in tree hollows. *Biodiversity and Conservation* **11**, 931–41.

Ranius, T. and Nilsson, S. G. (1997) Habitat of *Osmoderma eremita* Scop. (Coleoptera: Scarabaeidae), a beetle living in hollow trees. *Journal of Insect Conservation* **1**, 193–204.

Ranius, T. and Jansson, N. (2002) A comparison of three methods to survey saproxylic beetles in hollow oaks. *Biodiversity and Conservation* **11**, 1759–71.

Ratsirarson, H., Robertson, H. G., Picker, M. D. and van Noort, S. (2002) Indigenous forests versus exotic eucalypt and pine plantations: a comparison of leaf-litter invertebrate communities. *African Entomology* **10**, 93–9.

Reid, T. (2000) *Butterfly Gardening in South Africa*. Cape Town, South Africa: Briza.

Rejmánek, M. (1999) Invasive plant species and invasible systems. In Sandlund, O. T., Schei, P. J. and Viken, Å. (eds.), *Invasive Species and Biodiversity Management*, pp. 79–102. Dordrecht, The Netherlands: Kluwer.

Rentz, D. C. F. (1993) Orthopteroid insects in threatened habitats in Australia. In Gaston, K. J., New, T. R. and Samways, M. J. (eds.), *Perspectives on Insect Conservation*, pp. 125–38. Andover: Intercept.

Resh, V. H., Brown, A. V., Covich, A. P. *et al.* (1988) The role of disturbance in stream ecology. *Journal of the North American Benthological Society* **7**, 433–55.

Resh, V. H. and Jackson, J. K. (1993) Rapid assessment approaches to biomonitoring using benthic macroinvertebrates. In Rosenberg, D. M. and Resh, V. H. (eds.), *Freshwater Biomonitoring and Benthic Macroinvertebrates*, pp. 195–233. New York, NY: Chapman and Hall.

Reyers, B., Fairbanks, D. H. K., Wessels, K. J. and van Jaarsveld, A. S. (2002a) A multicriteria approach to reserve selection: addressing long-term biodiversity maintenance. *Biodiversity and Conservation* **11**, 769–93.

Reyers, B., Wessels, K. J. and van Jaarsveld, A. S. (2002b) An assessment of biodiversity surrogacy options in the Limpopo Province of South Africa. *African Zoology* **37**, 185–95.

Ricketts, T. H., Daily, G. C., Ehrlich, P. R. and Fay, J. P. (2001) Countryside biogeography of moths in a fragmented landscape: biodiversity in native and agricultural habitats. *Conservation Biology* **15**, 378–88.

Riley, J. R., Reynolds, D. R., Mukhopadhyay, S., Ghosh, M. R. and Sarkar, T. K. (1995). Long-distance migration of aphids and other small insects in northeast India. *European Journal of Entomology* **92**, 639–53.

Risch, S. J. and Carroll, C. R. (1982) Effect of a keystone predaceous ant, *Solenopsis geminata*, on arthropods in a tropical agroecosystem. *Ecology* **63**, 1979–83.

Rivers-Moore, N. A. and Samways, M. J. (1996) Game and cattle trampling, and impacts of human dwellings on arthropods at a game park boundary. *Biodiversity and Conservation* **5**, 1545–56.

Roberts, D. L., Cooper, R. J. and Petit, L. J. (2000) Use of premontane moist forest and shade coffee agroecosystems by army ants in western Panama. *Conservation Biology* **14**, 192–9.

Robinson, M. (1991) The Neolithic and Late Bronze Age insect assemblages. In S. Needham (ed.), *Excavation and Salvage at Runningmede Bridge, 1978: The Late Bronze Age Waterfront Site*, pp. 277–326. London: British Museum.

Roland, J. and Taylor, P. D. (1997) Insect parasitoid species respond to forest structure at different spatial scales. *Nature* **386**, 710–14.

Rolston, H. III (1994) *Conserving Natural Value*. New York, NY: Columbia University Press.

(2000) The land ethic at the turn of the millennium. *Biodiversity and Conservation* **9**, 1045–58.

Rogers, D. J. and Randolph, S. E. (2000) The global spread of malaria in a future, warmer world. *Science* **289**, 1763–6.

Rösch, M., Chown, S. L. and McGeoch, M. A. (2001) Testing a bioindicator assemblage: gall-inhibiting moths and urbanization. *African Entomology* **9**, 85–94.

Rosenberg, D. K., Noon, B. R. and Meslow, E. C. (1997) Biological corridors: form, function, and efficacy. *BioScience* **47**, 677–87.

Rothenberg, D. (1989) Introduction: Ecosophy T – from intuition to system. In Naess, A. (ed.), *Ecology, Community and Lifestyle*, pp. 1–22. Cambridge: Cambridge University Press.

Rothschild, M. (1998) Gardening with butterflies In *Butterfly Gardening*, created by The Xerces Society in association with The Smithsonian Institution, pp. 7–15. San Francisco, CA: Sierra Club.

Roy, D. B. and Sparks, T. H. (2000) Phenology of British butterflies and climate change. *Global Change Biology* **6**, 407–16.

Roy, D. B., Bohan, D. A., Haughton, A. J. *et al.* (2003) Invertebrates and vegetation of field margins adjacent to crops subject to contrasting herbicide regimes in the Farm Scale Evaluations of genetically modified herbicide-tolerant crops. *Proceedings of the Royal Society of London, B* **358**, 1879–98.

Roy, D. B., Rothery, P., Moss, D., Pollard, E. and Thomas, J. A. (2001) Butterfly numbers and weather: predicting historical trends in abundance and the future effects of climate change. *Journal of Animal Ecology* **70**, 201–17.

Rozema, J., van der Staaij, J., Björn, L. O. and Caldwell, M. (1997) UV-B as an environmental stress factor in plant life: stress and regulation. *Trends in Ecology and Evolution* **12**, 22–8.

Rozzi, R., Silander J., Jr, Armesto, J. J., Feinsinger, P. and Massardo, F. (2000) Three levels of integrating ecology with the conservation of South American temperate forests: the initiative of the Institute of Ecological Research Chiloé, Chile. *Biodiversity and Conservation* **9**, 1199–217.

Ruohomäki, K., Kaitaniemi, P., Kozlov, M., Tammaru, T. and Haukioja, E. (1996) Density and performance of *Epirrita autumnata* (Lepidoptera: Geometridae) along three air pollution gradients in northern Europe. *Journal of Applied Ecology* **33**, 773–85.

Ruszczyk, A. (1996) Spatial patterns in pupal mortality in urban palm caterpillars. *Oecologia* **107**, 356–63.

Rutherford, M. C., Midgley, G. F., Bond, W. J., Powrie, L. W., Roberts, R. and Allsopp, J. (2000) Plant biodiversity. In Kiker, G. (ed.), *Climate Change Impacts in Southern Africa*. Pretoria, South Africa: Department of Environmental Affairs and Tourism.

Rutherford, M. C., Powrie, L. W. and Schulze, R. E. (1999) Climate change in conservation areas of South Africa and the potential impact on floristic composition: a first assessment. *Diversity and Distributions* **5**, 253–62.

Ryder, O. A. (1986) Species conservation and systematics: the dilemma of subspecies. *Trends in Ecology and Evolution* **1**, 9–10.

Ryszkowski, L., Karg, J., Margarit, G., Paoletti, M. G. and Zlotin, R. (1993) Above-ground insect biomass in agricultural landscapes of Europe. In Bunce, R. G. H., Ryszkowski, L. and Paoletti, M. G. (eds.), *Landscape Ecology and Agroecosystems*, pp. 71–82. Boca Raton, FL: Lewis.

Saarinen, K., Lahti, T. and Marttila, O. (2003) Population trends of Finnish butterflies (Lepidoptera: Hesperioidea, Papilionoidea) in 1991–2000. *Biodiversity and Conservation* **12**, 2147–59.

Saccheri, I., Kuussaari, M., Kankare, M., Vikman, P., Fortelius, W. and Hanski, I. (1998) Inbreeding and extinction in a butterfly metapopulation. *Nature* **392**, 491–4.

Sahlén, G. and Ekestubbe, K. (2001) Identification of dragonflies (Odonata) as indicators of general species richness in boreal forest lakes. *Biodiversity and Conservation* **10**, 673–90.

Samways, M. J. (1983a) Asymmetrical competition and amensalism through soil dumping by the ant, *Myrmicaria natalensis*. *Ecological Entomology* **8**, 191–4.

(1983b) Community structure of ants (Hymenoptera: Formicidae) in a series of habitats associated with citrus. *Journal of Applied Ecology* **20**, 833–47.

(1988) Classical biological control and insect conservation: are they compatible? *Environmental Conservation* **15**, 349–54.

(1989a) Insect conservation and the disturbance landscape. *Agriculture, Ecosystems and Environment* **27**, 183–94.

(1989b) Insect conservation and landscape ecology: a case-history of bush crickets (Tettigoniidae) in southern France. *Environmental Conservation* **16**, 217–26.

(1989c) Farm dams as nature reserves for dragonflies (Odonata) at various altitudes in the Natal Drakensberg mountains, South Africa. *Biologial Conservation* **48**, 181–7.

(1990) Land forms and winter habitat refugia in the conservation of montane grasshoppers in southern Africa. *Conservation Biology* **4**, 375–82.

(1993a) A spatial and process sub-regional framework for insect and biodiversity research and management. In Gaston, K. J., New, T. R. and Samways, M. J. (eds.), *Perspectives on Insect Conservation*, pp. 1–27. Andover, UK: Intercept.

(1993b) Insects in biodiversity conservation: some perspectives and directives. *Biodiversity and Conservation* **2**, 258–82.

(1994) *Insect Conservation Biology*. London: Chapman and Hall.

(1995) Southern hemisphere insects: their variety and the environmental pressures upon them. In Harrington, R. and Stork, N. E. (eds.), *Insects in a Changing Environment*, pp. 297–320. London: Academic Press.

(1996a) Skimming and insect evolution. *Trends in Ecology and Evolution* **11**, 471.

(1996b) Insects on the brink of a major discontinuity. *Biodiversity and Conservation* **5**, 1047–58.

(1996c) Insects in the urban environment: pest pressures versus conservation concern. In Wildey, K. B. (ed.). *Proceedings of the 1996 International Conference on Insects Pests in the Urban Environment*, pp. 129–33. Edinburgh: International Conference on Insect Pests in the Urban Environment.

(1996d) The art of unintelligent tinkering. *Conservation Biology* **10**, 1307.

(1997a) Conservation biology of Orthoptera. In Gangwere, S. K., Muralirangan, M. C. and Muralirangan, M. (eds.), *The Bionomics of Grasshoppers, Katydids and their Kin*, pp. 481–496. Wallingford: CAB International.

(1997b) ESUs and conservation of pests. *Conservation Biology* **11**, 304.

(1997c) Reserves do not guarantee survival from the vagaries of El Niño. *Journal of Insect Conservation* **1**, 145.

(1997d) Classical biological control and biodiversity: what risks are we prepared to accept? *Biodiversity and Conservation* **6**, 1309–16.

(1998a) Establishment of resident Odonata populations on the formerly waterless Cousine island, Seychelles: an Island Biogeography Theory (IBT) perspective. *Odonatologica* **27**, 253–8.

(1998b) Insect population changes and conservation in the disturbed landscapes of Mediterranean-Type Ecosystems. In Rundel, P. W., Montenegro, G. and Jaksic, F. M. (eds.), *Landscape Disturbance and Biodiversity in Mediterranean-Type Ecosystems*, pp. 313–31. Berlin: Springer.

(1999a) Managing insect invasions by watching other countries. In Sandlund, O. J., Schei, P. J. and Viken, Å. (eds.), *Invasive Species and Biodiversity Management*, pp. 295–304. Dordrecht, The Netherlands: Kluwer.

(1999b) Landscape triage for conserving insect diversity. In Ponder, W. and Lunney, D. (eds.), *The Other 99%: The Conservation and Biodiversity of Invertebrates*, pp. 269–73. Mosman: The Royal Zoological Society of New South Wales.

(2000a) Can locust control be compatible with conserving biodiversity? In Lockwood, J. A., Latchininsky, A. V. and Sergeev, M. G. (eds.), *Grasshoppers and Grassland Health*, pp. 173–9. Dordrecht, The Netherlands: Kluwer.

(2000b) A conceptual model of ecosystem restoration triage based on experiences from three remote oceanic islands. *Biodiversity and Conservation* **9**, 1073–83.

(2002a) Caring for the multitude current challenges. *Biodiversity and Conservation* **11**, 341–3.

(2002c) Red Listed Odonata species of Africa. *Odonatologica* **31**, 117–28.

(2003a) Threats to the tropical island dragonfly fauna (Odonata) of Mayotte, Comoro archipelago. *Biodiversity and Conservation* **12**, 1785–92.

(2003b) Conservation of an endemic dragonfly fauna in the ancient tropical Seychelles archipelago. *Odonatologica* **32**, 177–82.

(2003c) Marginality and national Red Listing of species. *Biodiversity and Conservation* **12**, 2523–5.

(2005) National Red List of South African dragonflies (Odonata). *Odonatologica* (in Press).

Samways, M. J., Caldwell, P. M. and Osborn, R. (1996) Ground-living invertebrate assemblages in native, planted and invasive vegetation in South Africa. *Agriculture, Ecosystems and Environment* **59**, 19–32.

Samways, M. J. and Grech, N. M. (1986) Assessment of the fungus *Cladosporium oxysporum* (Berk. and Curt.) as a potential biocontrol agent against certain Homoptera. *Agriculture, Ecosystems and Environment* **15**, 231–9.

Samways, M. J. and Kreuzinger, K. (2001) Vegetation, ungulate and grasshopper interactions inside vs. outside an African savanna game park. *Biodiversity and Conservation* **10**, 1963–81.

Samways, M. J. and Lockwood, J. A. (1998) Orthoptera conservation: pests and paradoxes. *Journal of Insect Conservation* **2**, 143–9.

Samways, M. J. and Moore, S. D. (1991) Influence of exotic conifer patches on grasshopper (Orthoptera) assemblages in a grassland matrix at a recreational resort, Natal, South Africa. *Biological Conservation* **57**, 205–19.

Samways, M. J. and Osborn, R. (1998) Divergence in a transoceanic circumtropical dragonfly on a remote island. *Journal of Biogeography* **25**, 935–46.

Samways, M. J., Osborn, R. and Carliel, F. (1997) Effect of a highway on ant (Hymenoptera: Formicidae) species composition and abundance, with recommendation for roadside verge width. *Biodiversity and Conservation* **6**, 903–13.

Samways, M. J., Osborn, R., Hastings, H. and Hattingh, V. (1999) Global climate change and accuracy of prediction of species geographical ranges: establishment success of introduced ladybirds (Coccinellidae, *Chilocorus* spp.) worldwide. *Journal of Biogeography* **26**, 795–812.

Samways, M. J. and Sergeev, M. G. (1997) Orthoptera and landscape change. In Gangwere, S. K., Muralirangan, M. C. and Muralirangan, M. (eds.), *The Bionomics of Grasshoppers, Katydids and their Kin*, pp. 147–62. Wallingford: CAB International.

Samways, M. J. and Stewart, D. A. B. (1997) An aquatic ecotone and its significance in conservation. *Biodiversity and Conservation* **6**, 1429–44.

Samways, M. J. and Steytler, N. S. (1996) Dragonfly (Odonata) distribution patterns in urban and forest landscapes and recommendations for riparian corridor management. *Biological Conservation* **78**, 279–88.

Samways, M. J. and Taylor, S. (2004) Impacts of invasive alien plants on red-listed South African dragonflies (Odonata). *South African Journal of Science* **100**, 78–80.

Sandalow, D. B. and Bowles, I. A. (2001) Fundamentals of treaty-making on climate change. *Science* **292**, 1838–40.

Sands, D. P. A. (1997) The 'safety' of biological control agents: assessing their impact on beneficial and other non-target hosts. *Memoirs of the Museum of Victoria* **56**, 611–15.

Sax, D. F. and Gaines, S. D. (2003) Species diversity: from global decreases to local increases. *Trends in Ecology and Evolution* **18**, 561–6.

Schatz, G. E. (1992) The race between deep flowers and long tongues. *Wings*, Summer 1992, 19–21.

Scheffer, M. and Carpenter, S. R. (2003) Catastrophic regime shifts in ecosystems: linking theory to observation. *Trends in Ecology and Evolution* **18**, 648–56.

Schemske, D. W. and Horvitz, C. C. (1984) Variation among floral visitors in pollination ability: a precondition for mutualism specialization. *Science* **255**, 519–21.

Schlik-Steiner, B. C., Steiner, F. M. and Schödl, S. (2003) A case study to qualify the value of voucher specimens for invertebrate conservation: ant records in Lower Austria. *Biodiversity and Conservation* **12**, 2321–8.

Schmidt, K. (1995) A new ant on the block. *New Scientist* **148** (No. 2002), 28–31.

Schmitt, T. and Seitz, A. (2001) Allozyme variation in *Polyommatus coridon* (Lepidoptera: Lycaenidae): identification of ice-age refugia and reconstruction of post-glacial expansion. *Journal of Biogeography* **28**, 1129–36.

Scholtz, C. H. and Chown, S. L. (1993) Insect conservation and extensive agriculture: the savanna of southern Africa. In Gaston, K. J., New, T. R. and Samways, M. J. (eds.), *Perspectives on Insect Conservation*, pp. 75–95. Andover: Intercept.

Scholtz, C. H. and Krüger, K. (1995) Effects of invermectin residues in cattle dung on dung insect communities under extensive farming conditions in South Africa. In Harrington, R. and Stork, N. E. (eds.), *Insects in a Changing Environment*, pp. 465–71. London: Academic Press.

Schowalter, T. D. (2000) Insects as regulators of ecosystem development. In Coleman, D. C. and Hendrix, P. F. (eds.), *Invertebrates as Webmasters in Ecosystems*, pp. 99–114. Wallingford: CAB International.

Schowalter, T. D. and Ganio, L. M. (1999) Invertebrate communities in a tropical rain forest canopy in Puerto Rico following Hurricane Hugo. *Ecological Entomology* **24**, 191–201.

Schreiner, I. and Nafus, D. (1986) Accidental introductions of insect pests to Guam, 1945–1985. *Proceedings of the Hawaiian Entomological Society* **27**, 45–72.

Schultz, C. B. and Hammond, P. C. (2003) Using population viability analysis to develop recovery criteria for endangered insects: case study of the Fender's blue butterfly. *Conservation Biology* **17**, 1372–85.

Schutte, G., Reich, M. and Plachter, H. (1997) Mobility of the rheobiont damselfly *Calopteryx splendens* (Harris) in fragmented habitats (Zygoptera: Calopterygidae). *Odonatologica* **26**, 317–27.

Shackleton, N. J. (1977) Oxygen isotope stratigraphy of the Middle Pleistocene. In Shotton, F. W. (ed.), *Quaternary Studies: Recent Advances*, pp. 1–16. Oxford: Clarendon Press.

Shahabuddin, G., Herzner, G. A., Cesar, A. and Gomez, M. D. C. (2000) Persistence of a frugivorous butterfly species in Venezuelan forest fragments: the role of movement and habitat quality. *Biodiversity and Conservation* **9**, 1623–41.

Sharratt, W. J., Picker, M. D. and Samways, M. J. (2000) The invertebrate fauna of the sandstone caves of the Cape Peninsula (South Africa): patterns of endemism and conservation priorities. *Biodiversity and Conservation* **9**, 107–43.

Shea, K. and Chesson, P. (2002) Community ecology theory as a framework for biological invasions. *Trends in Ecology and Evolution* **17**, 170–6.

Shepherd, M., Buchmann, S. L., Vaughan, M. and Black, S. H. (2003) *Pollinator Conservation Handbook*. Portland, OR: The Xerces Society.

Shetler, S. G. (1998) Butterfly gardening and conservation. In *Butterfly Gardening*, created by The Xerces Society in association with The Smithsonian Institution, pp. 131–3. San Francisco, CA: Sierra Club.

Sherley, G. H. (1998) Translocating a threatened New Zealand giant Orthoptera, *Deinacrida* sp. (Stenopelmatidae): some lessons. *Journal of Insect Conservation* **2**, 195–9.

Shirt, D. B. (ed.) (1987) *British Red Data Books: 2. Insects*. Peterborough: Nature Conservancy Council.

Shreeve, T. G., Dennis, R. L. H. and Pullin, A. S. (1996) Marginality: scale determined processes and the conservation of the British butterfly fauna. *Biodiversity and Conservation* **5**, 1131–41.

Siegert, F., Ruecker, G., Heinrichs, A. and Hoffmann, A. A. (2001) Increased damage from fires in logged forests during droughts caused by El Niño. *Nature* **414**, 437–40.

Simberloff, D. (1998) Flagships, umbrellas and keystones: is single-species management passé in the landscape era? *Biological Conservation* **83**, 247–57.

Simonson, S. E., Opler, P. A., Stohlgren, T. J. and Chong, G. W. (2001) Rapid assessment of butterfly diversity in a montane landscape. *Biodiversity and Conservation* **10**, 1369–86.

Slone, T. H., Orsak, L. J. and Malver, O. (1997) A comparison of price, rarity and cost of butterfly specimens: implications for the insect trade and for habitat conservation. *Ecological Economics* **21**, 77–85.

Small, E. C., Sodler, J. P. and Telfer, M. G. (2003) Carabid assemblages on urban derelict sites in Birmingham, UK. *Journal of Insect Conservation* **6**, 233–46.

Smallidge, P. J., Lepold, D. J. and Allen, C. M. (1996) Community characteristics and vegetation management for Karner blue butterfly (*Lycaeides melissa samuelis*) habitats on rights-of-way in east-central New York, USA. *Journal of Applied Ecology* **33**, 1405–19.

Smart, S. M., Bunce, R. G. H., Firbank, L. G. and Coward, P. (2002) Do field boundaries act as refugia for grassland plant species diversity in intensively managed agricultural landscapes in Britain? *Agriculture, Ecosystems and Environment* **91**, 73–87.

Smit, J., Höppner, J., Hering, D. and Plachter, H. (1997) Spider and ground beetle communities (Araneae, Carabidae) on gravel bars along streams in Northern Hesse (Germany). *Verhandlungen der Gesellschaft für Ökologie* **27**, 357–64.

Smith, D. S. and Hellmund, P. C. (eds.) (1993) *Ecology of Greenways*. Minneapolis, MN: University of Minnesota Press.

Smith, R. (1999) "Bugging the media": TV broadcasting and the invertebrate agenda. In Ponder, W. and Lunney, D. (eds.), *The Other 99%: The Conservation of Biodiversity of Invertebrates*, pp. 413–17, Mosman: The Royal Zoological Society of New South Wales.

Smith, T. B., Kark, S., Schneider, C. J., Wayne, R. K. and Moritz, C. (2001) Biodiversity hotspots and beyond: the need for preserving environmental transitions. *Trends in Ecology and Evolution* **16**, 431.

Smith, V. R., Avenant, N. L. and Chown, S. L. (2002) The diet and impact of house mice on a sub-Antarctic island. *Polar Biology* **25**, 703–15.

Sole, R. and Goodwin, B. (2000) *Signs of Life*. New York, NY: Basic Books.

Soulé, M. E. (1989) Conservation biology in the twenty-first century: summary and outlook. In Western, D. and Pearl, M. C. (eds.), *Conservation for the Twenty-first Century*, pp. 297–303. New York, NY: Oxford University Press.

Sotherton, N. (1992) The environmental benefits of conservation headlands in cereal fields. *Outlook on Agriculture* **21**, 219–24.

Southwood, T. R. E. (1973) The insect/plant relationship – an evolutionary perspective. *Symposium of the Royal Entomological Society of London* **6**, 3–30.

Southwood, T. R. E., Brown, V. K. and Reader, P. M. (1979) The relationship of plant and insect diversities in succession. *Biological Journal of the Linnean Society* **12**, 327–48.

Southwood, T. R. E. and Henderson, P. A. (2000) *Ecological Methods, 3rd Edition*. Oxford: Blackwell Science.

Spagarino, C., Postur, G. M. and Peri, P. L. (2001) Changes in *Nothofagus pumilio* forest biodiversity during the forest management cycle. 1. Insects. *Biodiversity and Conservation* **10**, 2077–92.

Sparks, T. H. and Yates, T. J. (1997) The effect of spring temperature on the appearance dates of British butterflies 1883–1993. *Ecography* **20**, 368–74.

Spector, S. (2002) Biogeographic crossroads as priority areas for biodiversity conservation. *Conservation Biology* **16**, 1480–7.

Spence, J. R., Langor, D. W., Hammond, H. E. J. and Pohl, G. R. (1997) Beetle abundance and diversity in a boreal mixed-wood forest. In Watt, A. D., Stork, N. E. and Hunter, M. D. (eds.), *Forests and Insects*, pp. 287–301. London: Chapman and Hall.

Spitzer, K., Bezděk, A. and Jaroš, J. (1999) Ecological succession of a relict Central European peat bog and variability of its insect biodiversity. *Journal of Insect Conservation* **3**, 97–106.

Spitzer, K., Jaros, J., Havelka, J. and Leps, J. (1997) Effect of small-scale disturbance on butterfly communities of an Indo-Chinese montane rainforest. *Biological Conservation* **80**, 9–15.

Stanley, D. P. and Weinstein, P. (1996) Leaf litter traps for sampling orthopteroid insects in tropical caves. *Journal of Orthoptera Research* **5**, 51–2.

Steenkamp, H. E. and Chown, S. L. (1996) Influence of dense stands of an exotic tree, *Prosopis glandulosa* Benson, assemblage in southern Africa. *Biological Conservation* **78**, 305–11.

Steffan-Dewenter, I. and Leschke, K. (2003) Effects of habitat management on vegetation and above-ground nesting bees and wasps of orchard meadow in Central Europe. *Biodiversity and Conservation* **12**, 1953–68.

Steffan-Dewenter, I. and Tscharntke, T. (1997) Early succession of butterfly and plant communities on set-aside fields. *Oecologia* **109**, 294–302.

Stephens, P. A. and Sutherland, W. J. (1999) Consequences of the Allee effect for behaviour, ecology and conservation. *Trends in Ecology and Conservation* **14**, 401–5.

Stewart, D. A. B. (1998) Non-target grasshoppers as indicators of the side-effets of chemical locust control in the Karoo, South Africa. *Journal of Insect Conservation* **2**, 263–76.

Stewart, D. A. B. and Samways, M. J. (1998) Conserving dragonfly (Odonata) assemblages relative to river dynamics in a major African savanna game reserve. *Conservation Biology* **12**, 683–92.

Steytler, N. S. and Samways, M. J. (1995) Biotope selection by adult male dragonflies (Odonata) at an artificial lake created for insect conservation in South Africa. *Biological Conservation* **72**, 381–6.

Stock, W. D. and Allsopp, N. (1992) Functional perspective of ecosystems. In Cowling, R. (ed.) *The Ecology of Fynbos*, pp. 241–59. Cape Town: Oxford University Press.

Stockwell, C. A., Hendry, A. P. and Kennison, M. J. (2003) Contemporary evolution meets conservation biology. *Trends in Ecology and Evolution* **18**, 94–101.

Stoker, R. L., Grant, W. E. and Vinson, S. B. (1995) *Solenopsis invicta* (Hymenoptera: Formicidae) effect on invertebrate decomposers of carrion in Central Texas. *Environmental Entomology* **24**, 817–22.

Stork, N. E. (1987) Guild structure of arthropods from Bornean rain forest trees. *Ecological Entomology* **12**, 69–80.

(1988) Insect diversity: facts, fiction and speculation. *Biological Journal of the Linnean Society* **35**, 321–37.

(1999) Estimating the number of species on Earth. In Ponder, W. and Lunney, D. (eds.), *The Other 99%: The Conservation and Biodiversity of Invertebrates*, pp. 1–7. Chipping Norton, Australia: Surrey Beatty.

Stork, N. E. and Lyal, C. H. C. (1993) Extinction or 'co-extinction' rates? *Nature* **366**, 307.

Stork, N. E. and Samways, M. J. (co-ordinators) (1995) Inventorying and monitoring of biodiversity. In Heywood, V. H. (ed.), *Global Biodiversity Assessment*, pp. 453–543. Cambridge: Cambridge University Press.

Stork, N. E., Srivastava, D. S., Watt, A. D. and Larsen, T. B. (2003) Butterfly diversity and silvicultural practice in lowland rainforests in Cameroon. *Biodiversity and Conservation* **12**, 387–410.

Strong, D. R., Lawton, J. H. and Southwood, T. R. E. (1984) *Insects on Plants*. Oxford: Blackwell Scientific Publications.

Sudling, K. N., Gross, K. L. and Houseman, G. R. (2004) Alternative states and positive feedbacks in restoration ecology. *Trends in Ecology and Evolution* **19**, 46–53.

Suh, A. N. and Samways, M. J. (2001) Development of a dragonfly awareness trail in an African botanical garden. *Biological Conservation* **100**, 345–53.

Summerville, K. S., Boulwane, M. J., Veech, J. A. and Crist, T. O. (2003) Spatial variation in species diversity and composition of forest Lepidoptera in eastern deciduous forests of North America. *Conservation Biology* **17**, 1045–57.

Sutcliffe, O. L., Bakkestuen, V., Fry, G. and Stabbetorp, O. E. (2003) Modeling the benefits of farmland restoration: methodology and application to butterfly movement. *Landscape and Urban Planning* **63**, 15–31.

Sutcliffe, O. L. and Thomas, C. D. (1996) Open corridors appear to facilitate dispersal of ringlet butterflies (*Aphantopus hyperantus*) between woodland clearings. *Conservation Biology* **10**, 1359–65.

Sutcliffe, O. L., Thomas, C. D. and Peggie, D. (1997a) Area-dependent migration by ringlet butterflies generate a mixture of patchy population and metapopulation attributes. *Oecologia* **109**, 229–34.

Sutcliffe, O. L., Thomas, C. D., Yates, T. J. and Greatorex-Davis, J. N. (1997b) Correlated extinctions, colonization and population fluctuation in a highly connected ringlet butterfly metapopulation. *Oecologia* **109**, 235–41.

Sverdrup-Thygeson, A. (2001) Can 'continuity indicator species' predict species richness or red-listed species of saproxylic beetles? *Biodiversity and Conservation* **10**, 815–32.

Swengel, A. B. (1998) Effects of management on butterfly abundance in tallgrass prairie and pine barrens. *Biological Conservation* **83**, 77–89.

(2001) A literature review of insect responses to fire, compared to other conservation managements of open habitat. *Biodiversity and Conservation* **10**, 1141–69.

Swengel, A. B. and Swengel, S. R. (1997) Co-occurrence of prairie and barrens butterflies: applications to ecosystem conservation. *Journal of Insect Conservation* **1**, 131–44.

(2001) Effects of prairie and barrens management on butterfly faunal composition. *Biodiversity and Conservation* **10**, 1757–85.

Swinton, A. H. (1880) *Insect Variety: its Propagation and Distribution*. London: Cassell, Petter and Galpin.

Taylor, R. A. J. and Reling, D. (1986) Density/height profile and long-range dispersal of first-instar gypsy moth (Lepidoptera: Lymantriidae). *Environmental Entomology* **15**, 431–5.

Terblanche, R. E., Morgent, T. L. and Cilliers, S. S. (2003) The vegetation of three localities of the threatened butterfly species *Chrysoritis aureus* (Lepidoptera: Lycaenidae) *Koedoe* **46/1**, 73–90.

The Xerces Society and The Smithsonian Institution (1998) *Butterfly Gardening*. San Francisco, CA: Sierra Club.

Thomas, C. D. (1990) Fewer species. *Nature* **347**, 237.

(1995) Ecology and conservation of butterfly metapopulations in the fragmented British landscape. In Pullin, A. S. (ed.), *Ecology and Conservation of Butterflies*, pp. 46–63. London: Chapman and Hall.

(2000) Dispersal and extinction in fragmented landscapes. *Proceedings of the Royal Society of London, B* **267**, 139–45.

Thomas, C. D., Bodsworth, E. J., Wilson, R. J., Simmons, A. D., Davies, Z. G., Musche, M. and Conradt, L. (2001) Ecological and evolutionary processes at expanding range margins. *Nature* **411**, 577–81.

Thomas, J. A. (1991) Rare species conservation: case studies of European butterflies. In Spellerberg, I. F., Goldsmith, F. B. and Morris, M. G. (eds.), *The Scientific Management of Temperate Communities for Conservation*, pp. 149–97. Oxford: Blackwell Scientific Publications.

Thomas, J. A., Bourn, N. A. D., Clarke, R. T. *et al.* (2001) The quality and isolation of habitat patches both determine where butterflies persist in fragmented landscapes. *Proceedings of the Royal Society of London, B* **268**, 1791–6.

Thomas, J. A., Clarke, R. T., Elmes, G. W. and Hochberg, M. E. (1998b) Population dynamics in the genus *Maculinea*. In Dempster, J. P. and McLean, I. F. G. (eds.), *Insect Population Dynamics: In Theory and Practice*, pp. 261–90. London: Chapman and Hall.

Thomas, J. A. and Morris, M. G. (1995) Rates and patterns of extinction among British invertebrates. In Lawton, J. H. and May, R. M. (eds.), *Extinction Rates*, pp. 111–30. Oxford: Oxford University Press.

Thomas, J., Surry, R., Shreeves, B. and Steele, C. (1998a) *New Atlas of Dorset Butterflies*. Dorchester: Dorset Natural History and Archaeological Society.

Thomas, J. and Webb, N. (1984) *Butterflies of Dorset*. Dorchester, UK: Dorset Natural History and Archaeological Society.

Thompson, D. J. and Purse, B. V. (1999) A search for long-distance dispersal in the southern damselfly *Coenagrion mercuriale* (Charpentier). *Journal of the British Dragonfly Society* **15**, 46–50.

Thornton, I. W. B. (1996) *Krakatau: The Destruction and Reassembly of an Island Ecosystem*. Cambridge, MASS: Harvard University Press.

Thornton, I. W. B., New, T. R., MacLaren, D. D., Sudarman, H. K. and Vaughan, P. J. (1988) Air-borne arthropod fall-out on Anak Krakatau and a possible pre-vegetation pioneer community. *Philosophical Transactions of the Royal Society of London, B* **322**, 471–9.

Thrall, P. H., Burdon, J. J. and Murray, B. R. (2000). The metapopulation paradigm: a fragmented view of conservation biology. In Young, A. G. and Clarke, G. M. (eds.), *Genetics, Demography and Viability of Fragmented Populations*, pp. 75–95. Cambridge: Cambridge University Press.

Tilman, D. (1997) Community invisibility, recruitment limitation, and grassland biodiversity. *Ecology* **78**, 81–92.

Tilman, D., Fargione, J., Wolff, B. *et al.* (2001) Forecasting agriculturally driven global environmental change. *Science* **292**, 281–284.

Tilman, D., May, R. M., Lehman, C. L. and Nowak, M. A. (1994) Habitat destruction and the extinction debt. *Nature* **371**, 65–6.

Toft, R. J. and Rees, J. S. (1998) Reducing predation of orb-web spiders by controlling common wasps (*Vespula vulgaris*) in a New Zealand beech forest. *Ecological Entomology* **23**, 90–5.

Tolman, T. and Lewington, R. (1997) *Collins Field Guide: Butterflies of Britain and Europe*. London: Harper Collins.

Torrusio, S., Cigliano, M. M. and Wysiecki, M. L. de (2002) Grasshopper (Orthoptera: Acridoidea) and plant community relationships in the Argentine pampas. *Journal of Biogeography* **29**, 221–9.

Travis, J. M. J. (2003) Climate change and habitat destruction: a deadly anthropogenic cocktail. *Proceedings of the Royal Society of London, B* **270**, 467–473.

Travis, J. M. J. and Dytham, C. (1999) Habitat persistence, habitat availability and the evolution of dispersal. *Proceedings of the Royal Society of London, B* **266**, 723–8.

Trewick, S. A. (2000) Molecular evidence for dispersal rather than vicariance as the origin of flightless insect species on the Chatham Islands, New Zealand. *Journal of Biogeography* **27**, 1189–200.

Tribe, G. D. and Richardson, D. M. (1994) The European wasp, *Vespula germanica* (Fabricius) (Hymenoptera: Vespidae), in southern Africa, and its potential distribution as predicted by ecoclimatic modeling. *African Entomology* **2**, 1–6.

Trombulak, S. C. and Frissell, C. A. (2000) Review of ecological effects of roads on terrestrial and aquatic communities. *Conservation Biology* **14**, 18–30.

Trumbo, S. T. and Bloch, P. L. (2000) Habitat fragmentation and burying beetle abundance and success. *Journal of Insect Conservation* **4**, 245–52.

Tscharntke, T., Steffan-Dewenter, I., Kruess, A. and Thies, C. (2002) Contribution of small habitat fragments to conservation of insect communities of grassland-cropland landscapes. *Ecological Applications* **12**, 354–63.

Tsutsui, N. D., Suarez, A. V. and Grosberg, R. K. (2003) Genetic diversity, asymmetrical aggression, and recognition in a widespread invasive species. *Proceedings of the National Academy of Sciences* **100**, 1078–83.

Tsutsui, N. D., Suarez, A. V., Holway, D. A. and Case, T. J. (2000) Reduced genetic variation and the success of an invasive species. *Proceedings of the National Academy of Science, USA* **97**, 5948–53.

Ubukata, H. (1997) Impact of global warming on insects. In Domoto, A. and Iwatsuki, K. (eds.), *Threats of Global Warming to Biological Diversity*, pp. 273–307. Tokyo: Tukijishokan.

Ugland, K. I., Gray, J. S. and Ellingsen, K. E. (2003) The species-accumulation curve and estimation of species richness. *Journal of Animal Ecology* **72**, 888–97.

Ungerer, M. J., Ayres, M. P. and Lombardero, M. J. (1999) Climate and its northern distribution limits of *Dendroctonus frontalis* Zimmerman (Coleoptera: Scolytidae). *Journal of Biogeography* **26**, 1133–45.

Usher, M. B. and Keiller, S. W. J. (1998) The macrolepidoptera of farm woodlands: determinants of diversity and community structure. *Biodiversity and Conservation* **7**, 725–48.

Van Jaarsveld, A. S. (2003) *The Evolution of Conservation Planning*. Inaugural Address, University of Stellenbosch, South Africa.

Van Jaarsveld, A. S., Freitag, S., Chown, S. L. *et al.* (1998) Biodiversity assessment and conservation strategies. *Science* **279**, 2106–8.

Van Lenteren, J. C., Babendreier, D., Bigler, F. *et al.* (2003) Environmental risk assessment of exotic natural enemies used in inundative biological control. *BioControl* **48**, 3–38.

Van Wilgen, B. W., Cowling, R. M. and Burgers, C. J. (1996) Valuation of ecosystem services: a case study from South African Fynbos ecosystems. *BioScience* **46**, 184–9.

Vanderwoude, C., Andersen, A. N. and House, A. P. N. (1997) Community organization, biogeography and seasonality of ants in an open forest of south-eastern Queensland. *Australian Journal of Zoology* **45**, 523–37.

Vane-Wright, R. I., Humphries, C. J. and Williams, P. H. (1991) What to protect? Systematics and the agony of choice. *Biological Conservation* **55**, 235–54.

Vasconcelos, H. L. (1999) Effects of forest disturbance on the structure of ground-foraging ant communities in central Amazonia. *Biodiversity and Conservation* **8**, 409–20.

Veldtman, R., McGeoch, M. A. and Scholtz, C. H. (2002) Variability in cocoon size in southern African wild silk moths: implications for sustainable harvesting. *African Entomology* **10**, 127–36.

Verdú, J. R., Crespo, M. B. and Galante, E. (2000) Conservation strategy of a nature reserve in Mediterranean ecosystems: the effects of protection from grazing on biodiversity. *Biodiversity and Conservation* **9**, 1707–721.

Vermeulen, H. J. W. (1994) Corridor function of a road verge for dispersal of stenotopic heathland ground beetles Carabidae. *Biological Conservation* **69**, 339–49.

Via, S. and Hawthorne, D. J. (2002) The genetic architecture of ecological specialization: correlated gene effects on host use and habitat choice in pea aphids. *American Naturalist* **159**, S76–88.

Vickery, M. (1998) *Gardening for Butterflies*. Dedham, Essex, UK: Butterfly Conservation.

Vines, G. (1999) Local heroes. *New Scientist* 27 February 1999, 34–39.

Vinson, S. B. (1991) Effect of the red imported fire ant (Hymenoptera: Formicidae) on a small plant-decomposing arthropod community. *Environmental Entomology* **20**, 98–103.

Voelz, N. J. and McArthur, J. V. (2000) An exploration of factors influencing lotic insect species richness. *Biodiversity and Conservation* **9**, 1543–70.

Vogler, A. P. and DeSalle, R. (1994) Diagnosing units of conservation measurement. *Conservation Biology* **8**, 354–63.

Walpole, M. J. and Leader-Williams, N. (2002) Tourism and flagship species in conservation. *Biodiversity and Conservation* **11**, 543–7.

Ward, D. F., New, T. R. and Yen, A. L. (2002) The beetle larva of remnant eucalypt woodlands from the Northern Plains, Victoria Australia. *Journal of Insect Conservation* **1**, 131–44.

Wardle, D. A., Nicholson, K. S. and Rahman, A. (1995) Ecological effects of the invasive weed species *Senecio jacobaea* L. (ragwort) in a New Zealand pasture. *Agriculture, Ecosystems and Environment* **56**, 19–28.

Warren, M. S. (1985) Influence of shade on butterfly numbers in woodland rides with special reference to the wood white *Leptidea sinapis*. *Biological Conservation* **33**, 146–64.

Warren, M. S., Hill, J. K., Thomas, J. A. *et al.* (2001) Rapid responses of British butterflies to opposing forces of climate and habitat change. *Nature* **414**, 65–9.

Walther, G.-R., Post, E., Convvey, P. *et al.* (2002) Ecological responses to recent climate change. *Nature* **416**, 389–95.

Watt, A. D., Stork, N. E., Eggleton, P. *et al.* (1997) Impact of forest loss and regeneration on insect abundance and diversity. In Watt, A. D., Stork, N. E. and Hunter, M. D. (eds.), *Forests and Insects*, pp. 273–86. London: Chapman and Hall.

Watt, A. D., Whittaker, J. B., Docherty, M., Brooks, G., Lindsay, E. and Salt, D. T. (1995) The impact of elevated atmospheric CO_2 on insect herbivores. In Harrington, R. and Stork, N. E. (eds.), *Insects in a Changing Environment*, pp. 197–217. London: Academic Press.

Webb, M. R. and Pullin, A. S. (1998) Effects of submergence by winter floods on diapausing caterpillars of a wetland butterfly, *Lycaena dispar batavus*. *Ecological Entomology* **23**, 96–9.

Webb, N. R. (1989). Studies on the invertebrate fauna of fragmented heathland in Dorset, UK, and the implications for conservation. *Biological Conservation* **47**, 153–65.

Webb, N. R. and Thomas, J. A. (1993) Concerning insect habitats in heathland biotopes: a question of scale. In Edwards, P. J., May, R. M. and Webb, N. R. (eds.), *Large-scale Ecology and Conservation Biology*, pp. 129–152. Blackwell, Oxford, UK.

Weibull, A.-C., Östman, Ö. and Granqvist, Å. (2003) Species richness in agroecosystems: the effect of landscape, habitat and farm management. *Biodiversity and Conservation* **12**, 1335–1355.

Weidemann, G., Reich, M. and Plachter, H. (1996) Influence of roads on a population of *Psophus stridulus* L. 1758 (Saltatoria, Acrididae). *Verhandlungen der Gesellschaft für Ökologie* **26**, 259–268.

Weissmann, M. J., Lederhouse, R. C. and Elia, F. C. (1995) Butterfly gardening and butterfly houses and their influence on conservation in North America. In Scriber, J. M., Tsubaki, Y. and Lederhouse, R. C. (eds.), *Swallowtail Butterflies: their Ecology and Evolutionary Biology*, pp. 393–400. Scientific Publishers, Gainesville, USA.

Wells, S. M., Pyle, R. M. and Collins, M. N. (1983) *The IUCN Invertebrate Red Data Book*. IUCN, Gland, Switzerland.

Wetterer, J. K., Miller, S. E., Wheeler, D. E. *et al.* (1999) Ecological dominance by *Paratrechina longicornis* (Hymenoptera: Formicidae), an invasive tramp ant, in Biosphere 2. *Florida Entomologist* **82**, 381–8.

Wheeler, Q. D. (1990) Insect diversity and cladistic constraints. *Annals of the Entomological Society of America* **83**, 1031–47.

Whitford, W. G. (2000). Keystone arthropods as webmasters in desert ecosystems. In Coleman, D. C. and Hendrix, P. F. (eds.), *Invertebrates as Webmasters in Ecosystems*, pp. 25–41. Wallingford: CAB International.

Whitmore, T. C. (1999) Arguments on the forest frontier. *Biodiversity and Conservation* **8**, 865–8.

Whitmore, T. C., Crouch, T. E. and Slotow, R. H. (2002) Conservation of biodiversity in urban environments: invertebrates on structurally enhanced road islands. *African Entomology* **10**, 113–26.

Whitten, T., Holmes, D. and Mackinnon, K. (2001) Conservation biology: a displacement behaviour for academia? *Conservation Biology* **15**, 1–3.

Wiens, J. A. (1989) Spatial scaling in ecology. *Functional Ecology* **3**, 385–97.

(1995) Landscape mosaics and ecological theory. In Hansson, L., Fahrig, L. and Merriam, G. (eds.), *Mosaic Landscapes and Ecological Processes*, pp. 1–26. London: Chapman and Hall.

Wiens, J. A. and Milne, B. T. (1989) Scaling of 'landscapes' in landscape ecology, or, landscape ecology from a beetle's perspective. *Landscape Ecology* **3**, 87–96.

Wiens, J. A., Stenseth, N. C., Horne, B. van and Ims, R. A. (1993) Ecological mechanisms in landscape ecology. *Oikos* **66**, 369–80.

Wilhelm, R. (1964) *The I Ching of Book of Changes*. London: Routledge and Kegan Paul.

Wikars, L.-O. (2002) Dependence on fire in wood-living insects: an experiment with burned and unburned spruce and birch logs. *Journal of Insect Conservation* **6**, 1–12.

Wilkie, L., Cassis, G. and Gray, M. (2003) A quality control protocol for terrestrial invertebrate biodiversity assessment. *Biodiversity and Conservation* **12**, 121–46.

Williams, B. L. (2002) Conservation genetics, extinction, and taxonomic status: a case history of the Regal fritillary. *Conservation Biology* **16**, 148–57.

Williams, K. S. (1993) Use of terrestrial arthropods to evaluate restored riparian woodlands. *Restoration Ecology* **1**, 107–16.

Williams, P. H. and Araújo, M. B. (2000) Using probability of persistence to identify important areas for biodiversity conservation. *Proceedings of the Royal Society of London, B* **267**, 1959–66.

Willams, P. H and Gaston, K. J. (1994) Measuring more of biodiversity: can higher taxon richness predict wholesale species richness? *Biological Conservation* **67**, 211–17.

Williams, R. and Martinez, N. (2000) Simple rules yield complex food webs. *Nature* **404**, 180–3.

Willamson, M. (1996) *Biological Invasions*. London: Chapman and Hall.

Willott, S. J., Lim, D. C., Compton, S. G. and Sutton, S. L. (2000) Effects of selective logging on the butterflies of a Bornean rainforest. *Conservation Biology* **14**, 1055–65.

Wilson, E. O. (1971) *The Insect Societies*. Harvard University Press, Cambridge, Massachusetts, USA.

Wilson, E. O. (1988) *Biodiversity*. Washington DC: National Academy Press.

(1991) Ants. *Wings* **16**, 3–13.

(1996) Hawaii: a world without social insects. *Bishop Museum Occasional Papers* **45**, 3–7.

Wilsson, S. G. and Baranowski, R. (1997) Habitat predictability and the occurrence of wood beetles in old growth beech forests. *Ecography* **20**, 491–8.

Winchester, N. N. (1997) Arthropods of coastal old-growth sitka spruce forests: conservation of biodiversity with special reference to the Staphylinidae. In Watt, A. D., Stork, N. E. and Hunter, M. D. (eds.), *Forests and Insects*, pp. 365–79. London: Chapman and Hall.

Woiwod, I. P. (1991) The ecological importance of longterm synoptic monitoring. In Firbank, L. G., Carter, N., Darbyshire, J. F. and Potts, G. R. (eds.), *The Ecology of Temperate Cereal Fields*, pp. 275–304. Oxford: Blackwell Scientific Publications.

(1997) Detecting the effects of climate change on Lepidoptera. *Journal of Insect Conservation* **1**, 149–58.

(2003) Are common moths in trouble? *Butterfly Conservation News* **2**, 9–11.

(2004) Butterflies and GM doom. *Butterfly* **85**, 34.

Wood, B. C. and Pullin, A. S. (2002) Persistence of species in a fragmented urban landscape: the importance of dispersal ability and habitat availability for grassland butterflies. *Biodiversity and Conservation* **11**, 1451–68.

Wood, J. G. Rev. (1876) *Homes without Hands*. London: Longmans, Green, and Co.

Wood, P. A. and Samways, M. J. (1991) Landscape element pattern and continuity of butterfly flight paths in an ecologically landscaped botanic garden, Natal, South Africa. *Biological Conservation* **58**, 149–66.

World Commission of Forests and Sustainable Development (1999) *Our Forests, Our Future*. Cambridge: Cambridge University Press.

World Resources Institute (1996) *World Resources 1996–1997.* New York, NY: Oxford University Press.

Worm, B. and Duffy, J. E. (2003) Biodiversity, productivity and stability in real food webs. *Trends in Ecology and Evolution* **18**, 628–32.

Wratten, S. D. and Van Emden, F. (1995) Habitat management for enhanced activity of natural enemies of insect pests. In Glen, D. M., Greaves, M. P. and Anderson, H. M. (eds.), *Ecology and Integrated Farming Systems*, pp. 117–45. Chichester: John Wiley & Sons.

Wright, D. J. (2002) Transgenetic crops expressing Bt toxins: status, prospects and resistance management strategies. *Antenna* **26**, 93–4.

Wright, M. G. (1993) Insect conservation in the African Cape fynbos, with special reference to endophagous insects. In Gaston, K. J., New, T. R. and Samways, M. J. (eds.), *Perspectives on Insect Conservation*, pp. 97–110. Andover: Intercept.

Wright, M. G. (1994) Variability in seed set: a defence mechanism against seed predators in *Protea* species (Proteaceae). *Oecologia* **99**, 397–400.

Wright, M. G. and Samways, M. J. (1998) Insect species richness tracking plant species richness in a diverse flora: gall-insects in the Cape Floristic Region, South Africa. *Oecologia* **115**, 427–33.

Wright, M. G. and Samways, M. J. (1999) Plant characteristics determine insect borer assemblages on *Protea* species in the Cape Fynbos, and importance for conservation management. *Biodiversity and Conservation* **8**, 1089–100.

Yen, A. L. (1987) A preliminary assessment of the correlation between plant, vertebrate and Coleoptera communities in the Victorian mallee. In Majer, J. D. (ed.), *The Role of Invertebrates in Conservation and Biological Survey*, pp. 73–88. Perth Western Australian Department of Conservation and Land Management Report.

Yen, A. L. and Butcher, R. J. (1997) *An Overview of the Conservation of Non-marine Invertebrates*. Canberra: Environment Australia.

York, A. (1999) Ecologically sustainable management: the utility of habitat surrogates for assessing terrestrial invertebrate diversity in temperate forests. In Ponder, W. and Lunney, D. (eds.), *The Other 99%: The Conservation and Biodiversity of Invertebrates*, pp. 34–9. Mosman, Australia: Transactions of the Royal Zoological Society of New South Wales.

Zedler, J. B. (2000) Progress in wetland restoration ecology. *Trends in Ecology and Evolution* **15**, 402–7.

Zgomba, M., Petrovic, D. and Srdic, Z. (1986) Mosquito larvicide impact on mayflies (Ephemeroptera) and dragonflies (Odonata) in aquatic biotopes. *Odonatologica* **16**, 221–2.

Zonneveld, C., Longcore, T. and Mulder, C. (2003) Optional schemes to detect the presence of insect species. *Conservation Biology* **17**, 476–87.

Zschokke, S., Dolt, C., Rusterholz, H.-P., Oggier, P., Braschler, B., Thommen, G. H., Lüdin, E., Erhardt, A. and Baur, B. (2000) Short-term responses of plants and invertebrates to experimental small-scale grassland fragmentation. *Oecologia* **125**, 559–72.

Zulka, K. P., Milasowszky, N. and Lethmayer, C. (1997) Spider biodiversity potential of an ungrazed and a grazed inland salt meadow in the National Park 'Neusiedler See-Seewinkel' (Austria): implications for management (Arachnida: Araneae). *Biodiversity and Conservation* **6**, 75–88.

Index